Finiteness Theorems for Limit Cycles

Translations of
MATHEMATICAL MONOGRAPHS

Volume 94

Finiteness Theorems for Limit Cycles

Yu. S. Il'yashenko

American Mathematical Society
Providence, Rhode Island

Ю. С. ИЛЬЯШЕНКО

ТЕОРЕМЫ КОНЕЧНОСТИ ДЛЯ ПРЕДЕЛЬНЫХ ЦИКЛОВ

Translated from the Russian by H. H. McFaden

1991 *Mathematics Subject Classification.* Primary 34-02, 34C05;
Secondary 34C20, 34C35, 57R25, 14E15, 41A60.

Library of Congress Cataloging-in-Publication Data

Il′i͡ashenko, I͡U. S.
[Teoremy konechnosti dli͡a predel′nykh t͡siklov. English]
Finiteness theorems for limit cycles/Yu. S. Il′yashenko; [translated from the Russian by H. H. McFaden].
p. cm.—(Translations of mathematical monographs, ISSN 0065-9282; v. 94)
Translation of: Teoremy konechnosti dli͡a predel′nykh t͡siklov.
Includes bibliographical references.
ISBN 0-8218-4553-5 (alk. paper)
1. Limit cycles. 2. Vector fields. 3. Differential equations. I. Title. II. Series.
QA371.I4413 1991 91-27411
515′.35—dc20 CIP

Printed in the United States of America

Information on Copying and Reprinting can be found at the back of this volume.

The paper used in this book is acid-free and falls within the guidelines
established to ensure permanence and durability. ♾
This publication was typeset using $\mathcal{A}\!\mathcal{M}\!\mathcal{S}$-TEX,
the American Mathematical Society's TEX macro system.

10 9 8 7 6 5 4 3 2 1 96 95 94 93 92 91

To my family:
Lena
Serezha
Lizochka
Aleksandr

Contents

Foreword

This book is devoted to a proof of the following finiteness theorem:

A polynomial vector field on the real plane has a finite number of limit cycles.

Some related results are proved along with it.

At the time of the discovery of limit cycles more than one hundred years ago Poincaré posed the question of whether the number of these cycles is finite for polynomial vector fields. He proved that the answer is yes for fields not having polycycles (separatrix polygons).

In a series of papers between 1889 and 1923, Dulac, a student of Poincaré, advanced greatly the local theory of differential equations (his achievements were finally understood only in the 1970s and early 1980s), and he presented a proof of the finiteness theorem in the memoir "Sur les cycles limites" (1923). In 1981 an error was found in this proof. The 1923 memoir practically concluded the mathematical creativity of Dulac. In the next thirty-two years (he died in 1955) he published only one survey (1934). Did he discover the error in his paper? Did he attempt to correct it during all his last years? These questions will surely remain forever unanswered.

To prove the finiteness theorem it suffices to see that limit cycles cannot accumulate on a polycycle of an analytic vector field (the nonaccumulation theorem). For this it is necessary to investigate the monodromy transformation (also called the Poincaré return mapping or the first return mapping) corresponding to this cycle. The investigation in this book uses the following five sources.

1. The theory of Dulac. This theory enables us to investigate the power asymptotics of the monodromy transformation. However, there exists a polycycle of an analytic vector field whose monodromy transformation has a nonidentity flat correction which thus decreases more rapidly than any power (the correction of a mapping is the difference between it and the identity). Therefore, power asymptotics are clearly insufficient for describing monodromy transformations.

2. Going out into the complex domain. The first systematic investigation of the global theory of analytic differential equations on the complex projective plane was undertaken by Petrovskiĭ and Landis in 1955. By extending

the solutions of an analytic differential equation into a neighborhood of a polycycle in the complex plane, the author was able to prove the nonaccumulation theorem for a polycycle whose vertices are nondegenerate saddles (1984). This step was taken under the influence of the work of Petrovskiĭ and Landis.

3. Resolution of singularities. This procedure, which reduces in its simplest variant to a finite series of polar blowing-ups (transitions from Cartesian coordinates to polar coordinates), enables us to essentially simplify the behavior of the solutions in a neighborhood of singular points of a vector field. The theorem on resolution of singularities asserts that in finitely many polar blowing-ups a compound singular point of an analytic vector field can be replaced by finitely many elementary singular points. The latter is the name for singular points at which the linearization of the field has at least one nonzero eigenvalue. The greatest complexity in the structure of the monodromy transformation is introduced by degenerate elementary singular points with one eigenvalue equal to zero and the other not equal to zero. They are investigated by methods of the geometric theory of normal forms.

4. The geometric theory of normal forms. Formal changes of variables enable us to reduce the germs of vector fields at singular points and the germs of diffeomorphisms at fixed points to comparatively simple so-called "resonant" normal forms (synonym: Poincaré-Dulac normal forms). As a rule, the normalizing series diverge when there are resonances, including the vanishing of an eigenvalue of the linearization (Bryuno, 1971; the author, 1981).

In this case the normal form is given not analytically as a series with a "relatively small number" of nonzero coefficients, but geometrically as a so-called "normalizing atlas." Namely, a punctured neighborhood of a singular point in a complex space is covered by finitely many domains of sector type that contain this singular point on the boundary. In each of these neighborhoods the vector field is analytically equivalent to its resonant normal form; a change of coordinates conjugating the original field with its normal form is said to be normalizing. A collection of normalizing substitutions is called a normalizing atlas. All the information about the geometric properties of the germ is contained in the transition functions from one normalizing substitution to another. The nontriviality (difference from the identity transformation) of these transition functions constitutes the so-called "nonlinear Stokes phenomenon." (A collection of papers by Elizarov, Shcherbakov, Voronin, Yakovenko, and the author will be devoted to this phenomenon.) It was first investigated for one-dimensional mappings by Ecalle, Malgrange, and Voronin in 1981. Normalizing atlases for germs of one-dimensional mappings are so-called functional cochains and play a fundamental role in the description of monodromy transformations of polycycles.

5. Superexact asymptotic series. These series are for use in describing asymptotic behavior with power terms and exponentially small terms simul-

taneously taken into account, and perhaps also iterated-exponentially small terms.

The structure of the book is as follows. In the Introduction we present all results about the Dulac problem obtained up to the writing of this book, with full proofs. An exception is formed by results in the local theory and theorems on resolution of singularities; their proofs belong naturally in textbooks, but such texts have unfortunately not yet been written. Superexact asymptotic series are discussed at the end of the Introduction and historical comments are given.

In the first chapter we give a complete description of monodromy transformations of polycycles of analytic vector fields and prove the nonaccumulation theorem. The main part of the chapter is the definition of regular functional cochains, which are used to describe monodromy transformations. This description is based on the group properties of regular functional cochains. Their verification recalls the proving of identities. However, since the definitions are very cumbersome, many details must be checked, and this takes a lot of space: Chapters II, IV, and, in part, V.

One of the most important properties of regular functional cochains is that they are uniquely determined by their superexact asymptotic series (STAR)[(1)] This is an assertion of the same type as the Phragmén-Lindelöf theorem for holomorphic functions of a single variable; it is used without proof in Chapter I and is proved in Chapter III and part of Chapter V.

Finally, the partial sums of STAR do not oscillate. This is established in §4.10. The proof of the nonaccumulation theorem is thus based on the following chain of implications.

A monodromy transformation has countably many fixed points $\Rightarrow$ the STAR for its correction is zero (since the partial sums of the nonzero series do not oscillate) $\Rightarrow$ the correction is zero by the Phragmén-Lindelöf theorem.

The introductory chapter overcomes all the difficulties connected with differential equations and reduces the finiteness theorem to questions of one-dimensional complex analysis.

The ideas for the proof presented were published by the author in the journal *Uspekhi Mathematicheskikh Nauk* **45** (1990), no. 2. This paper is the first part of the proposed work, of which the second part—the present book—is formally independent. In the first part the nonaccumulation theorem is proved for the case when the monodromy transformation has power and exponential asymptotics but not iterated-exponential asymptotics. The scheme of the paper is close to that of the book: the four sections of the paper are parallel to the first four chapters of the book and contain the same

---

([1]) This is the Russian abbreviation of Super Exact Asymptotic Series (Сверх Точные Асимптотические Ряды). It is chosen because it seems to sound better in English than the English abbreviation.

ideas, but there are essentially fewer technical difficulties in them. The reading of the paper should facilitate markedly the reading of the book. However, the text of the book is independent of the first part, and it is intended for autonomous reading.

Some words about the organization of the text. The book is divided into chapters and sections; almost all sections are divided into subsections. The numbering of the lemmas begins anew in each chapter, and the formulas in a chapter are labelled by asterisks; the labelling for formulas and propositions begins anew in each section. References to formulas and propositions in other sections are rare. In referring to a subsection of the same section we indicate only the letter before the heading of the subsection, and in referring to a subsection of another section of the same chapter we indicate the number of the chapter and the section; in referring to another chapter we indicate the chapter, section, and subsection.

I received a great deal of diverse help in writing this book from R. I. Bogdanov, A. L. Vol′berg, E. A. Gorin, V. P. Gurariĭ, P. M. Elizarov, A. S. Il′yashenko, E. M. Landis, A. Yu. and V. M. Nemirovich-Danchenko, and A. G. Khovanskiĭ. To all of them I express warm gratitude.

The several years spent writing this book were a heavy burden on my wife and children. Without their understanding, patience, and love the work would certainly never have been completed. I dedicate the book to my family.

Introduction

§0.1. Formulation of results: finiteness theorems and the identity theorem

In this section we formulate the main results: three finiteness theorems, the nonaccumulation theorem, and the identity theorem. Then we derive the first four theorems from the fifth.

A. Main theorems. This book is devoted to a proof of the following results.

THEOREM I. *A polynomial vector field on the real plane has only finitely many limit cycles.*

THEOREM II. *An analytic vector field on a closed two-dimensional surface has only finitely many limit cycles.*

THEOREM III. *A singular point of an analytic vector field on the real plane has a neighborhood free of limit cycles.*

These three theorems are called finiteness theorems.

As known from the times of Poincaré and Dulac ([10], [21]; a detailed reduction is carried out in §0.1D), the first two theorems are consequences of the following theorem.

THEOREM IV (nonaccumulation theorem). *An elementary polycycle of an analytic vector field on a two-dimensional surface has a neighborhood free of limit cycles.*

Recall that a polycycle of a vector field is a separatrix polygon; more precisely, it is a union of finitely many singular points and nontrivial phase curves of this field, with the set of singular points nonempty; solutions corresponding to nontrivial phase curves tend to singular points as $t \to +\infty$ and $t \to -\infty$; a polycycle is connected and cannot be contracted in itself to some proper subset of itself (Figure 1, next page).

A polycycle is said to be elementary if all its singular points are elementary, that is, their linearizations have at least one nonzero eigenvalue.

The monodromy transformation of a polycycle is defined in the same way as for an ordinary cycle, except that a half-open interval is used in place of

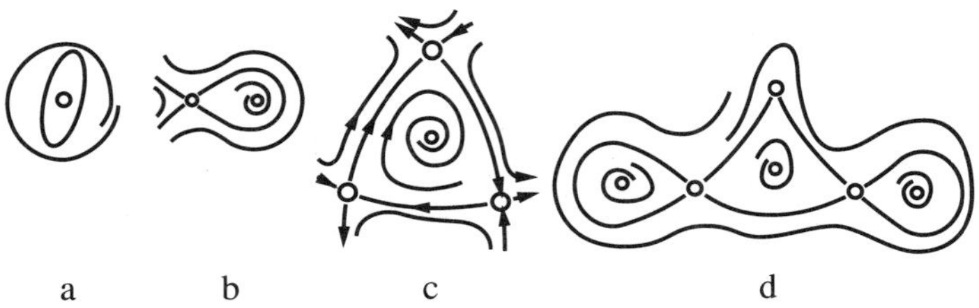

FIGURE 1

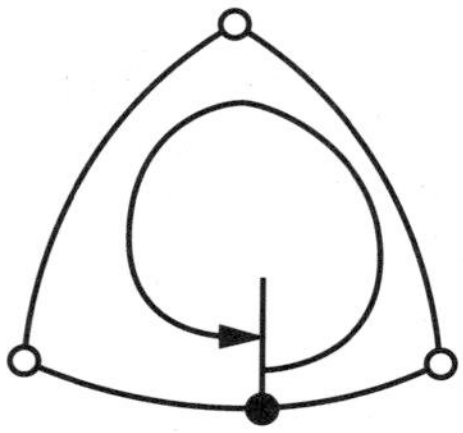

FIGURE 2

an open interval (Figure 2). It is convenient to regard monodromy transformations as germs of mappings $(\mathbf{R}^+, 0) \to (\mathbf{R}^+, 0)$.

THEOREM V (identity theorem). *Suppose that the monodromy transformation of polycycle of an analytic vector field on a two-dimensional surface has countably many fixed points. Then it is the identity.*

B. Reductions: A geometric lemma. The logical connections between Theorems I to V are shown in Figure 3. We strengthen the nonaccumulation Theorem IV by replacing the word "elementary" by the word "any." At the end of the subsection this new theorem will be derived from the identity theorem. In turn, it immediately yields Theorem III: a singular point is a one-point polycycle. Theorem III is singled out because its assertion has been widely used in the mathematical literature and folklore without reference to Dulac's memoir [10], the only source laying claim to a proof, though not containing one.

Theorem II also follows from the strengthened nonaccumulation theorem. Indeed, assume that it is false, and there exists an analytic vector field with countably many limit cycles on a closed two-dimensional surface.

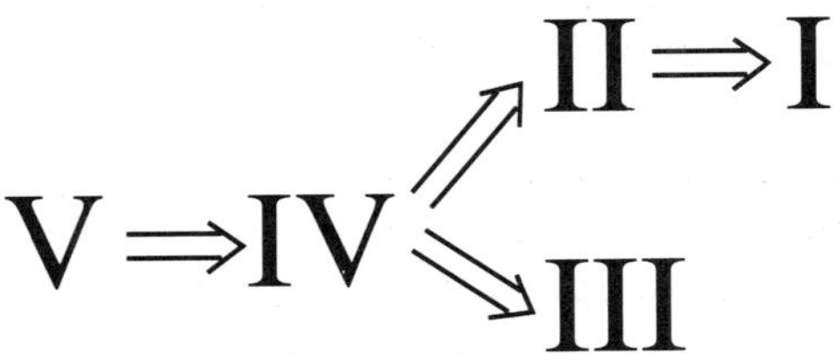

FIGURE 3

FIGURE 4

GEOMETRIC LEMMA. *Suppose that an analytic vector field on a closed two-dimensional surface has a sequence of closed phase curves. Then there exists a subsequence of this sequence accumulating on a closed phase curve or a polycycle* (*recall that a polycycle can degenerate into a point*).

REMARK. This lemma is true for smooth fields with singular points of finite multiplicity, and is false for arbitrary smooth fields: the limit set for such fields can be a "separatrix polygon with infinitely many sides" (Figure 4).

Everywhere below, "smoothness" means "infinite smoothness," and "diffeomorphism" means a C^{∞}-diffeomorphism.

PROOF. We prove the geometric lemma for the case when the surface in it is a sphere. Consider a disk on the sphere that does not intersect the countable set of curves in the given sequence. The stereographic projection with center at the center of the disk carries the original vector field into an analytic vector field on the plane that has a countable set of closed phase curves in some disk. It can be assumed without loss of generality that the number of singular points of the field in each disk is finite: otherwise the analytic functions giving the components of the field would have a common noninvertible factor, and by dividing both components by a suitable common analytic noninvertible factor we could make the resulting field have only finitely many singular points in each compact set. The closed phase curves of the original field remain phase curves of the new field.

Only finitely many curves in the sequence under consideration can be located pairwise outside each other: inside each of these curves is a singular point of the field, and the number of singular points is finite. Consequently, our sequence decomposes into finitely many subsets called nests: each curve of a nest bounds a domain with a countable set of curves of the same nest outside it or inside it. Take a sequence of curves of a nest; it is possible to take one point on each of them in such a way that the sequence of points converges. Then the chosen sequence of curves accumulates on a connected set γ. By the theorem on continuous dependence of the solutions on the initial conditions, this set consists of singular points of the equation and of phase curves. The considerations used in the Poincaré-Bendixson theorem enable us to prove that these curves go from some singular points to others if γ is not a cycle. Up to this point the argument has been for smooth vector fields. However, in the smooth case the set γ can contain countably many phase curves going out from a singular point and returning to it: a singular point of a smooth field can have countably many "petals" (Figure 4). This pathology is prevented by analyticity, as follows from the Bendixson-Dumortier theorem formulated below in §0.1C. This implies that γ is a closed phase curve or a polycycle.

The proof is analogous when the sphere is replaced by an arbitrary closed surface, but additional elementary topological considerations are needed, and we do not dwell on them. ▶

All the subsequent arguments are also given for the case when S is a sphere.

This concludes the derivation of Theorem II from the strengthened nonaccumulation theorem.

We derive this last theorem from the identity theorem. To do that is suffices to prove that a monodromy transformation is defined for the polycycle γ in the geometric lemma (the limit cycle corresponds to a fixed point of the monodromy transformation). For this, in turn, it is necessary to make more precise the definition given in §0.1A on an intuitive level.

DEFINITION 1. A semitransversal to a polycycle of a vector field on a surface S is defined to be a curve $\varphi\colon [0, 1) \to S$ satisfying the following conditions: the point $\varphi(0)$, called the vertex of the semitransversal, lies on the cycle; the curve φ is transversal to the field at all points except perhaps the vertex.

Using the word loosely, we also call the image of φ a semitransversal.

DEFINITION 2. A polycycle γ of a vector field is said to be monodromic if for an arbitrary neighborhood $\mathscr{U}$ and an arbitrary point P of the cycle there exist two semitransversals Γ and Γ' with vertex on γ, one belonging to the other, that have the following properties: the positive semitrajectory beginning at an arbitrary point q on Γ intersects Γ' at a positive time, with the first such point of intersection denoted by $\Delta(q)$; the arc of the semitrajectory with initial point q and endpoint $\Delta(q)$ lies in the neighborhood $\mathscr{U}$.

The germ of the mapping $\varphi^{-1} \circ \Delta \circ \varphi\colon (\mathbf{R}^+, 0) \to (\mathbf{R}^+, 0)$ is called the monodromy transformation of the cycle γ and denoted by Δ_γ.

We now take a transversal to the polycycle γ constructed in the geometric lemma; let γ_n be a sequence of closed phase curves accumulating on γ. One of the semitransversals of this transversal intersects countably many curves in this sequence. The curves γ_n and γ_{n+1} bound a domain homeomorphic to an annulus; let Γ_n be the intersection of this "annulus" with Γ. It can be assumed without loss of generality that there are no singular points inside this annulus, since there are only finitely many such points. Consequently, by the theorem on extension of phase curves, the monodromy transformation $\Gamma_n \to \Gamma_n$ is defined. This implies that the polycycle γ is monodromic.

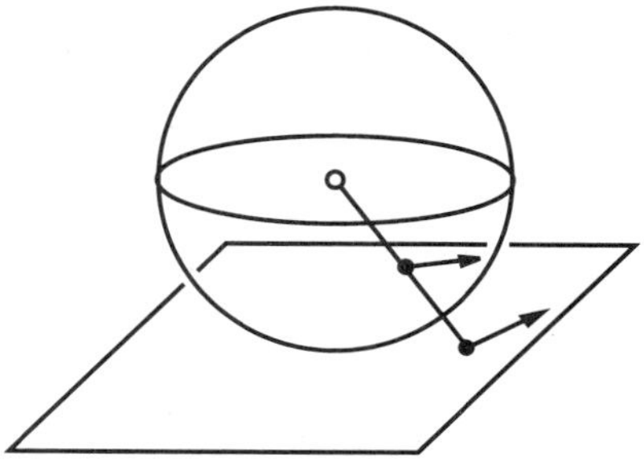

FIGURE 5

The strengthened nonaccumulation theorem thereby follows from the identity theorem.

Finally, Theorem I is a simple consequence of Theorem II. The reduction is carried out with the help of a well-known construction of Poincaré (Figure 5). Consider a sphere tangent to the plane at its South Pole, and a polynomial vector field on the plane. We project the sphere from the center onto this plane. Everywhere off the equator of the sphere there arises an analytic vector field that is "lifted from the plane" and tends to infinity on approaching the equator. Multiplying the constructed field by a suitable power of the analytic function "distance to the equator," we get a new field with finitely many singular points on the equator, and hence on the entire sphere. By Theorem II, it has finitely many limit cycles. Thus, the original field on the plane also has finitely many limit cycles. This proves Theorem I.

We remark that the identity theorem is obvious in the case when γ is a closed phase curve. In this case the monodromy transformation Δ_γ is the germ of an analytic mapping $(\mathbf{R}, 0) \to (\mathbf{R}, 0)$. If the limit cycles accumulate on γ, then the mapping Δ_γ has a countable set of fixed points that accumulate at an interior point of the domain of definition, which contradicts the uniqueness theorem for analytic functions.

The proof of the identity theorem in the general case goes according to the same scheme. The only difficulty is that isolated fixed points of the monodromy transformation accumulate not at an interior point, but at a boundary point of the domain of definition, and this is not forbidden for a biholomorphic mapping.

In the next subsection the strengthened nonaccumulation theorem is derived from Theorem IV.

C. The reduction: Resolution of singularities. We recall the definition of resolution of singularities (otherwise known as the σ-process or blowing-up), following Arnol′d [1]. Consider the natural mapping of the punctured real plane $\mathbf{R}^2\backslash\{0\}$ onto the projective line $\mathbf{R}P^1$: with each point of the punctured plane we associate the line joining this point to zero. The graph of this mapping is denoted by M; its closure $\overline{M}$ in the direct product $\mathbf{R}^2 \times \mathbf{R}P^1$

is diffeomorphic to the Möbius strip. The projection $\pi\colon \mathbf{R}^2\times\mathbf{R}P^1 \to \mathbf{R}^2$ along the second factor carries $\overline{M}$ into $\mathbf{R}^2$; the projective line $L=\mathbf{R}P^1$ (called a glued-in line below) is the complete inverse image of zero under this mapping; the projection $\pi\colon M\to\mathbf{R}^2\backslash\{0\}$ is a diffeomorphism.

The germ of an analytic vector field at an isolated singular point becomes the germ of an analytic field of directions with finitely many singular points on a glued-in line, as shown by the lemma stated below.

LEMMA (see, for example, [11], [2]). *To an analytic vector field* v *given in a neighborhood of the isolated singular point* 0 *in* $\mathbf{R}^2$ *there corresponds an analytic field of directions* α *defined in some neighborhood of a glued-in line* L *on the surface* $\overline{M}$ *everywhere except for finitely many points located on* L *and called singular points. Under the projection* $\pi\colon M\to\mathbf{R}^2\backslash\{0\}$ *the field* α *passes into the field of directions generated by the field* v. *In a neighborhood of each singular point the field* α *is generated by the analytic vector field* $\tilde{v}$.

The last assertion allows the σ-process to be continued by induction.

A singular point of the field of directions is elementary if the germ of the field at this point is generated by the germ of the vector field at an elementary singular point.

THE BENDIXSON-DUMORTIER THEOREM [5], [11], [29]. *By means of finitely many* σ*-processes a real-analytic vector field given in a neighborhood of a real-isolated singular point on the plane* $\mathbf{R}^2$ *can be carried into an analytic field of directions given in a neighborhood of a union of glued-in projective lines and having only finitely many singular points, each of them elementary and different from a focus or a center.*

The composition of σ-processes described in the Bendixson-Dumortier theorem is called a nice blowing-up. A nice blowing-up enables us to turn an arbitrary polycycle of an analytic vector field on the plane into an elementary polycycle with the same monodromy transformation. This gives a reduction of the strengthened nonaccumulation theorem to Theorem IV and concludes the proof of the chain of implications represented in Figure 3.

§0.2. The theorem and error of Dulac

In this section we present a proof of the main true result in Dulac's memoir [10] and point out the error in his proof of the finiteness theorem. The scheme of his argument lies at the basis of the proof of the identity theorem given below (see subsection D).

A. Semiregular mappings and the theorem of Dulac.

DEFINITION 1. A Dulac series is a formal series of the form

$$\Sigma = cx^{\nu_0}+\sum_1^\infty P_j(\ln x)x^{\nu_j},$$

where $c>0$, $0<\nu_0<\cdots<\nu_j<\cdots$, $\nu_j\to\infty$, and the P_j are polynomials.

DEFINITION 2. The germ of a mapping $f\colon (\mathbf{R}^+, 0) \to (\mathbf{R}^+, 0)$ is said to be semiregular if it can be expanded in an asymptotic Dulac series. In other words, for any N there exists a partial sum S of the above series such that $f(x) - S(x) = o(x^N)$.

REMARK. The concept of a semiregular mapping is invariant: semiregularity of a germ is preserved under a smooth change of coordinates in a full neighborhood of zero on the line. This follows from

LEMMA 1. *The germs of the semiregular mappings form a group.*

The lemma follows immediately from the definition. The main true result in [10] is

DULAC'S THEOREM. *A semitransversal to a monodromic polycycle of an analytic vector field can be chosen in such a way that the corresponding monodromy transformation is a flat, or vertical, or semiregular germ.*

The proof of this theorem is presented in subsections E to H.

B. The lemma of Dulac and a counterexample to it.

Dulac derived the finiteness Theorem I from the preceding theorem and a lemma.

LEMMA. *The germ of a semiregular mapping* $f\colon (\mathbf{R}^+, 0) \to (\mathbf{R}^+, 0)$ *is either the identity or has the isolated fixed point zero.*

This lemma is proved in §23 of [10] with the help of the following argument. The fixed point of the germ of f is found from the equation $f(x) = x$. If the principle term in the Dulac series for f is not the identity, then this equation has the isolated solution 0. If the principal term in the series is the identity but f itself is not the identity, then the equation $f(x) = x$ is equivalent to the equation

$$x^{\nu_1} P_1(\ln x) + o(x^{\nu_1}) = 0, \qquad (*)$$

where P_1 is a nonzero polynomial. This equation has the isolated root 0. Namely, dividing the equation by x^{ν_1}, we get an equation not having a solution in a sufficiently small neighborhood of zero. Indeed, the first term on the right-hand side of the new equation has a nonzero (perhaps infinite) limit as $x \to 0$, while the second term tends to zero. This concludes the proof of the lemma.

The lemma is false: a counterexample is supplied by the semiregular mapping $f\colon x \mapsto x + e^{-1/x} \sin \frac{1}{x}$, which has countably many fixed points accumulated at zero. The asymptotic Dulac series for f consists in the single term x. The error in the proof above amounts to the fact that the Dulac series for a semiregular mapping can be "trivial"—it may not contain terms other than x. Then the left-hand side of equation $(*)$ is equal to $o(x^{\nu_1})$, and we cannot investigate its zeros.

Actually, we proved here

THE CORRECTED LEMMA OF DULAC. *The germ of a semiregular mapping* $f: (\mathbf{R}^+, 0) \to (\mathbf{R}^+, 0)$ *has either a trivial* (*equal to* x) *Dulac series or an isolated fixed point at zero.*

The difficulty of the problem of finiteness is due to the fact that triviality of the Dulac series for a monodromy transformation does not imply that this transformation is the identity, as shown by the example in the next subsection.

C. Monodromy transformations with nonzero flat correction. It may seem likely that nonzero flat functions cannot arise in the theory of analytic differential equations. The example below destroys this illusion. The construction is carried out with the help of gluing, which is a powerful tool in the nonlocal theory of bifurcations and differential equations. As a result we obtain on a two-dimensional analytic surface an analytic vector field having a polycycle with two vertices—a separatrix lune—whose monodromy transformation has a nonzero flat correction [19].

This analytic surface is obtained by gluing together two planar domains with vector fields on them; we proceed to describe the latter.

In the rectangle $\mathscr{U} : |x| \le 1, |y| \le e^{-1}$ on the plane (e is the base of natural logarithms) consider a vector field giving a standard saddle node:

$$v(x, y) = x^2 \partial/\partial x - y \partial/\partial y$$

(Figure 6). In the same rectangle consider the field w obtained from v by symmetry with respect to the vertical axis and by time reversal:

$$w(x, y) = x^2 \partial/\partial x + y \partial/\partial y.$$

Take two copies of the rectangle $\mathscr{U} : \mathscr{U}_0 = \mathscr{U} \times \{0\}$ and $\mathscr{U}_1 = \mathscr{U} \times \{1\}$. We glue together points on two pairs of boundary segments of each of the rectangles (Figure 6; the arrows outside the rectangles indicate the gluing maps). Namely, we identify the points of the segments

$$\Gamma_0^+ = [-\tfrac{1}{2}, \tfrac{1}{2}] \times \{e^{-1}\} \times \{0\},$$
$$\Gamma_1^- = [-\tfrac{1}{2}, \tfrac{1}{2}] \times \{e^{-1}\} \times \{1\}$$

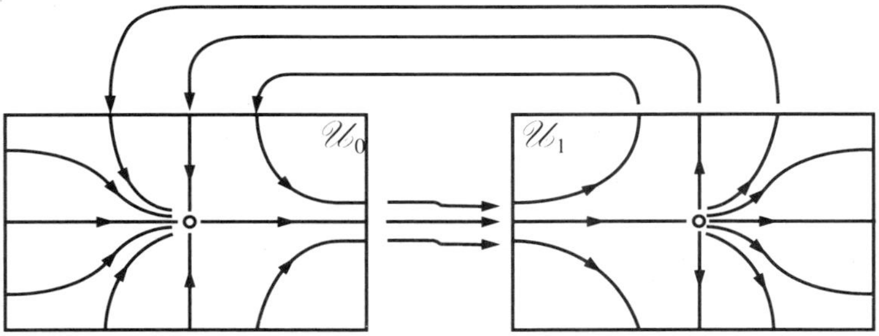

FIGURE 6

with the abscissae x and $-x$, and the points of the segments

$$\Gamma_0^- = 1 \times [-\tfrac{14}{4}, \tfrac{1}{4}] \times \{0\},$$
$$\Gamma_1^+ = \{-1\} \times [-\tfrac{5}{16}, \tfrac{3}{16}] \times \{1\}$$

with the ordinates y and $f(y) = y - y^2$, respectively. In the notation for the transversals $\Gamma_l^+, \Gamma_l^- \subset \mathscr{U}_l$ the plus sign indicates entry of the trajectories into the domain $\mathscr{U}_l$ across the transversal, while the minus sign indicates exit.

As a result of the gluings we get from the rectangles $\mathscr{U}_0$ and $\mathscr{U}_1$ a two-dimensional manifold homeomorphic to an annulus. By means of a well-known construction [19] we can introduce on it an analytic structure coinciding with the original structure on $\mathscr{U}_0$ and $\mathscr{U}_1$ and such that the vector field coinciding with v on $\mathscr{U}_0$ and with w on $\mathscr{U}_1$ is analytic on the whole surface. This vector field will have the separatrix lune γ obtained from segments of the coordinate axes in the gluings.

It is easy to compute the monodromy transformation for this polycycle γ. First we compute the correspondence mapping $\Delta\colon \Gamma_0^+ \to \Gamma_0^-$ along the trajectories of the field v where it is defined (Figure 6). The function $ye^{-1/x}$ is a first integral of v. Consequently, for $x > 0$,

$$e^{-1} \cdot e^{-1/x} = \Delta(x) \cdot e^{-1}, \qquad \Delta(x) = e^{-1/x}.$$

Similarly, the germ of the correspondence mapping at the point $y = 0$ of the semitransversal $\Gamma_1^+ \cap \{y > 0\}$ onto Γ_1^- is equal to

$$-\Delta^{-1}(y) = 1/\ln y.$$

By construction of the gluing mappings, the germ Δ_γ has the form

$$\Delta_\gamma = \Delta^{-1} \circ f \circ \Delta, \qquad f(y) = y - y^2$$

and acts according to the formula

$$x \overset{\Delta}{\mapsto} e^{-1/x} \overset{f}{\mapsto} e^{-1/x}(1 - e^{-1/x}) \overset{\ln}{\mapsto} \left(-\frac{1}{x} + \ln(1 - e^{1/x})\right)$$
$$\overset{-1/y}{\mapsto} x[1 - x\ln(1 - e^{-1/x})]^{-1}.$$

This mapping has a nonzero flat correction at zero, which is what was required. Thus, Dulac series do not suffice for describing the asymptotic behavior of monodromy transformations.

D. The scheme for proving the identity theorem. For the proof we construct a set of germs mapping $(\mathbf{R}^+, \infty) \to (\mathbf{R}^+, \infty)$ that contains the monodromy transformations of polycycles of analytic vector fields written in a suitable map, but is broader. The germs in this new set have two properties: they can be expanded and they can be extended.

That they can be expanded means that to each germ corresponds an asymptotic series containing information not only about power asymptotics but also about exponential asymptotics. Such series are called STAR—superexact asymptotic series; see §0.5 for more details. The triviality of such a series, that is, the condition that it equals the identity, means that the correction of this series is very rapidly decreasing as $x \to \infty$. In turn, the terms of the series do not oscillate, and hence the existence of a countable number of fixed points for the germ implies that the corresponding series is trivial.

That they can be extended means that a germ can be extended into the complex domain to be a map-cochain—a piecewise continuous mapping holomorphic off the lines of discontinuity. The Phragmén-Lindelöf theorem holds for the map-cochains arising upon extension of a monodromy transformation: if the correction of a germ decreases too rapidly, then it is identically equal to zero. The triviality of the STAR ensures precisely a "too rapid" decrease of the correction.

The following implication is obtained (Δ_γ is the monodromy transformation of the polycycle γ, $\widehat{\Delta}_\gamma$ is the STAR for Δ_γ, and Fix_∞ is the set of germs with countably many fixed points):

$$\Delta_\gamma \in \mathrm{Fix}_\infty \Rightarrow \widehat{\Delta}_\gamma = x \Rightarrow \Delta_\gamma - x \equiv 0\,.$$

If all the singular points on an elementary polycycle are hyperbolic saddles, then it is said to be hyperbolic, otherwise it is said to be nonhyperbolic. In the hyperbolic case the program presented was carried out in [19] with the use of ordinary and not superexact asymptotic series; see §0.3 below. In the general case the geometric theory of normal forms of resonant fields and mappings is used to describe the monodromy transformation (§0.4).

We return to the presentation of the proof of Dulac's theorem.

E. The classification theorem. Dulac's theorem describes the power asymptotics of a monodromy transformation. To get these asymptotic expressions it suffices to use the theory of smooth, and not analytic, normal forms.

DEFINITION 3. Two vector fields are smoothly (analytically) orbitally equivalent in a neighborhood of the singular point 0 if there exists a diffeomorphism (an analytic diffeomorphism) carrying one neighborhood of zero into another that leaves 0 fixed and carries phase curves of one field into phase curves of the other (perhaps reversing the direction of motion along the phase curves).

REMARK. In the definition of orbital equivalence it is usually required that the directions of motion along phase curves be preserved; to simplify the table below we do not require this.

THEOREM. *An analytic vector field in some neighborhood of an isolated elementary singular point on the real plane is smoothly orbitally equivalent to one of the vector fields in the table.*

Here the numbers k, m, *and* n *are positive integers,* a *is a real number,* $\underline{x} \in \mathbf{R}^2$, $\underline{x} = (x, y)$, $r^2 = x^2 + y^2$, I *is the operator of rotation through the angle* $\pi/2$, *the fraction* m/n *is irreducible, and* $\varepsilon \in \{0, 1, -1\}$.

Type of singular point	Normal form
1. A field with nonresonant linear part $v(\underline{x}) = \Lambda x + \cdots$	$w(\underline{x}) = \Lambda \underline{x}$
2. A center with respect to the linear terms	$w(\underline{x}) = I\underline{x} + \varepsilon(r^{2k} + ar^{4k})x$
3. A resonant node	$w(x, y) = (kx + \varepsilon y^k)\frac{\partial}{\partial x} + y\frac{\partial}{\partial y}$
4. A resonant saddle with the eigenvalue ratio $-\lambda = \lambda_2/\lambda_1 = -m/n$	$w(x, y) = x[1 + \varepsilon(u^k + au^{2k})] \cdot\frac{\partial}{\partial x} - \lambda y\frac{\partial}{\partial y}$ $u = x^m y^n$ the resonant monomial
5. A degenerate elementary singular point	$w(x, y) = x^{k+1}(\pm 1 + ax^k)^{-1}\frac{\partial}{\partial x} - y\frac{\partial}{\partial y}$

A closely related assertion was formulated as a conjecture by Bryuno [7]. A large part of the list given above is contained in the article [6] of Bogdanov. The classification theorem in its present form was formulated in [18]; fragments of the proof are contained in [6], [8], and [34]; a complete proof is given in [21].

COROLLARY (topological classification of elementary singular points). *An elementary singular point of an analytic vector field is one of five topological types: a saddle, a node, a focus, a center, and a saddle node.*

PROOF. Points of types 1, 3, or 4 in the table are among the saddles, nodes, or foci and are identified according to the linear part. For $\varepsilon = 1$ the points of type 2 are unstable foci; for $\varepsilon = -1$ they are stable; for $\varepsilon = 0$ they are centers. Degenerate elementary singular points reduce to normal forms with separating variables; depending on the sign $+$ or $-$ in front of the 1 in parentheses and the parity of k, these can be saddles, nodes, or saddle nodes (Figure 7, next page).

REMARK. This corollary was known as far back as Bendixson [5]. Another proof of it is a derivation from the reduction principle by Shoshitaĭshvili [32], [21].

As noted by Bogdanov, the normal forms given by the classification theorem can be integrated in elementary functions. The proof of Dulac's theorem is based on this remark.

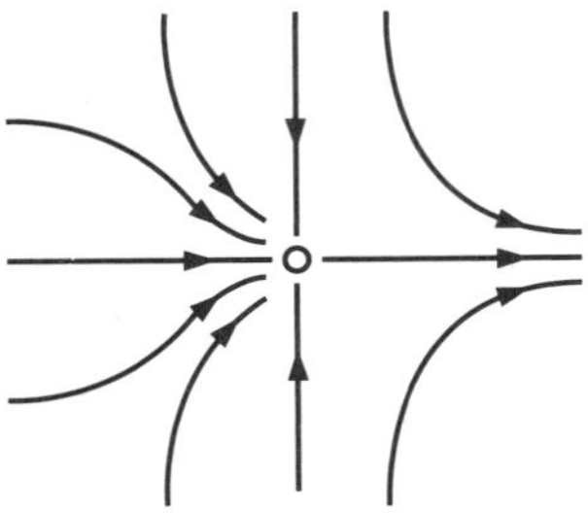

FIGURE 7

F. The scheme for proving Dulac's theorem, and the correspondence mappings. For a one-point elementary monodromic polycycle Dulac's theorem is trivial: the corresponding singular point is a focus or a center. Its monodromy transformation extends analytically to a full neighborhood of zero on the line, and hence can be expanded in a convergent (and not just asymptotic) Taylor series (a special case of a Dulac series). Everywhere below we consider an elementary polycycle with more than one point.

Note that if an elementary polycycle with more than one point is monodromic, then all the singular points on it have the topological type of a saddle or saddle node.

This follows immediately from the preceding corollary.

The monodromy transformation of an elementary polycycle can be decomposed in a composition of correspondence mappings for hyperbolic sectors of elementary singular points (Figures 8 and 9). A hyperbolic sector is represented in Figure 9; the correspondence mapping carries a semitransversal across which phase curves enter the sector into a semitransversal across which they leave the sector; the image and inverse image belong to a single phase curve. The inverse mapping is also called a correspondence mapping.

The first part of the proof of Dulac's theorem consists in a computation of the correspondence mappings for hyperbolic sectors of saddles and saddle nodes (rows 1, 4, and 5 of the table in subsection E). The second part is

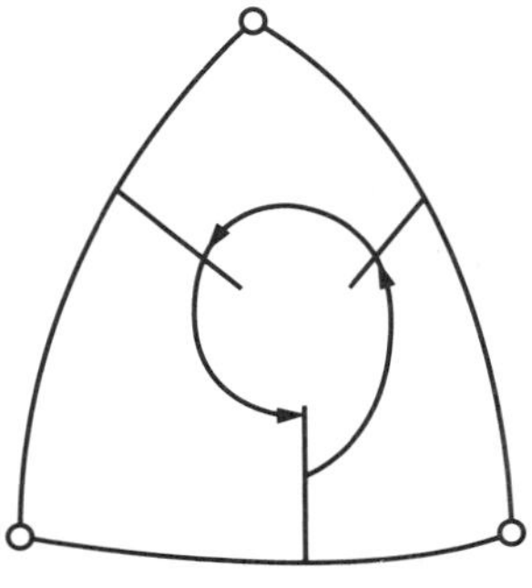

FIGURE 8

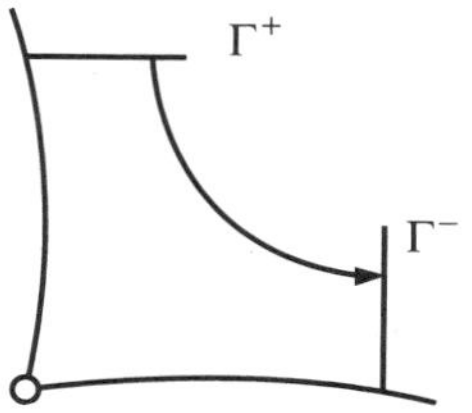

FIGURE 9

an investigation of compositions of these mappings with smooth changes of coordinates and with each other.

LEMMA 2. *A correspondence mapping for a hyperbolic sector of a nondegenerate saddle of a smooth vector field is semiregular.*

REMARK. It is natural to prove the lemma for smooth vector fields, and to use it for analytic vector fields.

PROOF. This lemma was proved for analytic vector fields in the first part of the memoir [10]. The proof below uses the classification theorem in subsection E.

By Lemma 1 in subsection A, the germ of a mapping smoothly equivalent to a semiregular mapping is also semiregular; therefore, it suffices to prove Lemma 2 for fields written in normal form (rows 1 and 4 of the table in E).

Suppose that the saddle under consideration is nonresonant: the eigenvalue ratio is irrational. Then the corresponding normal form has the form

$$w(x, y) = x\frac{\partial}{\partial x} - \lambda y\frac{\partial}{\partial y}, \qquad \lambda > 0$$

(the normal form in the first row of the table, multiplied by λ_1^{-1}, where $\Lambda = \mathrm{diag}(\lambda_1, \lambda_2))$. In this and the next subsections Γ^- is the semitransversal $\{x = 1, y \in [0, 1]\}$ with chart y, and Γ^+ is the semitransversal $\{y = 1, x \in [0, 1]\}$ with chart x. The correspondence mapping $\Gamma^+ \to \Gamma^-$ can be computed in an elementary way and has the form

$$x \mapsto y = \Delta(x) = x^\lambda;$$

this mapping is semiregular. Lemma 2 is proved in the nonresonant case.

Suppose now that the saddle is resonant, $\lambda = m/n$. The corresponding normal form has a form equivalent to formula 4 of the table in subsection E:

$$w(x, y) = x\frac{\partial}{\partial x} - y(\lambda + \tilde{\varepsilon}(u^k + \tilde{a}u^{2k}))\frac{\partial}{\partial y},$$

where $u = x^m y^n$ is the resonant monomial, $\tilde{\varepsilon} \in \{0, 1, -1\}$, and $\tilde{a} \in \mathbf{R}$. The case $\varepsilon = 0$ is treated as above. Let $\varepsilon \neq 0$. The proof of the lemma is based on the fact that the equation $\dot{\underline{x}} = w(\underline{x})$ can be integrated. The integration is carried out as follows. The factor system is written with respect to the resonant monomial:

$$\dot{u} = f(u).$$

Here $\dot{u} \equiv L_w u$ and $f = -\tilde{\varepsilon} n u^{k+1}(1 + \tilde{a} u^k)$. The factor system can be integrated in quadratures, and the same is true for the equation $\dot{x} = x$. Then y is found as a function of t.

To compute the correspondence mapping for the field w it is not necessary to carry these computations to completion. Denote by g_v^t the transformation of the phase flow of the vector field v over the time t (where it is defined). Let

$$\xi = (x, 1) \in \Gamma^+, \qquad \eta = (1, \Delta(x)) \in \Gamma^-.$$

The phase curve of the field w passes from the point ξ to the point η in the time $t = -\ln x$. Therefore

$$u(\eta) = g_f^{-\ln x} u(\xi),$$

where f is the right-hand side of the factor system with respect to u. But

$$u(\xi) = x^m, \qquad u(\eta) = (\Delta(x))^n.$$

Finally,

$$\Delta(y) = [g_f^{-\ln x}(x^m)]^{1/n}.$$

The last formula gives a semiregular mapping. We prove this first for $m = n = 1$. The local phase flow of the field $f(\partial/\partial u)$ at the point $(0, 0)$ in (t, u)-space is given by the germ of an analytic function of two variables; denote it by F. We extend the germ F into the complex domain and prove that for sufficiently small δ the Taylor series for F converges on the curve $L : t = -\ln u$, $(t, u) \in \mathbf{R}$, $u \in (0, \delta)$. For this we consider the equation $\dot{u} = f(u)$ with the complex phase space $\{u\}$ and with complex time. The solution φ of this equation with the initial condition $\varphi(0) = u$ is holomorphic in a disk with radius of order $|u|^{-k}$. For $a = 0$ $(f = -\varepsilon n u^{k+1})$ this follows from the explicit formula $\varphi(t) = u(1 + \varepsilon n k t u^k)^{-1/k}$ for the solution; for $a \neq 0$ it can be proved by simple estimates. Hence, the Taylor series for F converges in the domain $|t| \leq A|u|^{-k}$, where A is some positive constant. This domain contains the curve L for sufficiently small δ. Consequently, the mapping $u \mapsto F(-\ln u, u)$ is semiregular. This shows that the mapping $y \mapsto \Delta(y)$ is semiregular for $m = n = 1$.

The semiregularity of the mapping $y \mapsto \Delta(y)$ for arbitrary m and n follows from Lemma 1.

Lemma 2 is proved.

G. The correspondence mapping for a hyperbolic sector of a saddle node or degenerate saddle. A hyperbolic sector of a degenerate elementary singular point includes in its boundary part of the stable (or unstable) and center manifolds. We recall that in this case the stable manifold is a smooth invariant curve of the field that passes through the singular point and is tangent there to an eigenvector of the linear part with nonzero eigenvalue. A center manifold is an analogous curve tangent at the singular point to the kernel of the linear part.

DEFINITION 4. For a hyperbolic sector of a degenerate elementary singular point a correspondence mapping whose image is a semitransversal to a center manifold is called the mapping TO the center manifold for brevity; its inverse is the mapping FROM the center manifold.

EXAMPLE. For a suitable choice of semitransversal the correspondence mapping of the hyperbolic sector of the standard saddle node $x^2(\partial/\partial x) - y(\partial/\partial y)$ has the form

$f_0(x) = e^{-1/x}$, TO the center manifold,

$f_0^{-1}(x) = -1/\ln x$, FROM the center manifold;

see C.

DEFINITION 5. The germ of the mapping $f\colon (\mathbf{R}^+, 0) \to (\mathbf{R}^+, 0)$ is said to be a flat semiregular germ if the composition $h = f_0^{-1} \circ f$ is semiregular.

REMARK. Below in subsection H it is proved that a germ smoothly equivalent to a flat semiregular germ is itself a flat semiregular germ (corollary to Lemma 4 in H).

LEMMA 3. *The germ of a mapping TO a center manifold for an analytic field is a flat semiregular germ.*

PROOF. By the preceding remark, it suffices to prove the lemma for the corresponding smooth orbital normal form w; see row 5 in the table in E. In this case the correspondence mapping is said to be standard and denoted by Δ_{st}. It can be assumed without loss of generality that the quadrant $x \geq 0$, $y \geq 0$ contains a hyperbolic sector of the field w. Therefore, the sign $+$ should be chosen in the indicated formula for w:

$$w(x, y) = x^{k+1}(1 + ax^k)^{-1}\frac{\partial}{\partial x} - y\frac{\partial}{\partial y}.$$

The corresponding differential equation has separating variables. It can be integrated, and the correspondence mapping can be computed explicitly. Namely, let the semitransversals Γ^+ and Γ^- be the same as in subsection F. Let $\Delta_{\mathrm{st}}\colon (\Gamma^+, 0) \to (\Gamma^-, 0)$ be the germ of the mapping TO the center manifold for the field w. Then the phase curve of the field falls from the point $(x, 1)$ to the point $(1, \Delta_{\mathrm{st}}(x))$ in the time

$$t = -\ln \Delta_{\mathrm{st}}(x) = \int_x^1 \frac{1 + a\xi^k}{\xi^{k+1}}\, d\xi.$$

An elementary computation yields:

$$\int_x^1 \frac{1 + a\xi^k}{\xi^{k+1}}\, d\xi = \frac{1}{h_{k,a}(x)} - \frac{1}{k},$$

where $h_{k,a}(x) = kx^k/(1 - akx^k \ln x)$ is a semiregular mapping. Consequently,

$$\Delta_{\mathrm{st}} = C\exp(-1/h_{k,a}), \qquad C = \exp 1/k,$$

and Δ_{st} is a flat semiregular mapping.

H. Conclusion of the proof of Dulac's theorem. By results in subsections E to G, it remains to prove that a composition of semiregular germs, flat semiregular germs, and inverses of them (perhaps after a cyclic permutation of the germs that corresponds to a proper choice of semitransversal) is a flat, vertical, or semiregular germ. Recall that the germs of semiregular mappings form a group; see Lemma 1 in A.

LEMMA 4. *Suppose that* f_1 *and* f_2 *are two flat semiregular germs, and* h *is a semiregular germ* $(\mathbf{R}^+, 0) \to (\mathbf{R}^+, 0)$. *Then the composition* $f = f_2^{-1} \circ h \circ f_1$ *is semiregular, and the asymptotic series for* f *depends on the principal term in the asymptotic expansion for* h *and does not depend on the remaining terms of this expansion.*

It can be said that the composition f *"forgets" all the terms of the asymptotic series for* h *except for the principal term.*

PROOF OF THE LEMMA. Suppose that $f_1 = f_0 \circ h_1$ and $f_2 = f_0 \circ h_2$, where f_0 is the standard flat mapping $x \mapsto \exp(-1/x)$, and the germs h_1 and h_2 are semiregular. Then

$$f = h_2^{-1} \circ f_0^{-1} \circ h \circ f_0 \circ h_1 .$$

By virtue of Lemma 1, it suffices to prove that the composition $f_0^{-1} \circ h \circ f_0$ is semiregular. Let

$$h(x) = cx^{\nu_0}(1 + \tilde{h}(x)),$$
$$\tilde{h}(x) = O(x^{\varepsilon}), \qquad \varepsilon > 0, \ c > 0 .$$

Then

$$\begin{aligned} f_0^{-1} \circ h \circ f_0(x) &= \left[\frac{\nu_0}{x} - \ln c + \ln(1 + \tilde{h} \circ f_0)\right]^{-1} \\ &= \frac{x}{\nu_0 - x \ln c + x \ln(1 + \tilde{h} \circ f_0)} \\ &= \frac{x}{\nu_0 - x \ln c} + \cdots , \end{aligned}$$

where the dots stand for an exponentially decreasing (as $x \to 0$) component.

Consequently, the mapping $f_0^{-1} \circ h \circ f_0$ is semiregular. Lemma 4 is proved.

COROLLARY. *A germ smoothly equivalent to a flat semiregular germ is itself a flat semiregular germ.*

PROOF. Let $f_0 \circ h$ be a flat semiregular germ, and f a germ smoothly equivalent to it. Here we have in mind so-called RL-equivalence: the substitutions in the image and the inverse image can be different. In other words, there exist germs of diffeomorphisms h_1 and h_2 such that

$$f = h_1 \circ f_0 \circ h \circ h_2 .$$

It must be proved that the composition $f_0^{-1} \circ f$ is semiregular. This follows immediately from Lemma 4 above. ►

We proceed to the proof of Dulac's theorem. Let γ be an elementary polycycle. The monodromy transformation Δ_γ can be decomposed into a composition of the correspondence mappings Δ_j described in subsections F and G:

$$\Delta_\gamma = \Delta_N \circ \cdots \circ \Delta_1 . \qquad (*)$$

DEFINITION 6. An elementary polycycle is said to be balanced if in $(*)$ the number of the mappings FROM the center manifold is equal to the number of mappings TO the center manifold. Otherwise the cycle is said to be unbalanced.

Useful for describing the composition $(*)$ is the function χ called the characteristic of this composition and defined on $[-N, 0]$ as follows. The function χ is continuous and linear on the closed interval between two adjacent integers, and $\chi(0) = 0$. If the mapping Δ_j in $(*)$ corresponds to a nondegenerate singular point, then let $\chi(-j) = \chi(-j+1)$. If Δ_j is the mapping FROM the center manifold, then $\chi(-j) = \chi(-j+1)+1$. If Δ_j is the mapping TO the center manifold, then $\chi(-j) = \chi(-j+1)-1$. Obviously, a cycle γ is balanced if and only if $\chi(0) = \chi(-N) = 0$. The characteristic is determined up to an additive constant and a "cyclic shift of the argument": $j \to j+k \pmod N$, both of which depend on the choice of the semitransversal.

DEFINITION 7. A semitransversal to a polycycle is said to be properly chosen if the cycle characteristic defined with the help of the decomposition $(*)$ for the corresponding monodromy transformation is nonpositive.

LEMMA 5. *For a suitable choice of semitransversal the monodromy transformation for a balanced polycycle is semiregular, while for an unbalanced polycycle it is flat or vertical.*

PROOF. In the composition $(*)$ we put a left parenthesis before each mapping TO the center manifold and a right parenthesis after each mapping FROM the center manifold (Figure 10). If the cycle is balanced and its characteristic is nonpositive, then the parentheses turn out to be placed correctly: the number of left parentheses is equal to the number of right parentheses; the first parenthesis is a left one, and the last is a right one. In this case all the flat and vertical mappings fall in parentheses. Inside all the parentheses the products are semiregular in view of Lemma 4 (this is demonstrated intuitively in Figure 10; the obvious general argument is omitted). Lemma 5

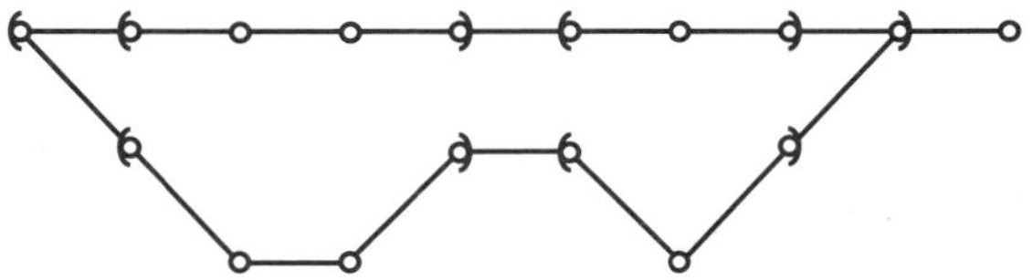

FIGURE 10

now follows from Lemma 1 for balanced cycles. The proof is analogous for unbalanced cycles. ►

Dulac's theorem follows immediately from Lemma 5.

The classification theorem, and with it Dulac's theorem, admits a generalization to the smooth case: it is necessary only to require that the vector field satisfy at all singular points a Łojasiewicz condition—upon approach of the singular point the modulus of the vector of the field decreases no more rapidly than some power of the distance to the singular point. The finiteness theorem is false, of course, for such fields.

To prove the finiteness theorem it turns out to be necessary to go out into the complex domain. In the next section the nonaccumulation theorem (the Theorem IV in §0.1A) is proved for a polycycle with hyperbolic singular points. Here essential use is made of the corrected lemma of Dulac (subsection B)—the strongest of the results in [10].

§0.3. Finiteness theorems for polycycles with hyperbolic vertices

In this section we introduce the class of almost regular germs, which contains the correspondence mappings of hyperbolic saddles in the analytic case, and we prove Theorems I, II, and IV for vector fields with nondegenerate singular points. Beginning with this section, all the vector fields under consideration are analytic; explicit mention of analyticity is often omitted. The presentation follows [21] and [24].

A. Almost regular mappings.

THEOREM IV BIS. *The limit cycles of an analytic vector field with nondegenerate singular points cannot accumulate on a polycycle of this field.*

Theorems I and II for fields with nondegenerate singular points (including singular points at infinity in Theorem I) can be derived from this as in §0.1.

A chart x that is nonnegative on a semitransversal, equal to zero at the vertex, and can be extended analytically to a full neighborhood of the vertex on the transversal containing the semitransversal, is said to be a natural chart. The chart $\xi = -\ln x$ is called a logarithmic chart.

It is convenient to write the correspondence mapping for a hyperbolic sector of an elementary (not necessarily hyperbolic) singular point in a logarithmic chart. In a natural chart this is the germ of the mapping $\Delta: (\mathbf{R}^+, 0) \to (\mathbf{R}^+, 0)$, and in a logarithmic chart it is the germ of $\widetilde{\Delta}: (\mathbf{R}^+, \infty) \to (\mathbf{R}^+, \infty)$; passage to the logarithmic chart is denoted by a tilde.

In the hyperbolic case the correspondence mapping, written in the logarithmic chart, extends to special domains similar to a half-plane and called quadratic domains.

DEFINITION 1. A quadratic standard domain is an arbitrary domain of the form

$$\varphi(\mathbf{C}^+ \backslash K), \qquad \varphi = \zeta + C(\zeta+1)^{1/2}, \ C > 0, \ K = \{|\zeta| \le R\}.$$

DEFINITION 2. A Dulac exponential series is a formal series of the form

$$\Sigma = \nu_0\zeta + c + \sum P_j(\zeta)\exp\nu_j\zeta,$$

where $\nu_0 > 0$, $0 > \nu_j \searrow -\infty$, and the P_j are polynomials; the arrow $\searrow$ means monotonically decreasing convergence.

COROLLARY 3. *An almost regular mapping is a holomorphic mapping of some quadratic standard domain* Ω *in* $\mathbf{C}$ *that is real on* $\mathbf{R}^+$ *and can be expanded in this domain as an asymptotic real Dulac exponential series. Expandability means that for any* $\nu > 0$ *there exists a partial sum approximating the mapping to within* $o(\exp(-\nu\xi))$ *in* Ω.

THEOREM 1. *An almost regular mapping is uniquely determined by an asymptotic series of it. In particular, an almost regular mapping with asymptotic series* ζ *is the identity.*

REMARK. This theorem explains the term "almost regular." At the beginning of the century "regularity" was often used as a synonym for "analyticity." Apparently, Dulac called the mappings he introduced "semiregular" because of their similarity to regular mappings: Dulac series are similar to Taylor series. However, semiregular mappings, in contrast to almost regular mappings, are not determined by their asymptotic series. On the other hand, Dulac exponential series for almost regular mappings diverge in general. "Almost regular mappings" are thus more regular than "semiregular mappings," but are still not exactly regular.

PROOF. The difference between two semiregular mappings with a common Dulac series is a holomorphic function f defined and bounded in some standard domain and decreasing on $(\mathbf{R}^+, \infty)$ more rapidly than any exponential ($\mathbf{R}^+$ is the positive semi-axis). By a theorem of Phragmén-Lindelöf type, this function is identically equal to zero. We prove this: the theorem is obtained immediately from it.

Let $\mathbf{C}^+$ be the right half-plane $\operatorname{Re}\zeta > 0$.

The following result is known to specialists and will be proved in §3.1C.

THEOREM 2. *If a function* g *is holomorphic and bounded in the right half-plane and decreases on* $(\mathbf{R}^+, \infty)$ *more rapidly than any exponential* $\exp(-\nu\xi)$, $\nu > 0$, *then* $g \equiv 0$.

If instead of a quadratic standard domain the function f were holomorphic in the right half-plane, then Theorem 1 would follow at once from Theorem 2. To exploit Theorem 2 we note that there exists a conformal mapping $\psi: \mathbf{C}^+ \to \Omega$ with the form $\psi(\xi) = \xi + O(\xi^{1/2})$ on $(\mathbf{R}^+, \infty)$. Consequently, the function $g = f \circ \psi$ satisfies the conditions of Theorem 2. From this, $g \equiv f \equiv 0$. ▶

B. Going out into the complex plane, and the proof of Theorem IV bis.

THEOREM 3. *The correspondence mapping of a hyperbolic saddle, written in a logarithmic chart, extends to a semiregular mapping in some quadratic domain.*

This theorem will be proved in subsections C and D. We derive Theorem IV bis from it.

DEFINITION 4. Two almost regular mappings are equivalent if they coincide in some quadratic standard domain. An equivalence class of such mappings is called an almost regular germ.

It follows from the definition of almost regular germs that these germs form a group with the operation of "composition." Therefore, it follows from Theorem 3 that the monodromy transformation Δ_γ of a polycycle with hyperbolic vertices, written in the logarithmic chart, extends to an almost regular germ. We now prove the chain of implications in §2D. Let $\Delta_\gamma \in \mathrm{Fix}_\infty$. Then by the corrected lemma of Dulac in §2B, the corresponding Dulac series is equal to id: $\widehat{\Delta}_\gamma = \mathrm{id}$. This implies that the mapping $\widetilde{\Delta}_\gamma$—the monodromy transformation Δ_γ written in the logarithmic chart—expands in an asymptotic Dulac exponential series equal to id. It follows from Theorem 1 and the almost regularity of $\widetilde{\Delta}_\gamma$ that $\widetilde{\Delta}_\gamma = \mathrm{id}$, which proves Theorem IV bis.

C. Hyperbolicity and almost regularity. Here we prove Theorem 3 in subsection B. By the definition of an almost regular germ, it must be proved that the mapping under investigation, written in the logarithmic chart (it is denoted by $\widetilde{\Delta}$): (a) extends biholomorphically to some quadratic standard domain Ω (regularity); (b) can be expanded in an asymptotic Dulac series in this domain (expandability).

The proof is broken up into four steps.

STEP 1. GEOMETRY AND ANALYTIC EXTENSION. Let us begin with an example. Consider the correspondence mapping of the linear saddle given by the field $v = z(\partial/\partial z) - \lambda w(\partial/\partial w)$ in $\mathbf{C}^2$. Let Γ^- and Γ^+ be the intervals $[0, 1] \times \{1\}$ and $\{1\} \times [0, 1]$. The correspondence mapping $\Delta\colon \Gamma^+ \to \Gamma^-$ has the form $z \mapsto z^\lambda$ in suitable natural coordinates. This mapping extends to the Riemann surface of the logarithm over the punctured disk (delete the center $z = 0$) on the line $w = 1$. On the real plane the image and inverse image of the mapping Δ are joined by phase curves of the field v. Which lines on the complex phase curves of v join the image and inverse image of the extended correspondence mapping?

The construction answering this question is easily analyzed in the linear case and is used in the general case.

Let (Figure 11)

$$B = \{|z| \le 1\} \times \{|w| \le 1\},$$
$$\mathscr{D}_0 = \{0 < |z| \le 1\} \times \{0\}, \quad \mathscr{D}_1 = \{0 < z \le 1\} \times \{1\}.$$

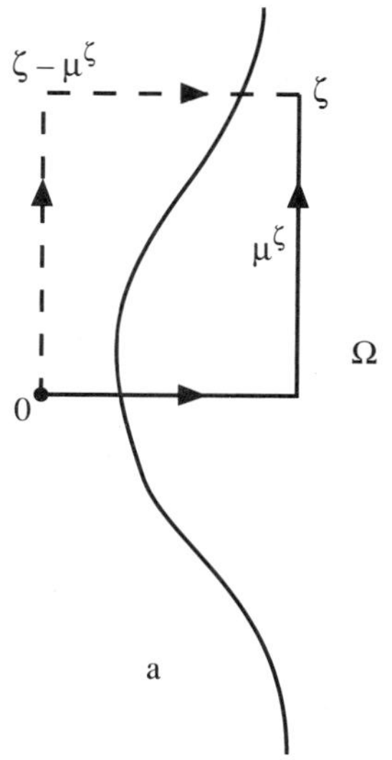

FIGURE 11(a)

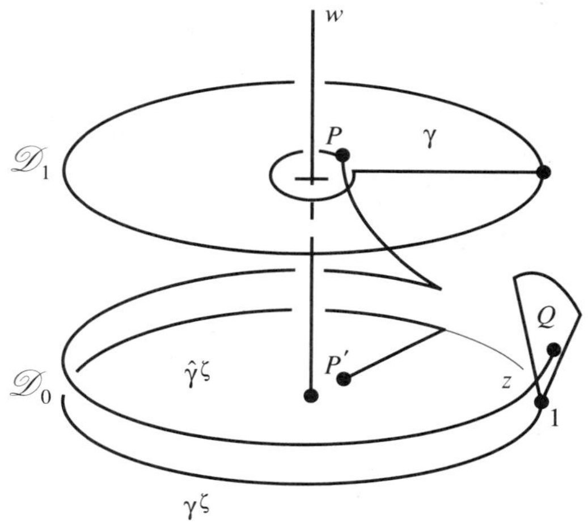

FIGURE 11(b)

For each $\zeta \in \mathbf{C}^+$, $\zeta = \xi + i\eta$, denote by μ^ζ (ζ is an index, not a power) the curve with initial point 0 and endpoint ζ consisting of the two intervals $[0, \xi]$ and $[\xi, \zeta]$, parametrized by the arclength s, and let $S = s(\zeta) = |\xi| + |\eta|$. The point corresponding to the parameter s is denoted by $\mu^\zeta(s)$; this defines a mapping $\mu^\zeta \colon [0, S] \to \mathbf{C}$, $s \mapsto \mu^\zeta(s)$.

We define the following curves (Figure 11):

$$\gamma = \gamma^{\zeta, 1} \colon [0, S] \to \mathscr{D}_1, \qquad s \mapsto (\exp(-\mu^\zeta(s)), 1),$$
$$\gamma_0 = \gamma^\zeta \colon [0, S] \to \mathscr{D}_0, \qquad s \mapsto (\exp(\mu^\zeta(s) - \zeta), 0).$$

Let $\gamma_T^\zeta = \gamma^\zeta|_{[0,T]}$. Denote by φ_p the complex phase curve of v passing through the point p. On the Riemann surface φ_p let $\hat{\gamma}^\zeta$ be an arc with initial point p that covers the curve γ^ζ under the projection $\pi_z : (z, w) \mapsto z$.

We prove that the endpoint of the arc $\hat{\gamma}^\zeta$ is the result of analytic extension of the correspondence mapping Δ (see the beginning of Step 1) along the curve $\gamma^{\zeta,1}$ with initial value $\Delta(1) = 1$.

We first prove that the covering $\hat{\gamma}^\zeta$ is defined and belongs to B. Indeed, the curve γ^ζ consists of two parts: a segment of a radius of the disk $\mathscr{D}_0$, and an arc of the corresponding circle. For $p = \gamma^{\zeta,1}(S) = (\exp(-\zeta), 1)$ the Riemann surface φ_p belongs to the real hypersurface $|w||z|^\lambda = \exp(-\lambda\xi)$, where $\xi = \operatorname{Re}\zeta$. Along the arc of the curve $\hat{\gamma}^\zeta$ lying over the radius, $|z|$ is increasing and $|w|$ is decreasing. The arc of $\hat{\gamma}^\zeta$ lying over the arc of the circle $|z| = 1$ belongs to the torus $|z| = 1$, $|w| = \exp(-\lambda\xi)$. Therefore, the whole curve $\hat{\gamma}^\zeta$ belongs to B.

Further, the endpoint of $\hat{\gamma}^\zeta$ depends continuously on ζ by construction. It depends on ζ analytically according to the theorem on analytic dependence of a solution on the initial conditions. Consequently, it is the result of the analytic extension under investigation.

We proceed to an investigation of the general case.

STEP 2. NORMALIZATION OF JETS ON A COORDINATE CROSS. It is proved in the memoir [10] that for any positive integer N a vector field is orbitally analytically equivalent in a neighborhood of a hyperbolic singular point to a field giving the equation

$$\dot{z} = z, \qquad \dot{w} = -w(\lambda + z^N w^N f(z, w))$$

for an irrational ratio $-\lambda$ of the eigenvalues of the singular point, and giving the equation

$$\dot{z} = z, \qquad \dot{w} = -w(\lambda + P(u) + u^{N+1} f(z, w)) \tag{$*$}$$

for $\lambda = m/n$. Here m and n are positive integers, m/n is an irreducible fraction, $u = z^m w^n$ is the resonant monomial, P is a real polynomial without free term and of degree at most N, and f is a function holomorphic at zero. It can be assumed without loss of generality that in the bidisk B the function $|f|$ is less than an arbitrarily preassigned constant, and the correspondence mapping $\Gamma^- \to \Gamma^+$ is defined, where Γ^- and Γ^+ are the same as in Step 1; this can be achieved by a change of scale.

The case of rational λ will be treated below; the proof of Theorem 3 for irrational λ is similar, only simpler.

STEP 3. THE CORRESPONDENCE MAPPING OF THE TRUNCATED EQUATION. This equation is obtained by discarding the last term in parentheses in $(*)$:

$$\dot{z} = z, \qquad \dot{w} = -w(\lambda + P(u)). \tag{$**$}$$

We prove that the correspondence mapping of this equation, written in the logarithmic chart, is almost regular. This was actually already done in §2F. It was proved there that

$$\Delta(z) = [g_{\tilde{f}}^{\ln z}(z^n)]^{1/m}, \qquad \dot{u} = \tilde{f}(u)$$

is the factor system for the truncated equation, $\tilde{f} = nwP(u)$.

We prove first that the mapping $u \mapsto g_{\tilde{f}}^{(\ln u)/n} u$, written in the logarithmic chart, is almost regular. Consider it first in the chart u. As in §2F, let $F(t, u)$ be the local phase flow for $\tilde{f}(\partial/\partial u)$ at the point $(0, 0)$ of (t, u)-space. The Taylor series for F converges in the domain $|t| \le A|u|^{-1}$, as proved in §2F. Under the substitution $t = (\ln u)/n$ this series becomes a Dulac series in the variable u. It converges in the domain where

$$|\ln u| \le A|u|^{-1}.$$

Here and below we consider the branch of the logarithm that is real on the positive semi-axis.

On the disk $\mathscr{D}_1$ containing the semitransversal Γ^-, the natural chart z, the function u, and the logarithmic chart ζ are connected by the relations:

$$u = z^m, \qquad \zeta = -\ln z = -\frac{1}{m}\ln u.$$

The previous inequality becomes the inequality

$$|\zeta| \le m^{-1} A|\exp(m\zeta)| = m^{-1} A \exp m\xi.$$

For every $A > 0$ there exists a C such that this inequality holds in the quadratic standard domain Ω_C. This proves that the correspondence mapping of the truncated equation is almost regular.

D. The second geometric lemma. The rest of the proof goes according to the following scheme. An analytic extension of the correspondence mapping of $(*)$ is constructed in a way similar to what was done for the linear case in Step 1 (geometric lemma). This enables us to extend the mapping to a quadratic standard domain (regularity). It is then proved that the difference between the correspondence mappings of the original and the truncated equations is small if N is large. This proves that the mapping under investigation can be expanded in a Dulac series (expandability).

STEP 4. GEOMETRIC LEMMA. Suppose that the curves μ^ζ, γ^ζ, γ_T^ζ and $\gamma^{\zeta,1}$ are the same as in Step 1. Let φ_p be the phase curve of the field $(*)$ containing the point p, and let $\hat{\gamma}^\zeta$ $(\hat{\gamma}_T^\zeta)$ be a covering on φ_p over the curve γ^ζ (γ_T^ζ) with initial point $p = (\exp(-\zeta), 1)$.

GEOMETRIC LEMMA. *For any equation* $(*)$ *the scale can be chosen in such a way that in the bidisk* B *the following holds. There is a* $C > 0$ *such that*:

1. *the arc* $\hat{\gamma}^\zeta$ *is defined for every* $\zeta \in \Omega_C$;

2. *the endpoint of this arc is the result of analytic extension of the correspondence mapping* $\Delta\colon \Gamma^+ \to \Gamma^-$ *of* $(*)$ *along the curve* γ_1^ζ *with initial value* $\Delta(1) = 1$;

3. *the first integral* $\tilde{u} = -(\lambda \ln z + \ln w)$ *of the linearized equation* $(*)$ *varies in modulus at most by* 1 *from the initial value* $\tilde{u}_0 = \tilde{u}(p)$, $p = (\exp(-\zeta), 1)$, *upon extension along the curve* $\hat{\gamma}^\zeta$;

4. *if* $\hat{\gamma}_0^\zeta$ *is the curve analogous to* $\hat{\gamma}^\zeta$ *on the phase curve of the truncated equation with initial point* p, *then the values of the function* $\tilde{u}$ *differ at most by* $\exp(-(N+1)\lambda\xi)$ *at the endpoints of the curves* $\hat{\gamma}_0^\zeta$ *and* $\hat{\gamma}^\zeta$.

REMARK. The first integral $\tilde{u}$ is chosen so that $\tilde{u}(1, \Delta(z)) = \tilde{\Delta}(\zeta)$ for $z \in \Gamma^+$, $z = \exp(-\zeta)$.

PROOF. Note that for arbitrary ζ and sufficiently small T the curve $\hat{\gamma}_T^\zeta$ is defined. We prove that it is defined for arbitrary $T \in [0, S]$ if $|z|$ is sufficiently small.

We prove first that over the part of γ^ζ belonging to a radius of the disk $\mathscr{D}_0$ the extensions of the solutions of equations $(*)$ and $(**)$ with initial point p are defined for sufficiently small values of $\exp(-\zeta)$. Consider the domain $\mathscr{U} = \{|u| \le \alpha\}$; smallness requirements are imposed on α below. Along the trajectories of equations $(*)$ and $(**)$ regarded as equations with real time, $\arg z$ does not change, but $|w|$ decreases in $\mathscr{U} \cap B$: for equation $(*)$

$$\tfrac{1}{2}(w\overline{w})^{\cdot} = \operatorname{Re}(\overline{w}\dot{w}) = -w\overline{w}(\lambda + \operatorname{Re}(P(u) + u^{N+1}f)).$$

In the domain $\mathscr{U} \cap B$, where u and f are sufficiently small, $|w|^{\cdot} < 0$ for $w \ne 0$. The derivative of $|w|$ in $\mathscr{U} \cap B$ along the field of the truncated equation can be estimated similarly. We must still prove that the curves under investigation do not leave $\mathscr{U} \cap B$.

We carry out the proof for the system $(*)$; it is analogous for $(**)$, only simpler. In the variables z, $\tilde{u}$ the system $(*)$ has the form

$$\dot{z} = z, \qquad \dot{\tilde{u}} = V(\tilde{u}) + R,$$
$$V(\tilde{u}) = -P(\exp(-n\tilde{u})),$$
$$R = \exp(-(N+1)\tilde{u}) f(z, w). \tag{$***$}$$

To conclude the passage to the new variables it would be necessary to express the function $f(z, w)$ in terms of z and $\tilde{u}$ in the expression for R, but this expression will not be used, since $|f|$ will be estimated from above by a constant. The solution of the system $(***)$ with initial point $(z_0, \tilde{u}_0) = (\exp(-\zeta), \lambda\zeta)$ is considered over the curve $\mu_T^\zeta - \zeta$. When the time t runs through this curve, the point $(z(t), 0)$ runs through the curve γ_T^ζ, while the point $(z, \tilde{u})(t)$ runs through the curve $\hat{\gamma}_T^\zeta$. The latter curve is defined for small T. We prove that it is defined for $T = S$ (S is the length of the curve μ^ζ); this will mean that the curve $\hat{\gamma}^\zeta$ is defined.

In the domain $\mathscr{U} \cap B$

$$|V(\tilde{u}) + R| \le \beta \exp(-n \operatorname{Re} \tilde{u})$$

for some $\beta > 0$. If $|\exp(-\zeta)|$ is small, then $\operatorname{Re} \tilde{u}_0$ is large; for an arbitrary sufficiently large value of $\operatorname{Re} \tilde{u}_0$ and for arbitrary $\delta \in \mathbf{C}$ with $|\delta| < 1$,

$$\beta |2\lambda^{-1} \tilde{u}_0| \exp(-n \operatorname{Re}(\tilde{u}_0 + \delta)) < 1. \qquad ({}^{*}_{**})$$

The number $|2\lambda^{-1}\tilde{u}_0|$ exceeds the time S over which the curve γ^ζ runs. Consequently, if $\operatorname{Re}\zeta$ is sufficiently large, then the curve $\hat{\gamma}^\zeta_\xi$ lying over the radius $\arg z = \text{const}$ is defined and belongs to the intersection $\mathscr{U} \cap B$: for $T < \xi$ the curve $\hat{\gamma}^\zeta_T$ does not go out to the boundary of this intersection. Similarly, it follows from the inequality $({}^{*}_{**})$ that the curve $\hat{\gamma}^\zeta_T$ does not go out to the boundary of $\mathscr{U}$ for $T \in [\xi, \xi + |\eta|]$, and the curve $\hat{\gamma}^\zeta$ is defined by the theorem on extension of phase curves.

It can be proved similarly that the curve $\hat{\gamma}^\zeta_0$ is defined. Suppose now that $\tilde{\varphi}$ is a solution of $(***)$ with the initial condition $\tilde{\varphi}(-\zeta) = (\exp(-\zeta), \lambda\zeta)$. Then $\tilde{u}(\tilde{\varphi}(0)) = \widetilde{\Delta}(\zeta)$ (see the remark after the formulation of the lemma). Let $\tilde{\varphi}_0$ be the solution of the truncated equation $\dot{z} = z$, $\dot{\tilde{u}} = V(\tilde{u})$ with the same initial condition; then $\tilde{u}(\tilde{\varphi}_0(0)) = \widetilde{\Delta}_0(\zeta)$. It was proved above that the solution $\tilde{u} \circ \tilde{\varphi}|_{\mu^\zeta - \zeta}$ runs through values lying in the disk K with center $\tilde{u}_0$ and radius 1. Let $L = \max_K(1, |V'(\zeta)|)$. Then, by Gronwall's lemma,

$$|\tilde{u} \circ \tilde{\varphi}(0) - \tilde{u} \circ \tilde{\varphi}_0(0)| \le \max |R| \exp LS = o(\exp(-N\lambda\xi)) = o(\exp(-\nu\xi))$$

for any previously assigned $\nu > 0$ if N is sufficiently large. This proves the geometric lemma. ▶

Theorem 3 now follows from the fact that the germ $\widetilde{\Delta}_0$ is almost regular. ▶

This finishes the proof of the finiteness theorem for fields with hyperbolic singular points.

§0.4. Correspondence mappings for degenerate elementary singular points. Normalizing cochains

The correspondence mappings in the heading can be described with the help of the geometric theory of normal forms. According to this theory, the germ of a vector field or mapping in a punctured neighborhood of a fixed point gives an atlas of normalizing charts with nontrivial transition functions. The normalizing charts conjugate the germ with its formal normal form; the transition functions contain all the information about the geometric properties of the germ.

A. Formulations. The correspondence mappings in the heading decompose into a product of three factors, of which two must be defined; we proceed to do this. The germ of a holomorphic vector field at an isolated elementary singular point is formally orbitally equivalent to the germ

$$\dot{z} = z^{k+1}(1+az^k)^{-1}, \qquad \dot{w} = -w. \tag{$*$}$$

Here $k+1$ is the multiplicity of the singular point, and a is a constant that is real if the original germ is real. For a formal normal form the manifold $z=0$ is contractive, and the manifold $w=0$ is the center manifold. The correspondence mapping of a semitransversal to the first manifold onto a semitransversal to the second (briefly, the mapping TO the center manifold) is denoted by Δ_{st} for the normalized system, and it has the form (see §2G) $\Delta_{\mathrm{st}} = \exp(-1/h_{k,a}(z))$, where $h_{k,a}(z) = kz^k/(1-akz^k \ln z)$. The factors of this form introduce exponentially small terms into the asymptotic expression for the monodromy transformations.

The germ of a real holomorphic vector field at an isolated degenerate elementary singular point always has a one-dimensional holomorphic invariant manifold that is contractive after a suitable time change, and it does not as a rule have a holomorphic center manifold. Corresponding to a contractive manifold is a monodromy transformation that has the following form after a suitable scale change:

$$f\colon z \mapsto z - 2\pi i z^{k+1} + \cdots. \tag{$**$}$$

This transformation is formally equivalent to a time shift $-2\pi i$ along the trajectories of the vector field $v(z) = z^{k+1}/(1+az^k)$. Here k and a are the same as in the formal orbital normal form of the germ. The corresponding normalizing formal series diverge as a rule, but they are asymptotic series for the normalizing cochains; we proceed to the definition of the latter.

A nice k-partition of the punctured disk is defined to be a partition of this disk into $2k$ equal sectors, one of which has a boundary ray on the real axis.

THEOREM 1 ON SECTORIAL NORMALIZATION [36], [27]. *For an arbitrary germ* $(**)$ *there exists a tuple of holomorphic functions, called a cochain normalizing the germ, having the following properties.*

1. *The functions in the tuple are in bijective correspondence with the sectors of a nice* k*-partition of some disk with center zero and radius* R; *each function is defined in a corresponding sector.*
2. *Each function in the tuple extends biholomorphically to a sector* S_j *with the same bisector and a larger angle* $\alpha \in (\pi/k, 2\pi/k)$; *the radius of the sector depends on* α.
3. *All the functions in the tuple have a common asymptotic Taylor series at zero with linear part the identity.*
4. *In the intersections of the corresponding sectors the functions in the tuple differ by* $o(\exp(-c/z^k))$ *for some* $c > 0$.
5. *Each of the functions in the tuple conjugates the germ* $(**)$ *in the sector* S_j *with the time shift by* $-2\pi i$ *along the trajectories of the field* $v(z)$.

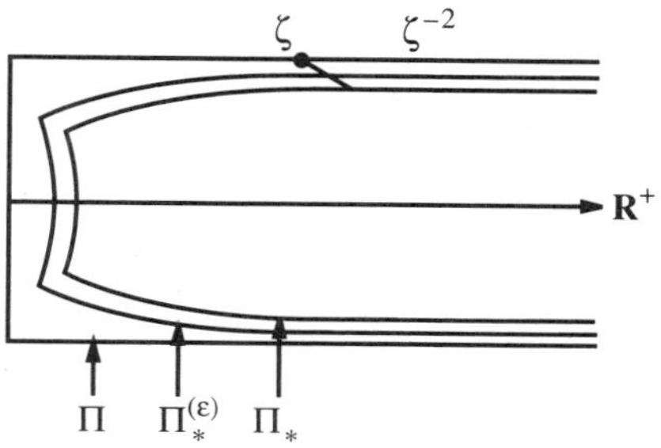

FIGURE 12

There is a unique normalizing cochain whose correction decreases more rapidly than the correction of the germ $(**)$, *that is, a cochain* $\mathrm{id}+o(z^{k+1})$.

DEFINITION 1. The set of all normalizing cochains described in the preceding theorem and its supplement below is denoted by $\mathscr{N}\mathscr{C}$ (= normalizing cochains); the set of mappings in the tuple corresponding to the sector adjacent from above to $(\mathbf{R}^+, 0)$ is denoted by $\mathscr{N}\mathscr{C}^u$ ($u=$ upper).

The following result is also known, but it will be proved below in C because it is contained "between the lines" in [36] and [27].

SUPPLEMENT. *A function in a tuple forming a normalizing cochain extends holomorphically to a domain broader than the sector* S_j. *For the sector* S^u *adjacent from above to the positive semi-axis in a nice k-partition, this domain has the form* $k^{-1}\Pi_*^{(\varepsilon)}$ *in the logarithmic chart* $\zeta = -\ln z$ ($\xi = -\ln x$, $\xi = \operatorname{Re}\zeta$, $x = \operatorname{Re} z$), *where*

$$\Pi_*^{(\varepsilon)} = \Phi_{1-\varepsilon}\Pi, \qquad \Pi = \{\xi \geq a, |\eta| \leq \pi/2\},$$
$$\Phi_{1-\varepsilon} = \zeta + (1-\varepsilon)\zeta^{-2}, \quad a = a(\varepsilon), \quad \varepsilon \in [0, 1).$$

An analogous result is valid for the remaining functions in the tuple. The mapping corresponding to S^u in the normalizing cochain is denoted by F^u_{norm}.

REMARK. Let $\Pi_*^{(0)} = \Pi_*$. The domains $\Pi_*^{(\varepsilon)}$ are ordered by inclusion: $\Pi_*^{(\varepsilon)} \subset \Pi_*^{(\varepsilon')}$ for $\varepsilon < \varepsilon'$. The domain $\Pi_*^{(\varepsilon)}$ is called the generalized ε-neighborhood of the curvilinear half-strip Π_* (Figure 12).

Everything is now ready for a description of the correspondence mappings in the section heading.

THEOREM 2 [22], [24]. *The correspondence mapping* $\Delta\colon (\mathbf{R}^+, 0) \to (\mathbf{R}^+, 0)$ *TO a center manifold of a degenerate elementary singular point of a real-analytic vector field is the restriction to* $(\mathbf{R}^+, 0)$ *of the composition*

$$\Delta = g \circ \Delta_{\mathrm{st}} \circ F^u_{\mathrm{norm}},$$

where F_{norm} *is the normalizing map-cochain (see the theorem on sectorial normalization) for the corresponding monodromy transformation;* F^u_{norm} *is that mapping in the tuple* F_{norm} *that is defined in the sector* S^u *adjacent from above to* $(\mathbf{R}^+, 0)$; *the mapping* Δ_{st} *was defined at the beginning of the subsection, and the germ* g *is holomorphic at zero.*

SUPPLEMENT. *The multiplier* $g'(0)$ *is positive.*

Theorem 2 will be proved in subsection B, and the supplement to it in D. The mapping F^u_{norm} is called the main mapping of the tuple F_{norm}.

REMARK. If F^u_{norm} is replaced by the mapping in F_{norm} corresponding to the sector adjacent to $\mathbf{R}^+$ from below, and the germ g is replaced by another holomorphic germ, then the product Δ does not change on $\mathbf{R}^+$. After choosing the main mapping in Theorem 2, we thereby dwelt upon one of two mutually symmetric variants, each one being by itself asymmetric.

Normalizing map-cochains make it necessary to use functional cochains for investigating monodromy transformations of polycycles.

The set of all germs described in the preceding theorem will be denoted by **TO**, the set of germs inverse to them by **FROM**, and the set of all almost regular germs by $\mathscr{R}$. The identity theorem follows from the next result.

THEOREM A. *A composition of germs in the classes* **TO**, $\mathscr{R}$, *and* **FROM** *either has no fixed points near zero, or is the identity.*

Chapters I–V are devoted to a proof of this theorem.

B. Proof of the theorem on the correspondence mapping. In this subsection we prove Theorem 2 in A.

For germs of vector fields at a degenerate elementary singular point there is a theorem on sectorial normalization analogous to Theorem 1 in A. To formulate it the germ must be reduced to a "preliminary normal form."

DULAC'S THEOREM [9]. *The germ of an analytic vector field at an isolated elementary singular point of multiplicity* $k+1$ *is orbitally analytically equivalent to the germ giving the equation*

$$\dot{z} = z^{k+1}, \qquad \dot{w} = -w + F(z, w), \qquad (***)$$
$$F(0, 0) = 0, \qquad dF(0, 0) = 0.$$

REMARK. An isolated singular point of an analytic vector field on the complex plane is always of finite multiplicity. If the original equation is real, then the normalizing substitution and equation $(***)$ are also real.

We proceed to formulate the theorem on sectorial normalization for equation $(***)$. A nice k-covering of the punctured disk $\mathscr{D}_0 : 0 < |z| < \varepsilon$ is a tuple of sectors S_j as described in the formulation of Theorem 1 in subsection A. Namely, we consider a nice partition of the punctured disk and replace each sector of this partition by a sector with the same bisector, the same radius, and a larger opening $\alpha \in (\pi/k, 2\pi/k)$. The resulting tuple of sectors S_j forms a nice k-covering of the punctured disk.

A nice k-covering of a neighborhood of zero in $\mathbf{C}^2$ with the w-axis deleted is defined to be a tuple $\{S_j \times \mathscr{D}\}$, where the S_j are the sectors of a nice k-covering of the disk $\mathscr{D}_0$, and $\mathscr{D} = \{|w| \leq \rho\}$ is a disk on the w-axis. The substitutions normalizing equation $(***)$ will be defined in the "sectors" $S_j \times \mathscr{D}$ and will have a common asymptotic expansion, to the definition of which we proceed.

DEFINITION 1. A semiformal z-preserving substitution is a substitution $\widehat{H}$ of the form $(z, w) \mapsto (z, w + \widehat{H}(z, w))$, $\widehat{H} = \sum_1^\infty H_n(w) z^n$; the functions H_n are holomorphic in one and the same disk $\mathscr{D}$; the series $\widehat{H}$ of powers of z is formal (a z-preserving substitution is denoted in the same way as the correction of its second component).

PROPOSITION [28]. *For an arbitrary equation* $(***)$ *there exists a unique substitution of the form* $h \circ \widehat{H}$, *where* $\widehat{H}$ *is a semiformal* z*-preserving substitution, and* h *is a holomorphic substitution of the form* $(z, w) \mapsto (h(z), w)$, $h(z) - z = o(z^{k+1})$, *carrying equation* $(***)$ *into the equation*

$$\dot{z} = z^{k+1}, \qquad \dot{w} = -w(1 + az^k) \qquad \left(\begin{smallmatrix} * \\ ** \end{smallmatrix}\right)$$

(*which can be reduced to equation* $(*)$ *by a change of time*).

THEOREM 3 ON SECTORIAL NORMALIZATION [15]. *For an arbitrary equation* $(***)$ *there exists in each sector* $S_j \times \mathscr{D}$ *of a nice* k*-covering of a neighborhood of zero in* $\mathbf{C}^2$ *with the* w*-axis deleted a unique biholomorphic mapping* $h \circ H_j$ *carrying equation* $(***)$ *into equation* $\left(\begin{smallmatrix} * \\ ** \end{smallmatrix}\right)$ *and such that the series* $\widehat{H}$ *is asymptotic for* H_j *in* $S_j \cap \mathscr{D}$ *as* $z \to 0$, $h(z) - z = o(z^{k+1})$.

The mappings $\widetilde{H}_j = h \circ H_j$ are said to be normalizing (equation $(***)$ in the sectors $S_j \times \mathscr{D}$). Let S_*^u (S_*^l) be the sector of a nice k-covering containing the sector S^u (S^l) adjacent to $(\mathbf{R}^+, \infty)$ from above (below) in a nice k-partition.

The corresponding normalizing substitutions in Theorem 3 are denoted by H^u and H^l. Each of them conjugates the germ of the correspondence mapping Δ of the original equation $(***)$ with the germ Δ_{st} of the correspondence mapping of the normalized system $(*)$. Moreover, these substitutions conjugate the monodromy transformations of the original system and of the normalized system in the sectors containing S^u and S^l, respectively. To verify this it suffices to see that a curve joining the image and inverse image of the monodromy transformation f_{st} on a phase curve of the standard equation belongs to the image of one of the sectors of a nice covering under the action of the normalizing substitution, and the transformation itself is imbeddable in a flow (this last means that it is a shift along the phase curves of some holomorphic vector field).

We compute the monodromy transformation of the normalized system $(*)$. Fix the transversal $\Gamma_A : w = A$, $0 < A < \rho$ with chart z. The inverse image z and the image $f_{\text{st}}(z)$ of the transformation f_{st} are by definition the z-coordinates of the initial point and the endpoint of the arc γ with initial point (z, A) covering on the solution of $(*)$ the loop $\{w = Ae^{i\varphi} | \varphi \in [0, 2\pi]\}$ which belongs to the w-axis. The system $(*)$ has separating variables; the desired arc γ has the form

$$\gamma = \{g^t(z, A) | t \in [0, -2\pi i]\}.$$

Here $\{g^t\}$ is the local phase flow of $(*)$; the arc γ is defined if $|z|$ is sufficiently small. Consequently,

$$f_{\text{st}} = g_{v(z)}^{-2\pi i}, \qquad v(z) = z^{k+1}/(1 + az^k),$$

and the monodromy transformation f_{st} is imbeddable.

If in addition z belongs to a sector between S^u and S^u_* and $|z|$ is sufficiently small, then the projection γ_z of the arc γ on the z-plane under the projection $(z, w) \mapsto z$ is entirely contained in the sector S^u_*. What is more, the arc γ belongs to the domain $H^u(S^u_* \times \mathscr{D})$—the image of the action of the normalizing substitution H^u. The curve $(H^u)^{-1}\gamma$ joins the image and inverse image of the monodromy transformation of the original system $(***)$ corresponding to the "curvilinear transversal" $\Gamma'_A = (H^u)^{-1}\Gamma_A$. Let π be the projection along the solutions $\Gamma'_A \to \Gamma_A$. Then the substitution $F^u = H^u \circ \pi^{-1}$ conjugates the monodromy transformation f with the imbeddable transformation f_{st}. The mapping F^u extends to a cochain F_{norm} normalizing the transformation f.

The restriction of the normalizing transformation H^u to the disk $\mathscr{D}_z = \{z\} \times \mathscr{D}$ carries the germ of this disk at the center into itself; denote the corresponding germ by g^{-1}: $g^{-1} = H^u|_{\mathscr{D}_z}$. We get that the correspondence mapping Δ is conjugated by means of the substitution $F^u = \pi \circ H^u$ in the inverse image and g^{-1} in the image with the mapping Δ_{st}:

$$g^{-1} \circ \Delta = \Delta_{\text{st}} \circ F^u, \qquad \Delta = g \circ \Delta_{\text{st}} \circ F^u.$$

Similarly, choosing the normalizing substitution H^l instead of H^u, we get that

$$\Delta = \tilde{g} \circ \Delta_{\text{st}} \circ F^l, \qquad \tilde{g}^{-1} = H^l|_{\mathscr{D}_z}.$$

This proves Theorem 2. ▶

REMARK. The substitutions F^u and F^l, as well as g and $\tilde{g}$, are not real on $(\mathbf{R}^+, \infty)$, not even for real equations $(*)$, $(***)$. It is proved in subsection D that the multiplier $g'(0)$ is nevertheless positive.

C. Proof of the supplement to the theorem on sectorial normalization. We consider the extension of the normalization mapping from the sector S^u; the remaining ones are investigated similarly, but this investigation is unnecessary for us. Recall that S^l (l for lower) denotes the sector symmetric to S^u with

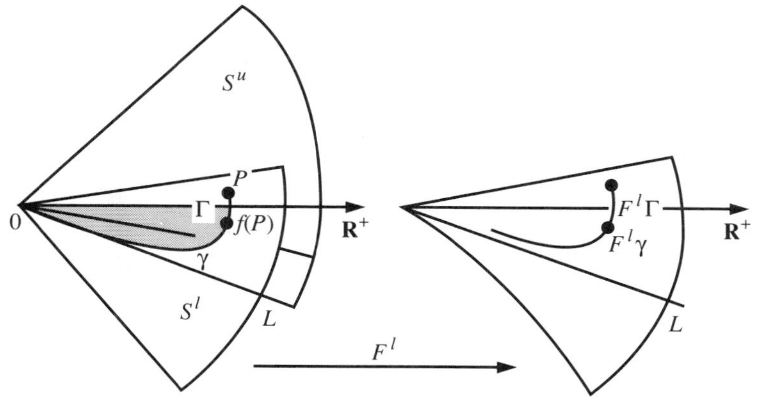

FIGURE 13

respect to the real axis (see Figure 13; it is convenient to represent the sectors on different copies of the plane $\mathbf{C}$). Let F^u and F^l be the functions in the tuple F_{norm} that conjugate the mapping $f = z - 2\pi i z^{k+1} + \cdots$ with the shift along the phase curves of the field $v(z) = z^{k+1}/(1+az^k)$ over the time $2\pi i$, or, what is the same, of the field $2\pi i v(z) = w(z)$ over the time 1. Denote this shift by g. The mapping F^u can be extended biholomorphically along the orbit of f by using the equality $F^u \circ f = g \circ F^u$. We first investigate the orbits of g and, for this, the phase curves of the field w. We prove that in a funnel of small radius containing an interval of the ray L: $\arg z = -\pi/2k$ these phase curves go into 0, having contact with L of order at least $k - 1/2$ (any $\varepsilon > 0$ could have been taken instead of $1/2$). Namely, we verify that on the lines

$$\gamma_\pm = r \exp i\left(-\frac{\pi}{2k} \pm r^{k-1/2}\right), \qquad r \in (0, r_0),$$

the field w is directed into the funnel bounded by these lines. A simple estimate gives us that

$$\arg\left(\frac{d\gamma_\pm}{dr}\right) - \arg w(z) = \pi \mp (\varepsilon + o(1)) r^{k-1/2}.$$

This implies the assertion about the contact of the phase curves of w with the ray L. Let p and $f(p)$ be two points in the sector $S^u \cap S^l$. We join them by an arc Γ such that $F^l\Gamma$ is an arc of the phase curve of the field w. The mapping F^u can be extended biholomorphically to the domain bounded by the lower radius of S^u and part of the curve γ given by

$$\gamma \stackrel{\text{def}}{=} \bigcup_0^\infty f^{[n]}\Gamma = (F^l)^{-1} \bigcup_{n=0}^\infty g^{[n]} F^l \Gamma.$$

This domain is shaded in Figure 13. The union on the right-hand side is a positive semitrajectory of the field w. It goes into 0, having contact of order at least $k-1/2$ with the ray L. The mapping F^l decomposes into an asymptotic Taylor series with linear part the identity in the sector S^l containing this semitrajectory. Therefore, the curve γ has contact of order at least $1/2$ with the ray L at zero (as a semicubic parabola contacts the x-axis). In passing to the logarithmic chart this curve goes into a line approaching the ray $\eta = -\pi/2k$ at the rate $\exp(-\xi/2)$. This yields the supplement to be proved.

D. The realness of the derivative $g'(0)$ in the expression for the correspondence mapping of a degenerate elementary singular point (a supplement to Theorem 2 in A). Since equation (***) is real, the complex conjugation involution $I\colon (z, w) \mapsto (\overline{z}, \overline{w})$ preserves it. The semiformal normalizing substitution in the proposition in subsection B passes into itself under this involution, by uniqueness. This implies that the normalizing substitution in Theorem 3, which is also uniquely determined, passes into itself under the involution of conjugation I: $H^u \circ I = I \circ H^l$. Then the substitutions F^u and F^l, which normalize the monodromy transformation, have the same property: $F^u(\overline{z}) = \overline{F^l(z)}$. Consequently, the Taylor series common for F^u and F^l is real. Accordingly, in the formula of Theorem 2 for the correspondence mapping,

$$\Delta = g \circ \Delta_{\text{st}} \circ F^u_{\text{norm}},$$

the first factor on the right-hand side can be expanded in a real asymptotic Taylor series. Recall that

$$\Delta_{\text{st}} = f_0 \circ h_{k,a}, \qquad f_0 = \exp(-1/z);$$

$$h_{k,a} = \frac{kz^k}{1 - akz^k \ln z}$$

can be expanded in a real Dulac series (subsection 2G). Thus, the Dulac series for the composition $\Delta_1 = h_{k,a} \circ F^u_{\text{norm}}$ is real. Suppose now that

$$g'(0) = \nu, \qquad \tilde{\nu} = \mathrm{Ad}(f_0)\nu\colon \zeta \mapsto \frac{\zeta}{1 - \zeta \ln \nu}.$$

Let $g = g_1 \circ \nu$; then $g_1'(0) = 1$. The Dulac series for the composition

$$f_0^{-1} \circ \Delta = \mathrm{Ad}(f_0) g_1 \circ \tilde{\nu} \circ \Delta_1$$

is real because Δ is real. The Dulac series for $\mathrm{Ad}(f_0)g_1$ is equal to x (see Lemma 4 in §2). Consequently, the Dulac series for $\tilde{\nu}$ is real, being a composition quotient of two real Dulac series. This implies that $\nu > 0$.

§0.5. Superexact asymptotic series

As shown in subsection 2C, ordinary asymptotic series are insufficient for the unique determination of monodromy transformations. Superexact asymptotic series are needed; the idea for constructing them goes as follows.

Suppose that a set M_1 of germs of mappings $(\mathbf{R}^+, \infty) \to (\mathbf{R}^+, \infty)$ is being investigated. Each of these germs can be expanded in an asymptotic series whose partial sums approximate the germ to within an arbitrary power of x, for example, in a Dulac series. Such series will be called ordinary series. However, it is desirable to expand the germs under study in series whose terms have not only a power order of smallness, but also an exponential order of smallness. At first glance this is impossible: an arbitrary remainder term of an ordinary series has power order of smallness, and it seems meaningless to take into account exponentially decreasing terms.

This difficulty is gotten around as follows. An intermediate class M_0 of functions is introduced; the functions in this class are expanded in ordinary series and are uniquely determined by them, that is, the zero function corresponds to the zero series. For example, M_0 can be taken to be the set of almost regular germs, given in a natural chart. Then the germs of the class M_1 are expanded in series of decreasing exponentials, and the coefficients in this series are no longer numbers but functions in the class M_0. The simplest example of a STAR has the appearance

$$\Sigma = a_0(x) + \sum a_j(x) \exp(-\nu_j/x),$$
$$a_j \in M_0, \quad 0 < \nu_j \nearrow \infty. \qquad (*)$$

A series Σ is said to be asymptotic for a germ f if for every $\nu > 0$ the series has a partial sum approximating the germ on $(\mathbf{R}^+, \infty)$ to within $o(\exp(-\nu/x))$. All information about the expansion of f in an ordinary series is included in the free term of the superexact series: the ordinary series for f and a_0 coincide.

By a simple example we show how to use superexact series to prove the identity theorem. Assume in addition to the preceding that the class M_0 (M_1) contains the germs of the functions 0 and x, and that the germs of this class can be expanded in Dulac series (respectively, in STAR $(*)$) and are uniquely determined by these series. Then we have the

THEOREM. $f \in M_1 \cap \mathrm{Fix}_\infty \Rightarrow f = \mathrm{id}$.

The theorem is proved according to the same scheme as the corrected Dulac lemma in §0.2B. Suppose that the theorem is false: there exists an $f \in M_1 \cap \mathrm{Fix}_\infty$, $f \neq \mathrm{id}$. Let $(*)$ be a STAR for f. Assume first that $a_0 \neq \mathrm{id}$. Then the corresponding Dulac series $\hat{a}_0 - \mathrm{id}$ is not 0. Consequently, the germ of $a_0 - \mathrm{id}$ is equal to the principal term of its Dulac series, multiplied by $1 + o(1)$; in particular, for some $\nu > 0$,

$$|a_0 - \mathrm{id}| > x^\nu.$$

Further, it follows from the expandability of f in a STAR $(*)$ that

$$|f - \mathrm{id}| \geq |a_0 - \mathrm{id}| + (|a_1| \exp(-\nu_1/x))(1 + o(1))$$
$$\geq x^\nu (1 + o(1)).$$

Consequently, $f - \mathrm{id} \neq 0$ for small x, and hence $f \notin \mathrm{Fix}_\infty$, a contradiction.

Suppose now that $a_0 = \mathrm{id}$, $f \neq \mathrm{id}$. Then the STAR $(*)$ is different from id; otherwise $f = \mathrm{id}$, since a germ in the class M_1 is uniquely determined by its series. We get from the definition of expandability that

$$f - \mathrm{id} = (a_1 \exp(-\nu_1/x))(1 + o(1)).$$

Arguing as in the preceding paragraph, we get that $a_1 \neq 0$ for small x. The two other factors in the formula for $f - \mathrm{id}$ also do not vanish near zero. Consequently, $f \notin \mathrm{Fix}_\infty$, a contradiction.

REMARK. Monodromy transformations of polycycles can be expanded in asymptotic series not only in simple but also in multiple exponentials of the type

$$\exp(-\nu \exp \circ \cdots \circ \exp 1/x).$$

The number of exponentials in this composition is the basic parameter used for proving the identity theorem by induction. In the first part [24] we take the case $n = 1$; in this book, which is the second part, we take arbitrary n. The second case is much more complicated technically, but all the basic ideas are used already in the case $n = 1$. An exception is Chapter V, an analogue of which is not needed for $n = 1$.

§0.6. Historical comments

For almost sixty years the finiteness problem was regarded as solved. Dulac's 1923 memoir [10] devoted to it was translated into Russian and published as a separate book in 1980. The first doubts as to the completeness of Dulac's proof were apparently expressed by Dumortier: in a report at the Bourbaki seminar [30] Moussu referred to a private communication from Dumortier in 1977. In the summer of 1981 Moussu sent to specialists letters in which he asked whether they regarded Dulac's assertion about finiteness of limit cycles as proved. Two month earlier the author of these lines had found a mistake in the memoir (see [17], [18]) and mentioned this in a reply to Moussu's letter. An up-to-date presentation of the main true result in Dulac's memoir and an analysis of his mistake are sketched briefly in [18] and [2] and given in detail in [21].

We mention that the greatest difficulties overcome in the memoir are related to the local theory of differential equations not for analytic vector fields, as might be assumed from the context, but for infinitely smooth vector fields, and these difficulties were connected with the description of correspondence mappings for hyperbolic sectors of elementary singular points. The investigation of compositions of these mappings that leads to the appearance of asymptotic Dulac series is then carried out in an elementary manner. The first part of Dulac's memoir concerns monodromy transformations of polycycles with nondegenerate elementary singular points, and the second part those of polycycles with arbitrary elementary singular points (degenerate ones are added). In the third part the application of resolution of singularities to the investigation of nonelementary polycycles is discussed; in the fourth part polycycles consisting of one singular point are considered. The complexity of the last two parts is due to the fact that they are based on a theorem on resolution of singularities that was proved only forty-five years later [31].

After Dumortier's detailed study of resolution of singularities of vector fields in 1977, the arguments in the last two parts of Dulac's memoir became commonplace; they are given a few lines in the survey [21] (see §0.1C). The difficulties connected with the first part of the memoir [10] are overcome by going out into the complex plane. Thus, all the papers [22], [24], [13], and [38] written in the last five years on the finiteness problem have overcome in more or less explicit form the difficulties that were not overcome in the second part of the memoir.

Correspondence mappings for degenerate elementary singular points were thoroughly studied in [20] and [22] from the point of view of their extension into the complex domain. The only difficulty consists in the investigation of compositions of these and almost regular mappings. To handle this difficulty it was necessary to develop a calculus of "functional cochains" and of "superexact asymptotic series." All subsequent work is devoted to the investigation of the indicated compositions. The main ideas in the present article are presented in [24], where Theorem A in subsection 4A is proved for compositions of germs in the classes $\mathscr{R}$, **TO**, and **FROM** in which germs in **TO** and **FROM** alternate.

The geometric theory of normal forms of resonant vector fields and mappings began to develop in parallel in Moscow and France with work of Ecalle [12] and Voronin [36] (see also [18], [27], [28]). The first steps in this theory were independent of the finiteness problem and taken before it was realized to be open.

Here I take the liberty of presenting a reminiscence that can be called a parable on the connection between form and substance. In December 1981 I made a report at a session of the Moscow Mathematical Society devoted to two questions that seemed to me to be independent of each other: the Dulac problem and the Ecalle-Voronin theory. Having to motivate the combination of the two parts in a single report, I improvised the following phrase: "Dulac's theorem shows what the smooth theory of normal forms gives for the investigation of the finiteness theorem. This theory cannot give a definitive proof. To obtain such a proof it is necessary to investigate the analytic classification of elementary singular points." In uttering the phrase, which originated there at the blackboard, I understood that this was not a pedagogical device, but a program of investigation. The first formulation of Theorem 3 in §0.4 was given in Leningrad at the International Topological Conference in 1982 [20], and a proof was published in [22].

The proofs below of the finiteness theorems make essential use of the geometric theory of normal forms (§0.4). The finiteness theorem was announced in [23].

Using the theory of resurgent functions he created in connection with local problems of analysis, Ecalle developed the approach of the four authors of [13] and obtained in parallel independent proofs of all the finiteness theorems stated in §0.1. At the time of writing this both proofs (those of Ecalle and of the author) exist as manuscripts.

CHAPTER I

Decomposition of a Monodromy Transformation into Terms with Noncomparable Rates of Decrease

In this chapter we define the basic concepts: germs of regular map-cochains (RROK) ([2]) and superexact asymptotic series (STAR). Their main properties are formulated, and all the rest of the text is devoted to proving them. The finiteness theorems are derived from these properties.

§1.1. Functional cochains and map-cochains

Let $\Omega \subset \mathbf{C}$ be an arbitrary domain, and Ξ a locally finite partition of it into analytic polyhedra: each domain of the partition is the closure of an open set given by finitely many inequalities of the form $\omega \leq 0$, where ω is a real-analytic function on a subdomain of $\mathbf{R}^2$. A tuple $f = \{f_j\}$ of functions is called a functional cochain corresponding to the partition Ξ if the functions in the tuple are in bijective correspondence with the domains of the partition, and each function extends holomorphically to some neighborhood of its domain of the partition. The partition corresponding to a functional cochain f is denoted by Ξ^f.

The coboundary δf of a functional cochain f is defined to be the tuple of holomorphic functions defined as follows on the boundary lines of the partition: corresponding to an ordered pair of domains of the partition Ξ^f that have the line $\mathscr{L}$ as common border is the germ on $\mathscr{L}$ of the holomorphic function $f_1 - f_2$, where f_1 and f_2 are the functions in f corresponding to the first and second domains of the pair. The tuple of these germs is called the coboundary of the cochain.

Map-cochains are constructed similarly: in the preceding definition it is necessary only to require that the functions f_j give biholomorphic mappings of the corresponding domains of the partition onto their images, and the difference $f_1 - f_2$ in the definition of the coboundary is replaced by the composition $f_1^{-1} \circ f_2$. It is required of the tuple f that for a pair of domains of the corresponding partition with common border along the line $\mathscr{L}$ the

([2]) Once more a Russian abbreviation: Ростки Регулярных Отображений-Коцепей.

composition $f_1^{-1} \circ f_2$ of the corresponding mappings in the tuple be defined in some neighborhood of $\mathscr{L}$.

The preceding definition gives the difference coboundary of a functional cochain, and the latter one gives the composition coboundary.

EXAMPLE. The tuple of functions giving the sectorial normalization of a mapping $\Delta\colon (\mathbf{C}, 0) \to (\mathbf{C}, 0)$ with linear part the identity (see §4 of the Introduction) forms a map-cochain, also called the normalizing cochain, denoted by F_{norm}. The role of the domain Ω is played by a punctured disk, and the domains of the partition are the sectors described in §0.4. The composition coboundary of the cochain forms the so-called Ecalle-Voronin modulus for the mapping Δ. This modulus completely characterizes the analytic equivalence class of Δ. The correction of the coboundary of the normalizing cochain for a nonidentity mapping decreases like $\exp(-C/|z|^k)$ for some $C > 0$; here $k+1$ is the degree of the lowest-order term of the Taylor expansion of the correction $\Delta - \text{id}$.

Functional cochains can be added, subtracted, and multiplied. Compositions are considered for map-cochains. The sum of two functional cochains f and g is the functional cochain denoted by $f+g$ and corresponding to the product of the partitions Ξ^f and Ξ^g. This means that to the intersection $\mathscr{D}_1 \cap \mathscr{D}_2$ of two domains of the respective partitions Ξ^f and Ξ^g there corresponds the function $f_1 + g_1$ equal to the sum of the functions in f and g corresponding to $\mathscr{D}_1$ and $\mathscr{D}_2$. Differences and products of functional cochains are defined analogously, as are compositions of map-cochains. Sums, differences, and products of functional cochains on one and the same domain Ω are always defined. The composition $f \circ g$ is not always defined; a sufficient condition for its existence is that there exists a positive number ε such that the mappings in f extend to the ε-neighborhoods of the corresponding domains of the partition, and the correction $g - \text{id}$ of g is less than ε in modulus.

§1.2. Transition to the logarithmic chart. Extension of normalizing cochains

A semitransversal to an elementary polycycle can always be chosen to belong to an analytic transversal: to an open interval transversal to the field. A chart on the semitransversal equal to zero at the vertex and analytically extendible to the transversal is said to be natural; its logarithm with the minus sign is called the logarithmic chart. A natural chart is denoted by x and the corresponding logarithmic chart by ξ; the transition function is $\xi = -\ln x$. In a natural chart the monodromy transformation of the polycycle is the germ of a mapping $(\mathbf{R}^+, 0) \to (\mathbf{R}^+, 0)$, and in the logarithmic chart it is the germ of a mapping $(\mathbf{R}^+, \infty) \to (\mathbf{R}^+, \infty)$. The notation $z = x + iy$, $\zeta = \xi + i\eta$ is used upon extension to the complex domain.

The following table contains examples of mappings used repeatedly in what follows.

	Mapping in a natural chart	The same mapping in the logarithmic chart
1	Power: $z \mapsto Cz^{\nu}$	Affine: $\zeta \mapsto \nu\zeta - \ln C$
2	Standard flat: $z \mapsto \exp(-1/z)$	Exponential: $\zeta \mapsto \exp\zeta$
3	A mapping defined in a sector with vertex 0 and expandable in a convergent or asymptotic Taylor series $\hat{f} = z(1 + \sum_1^\infty a_j z^j)$	A mapping defined in a horizontal half-strip and expandable in a convergent or asymptotic Dulac (exponential) series $\tilde{f} = \zeta + \sum_1^\infty b_j \exp(-j\zeta)$
4	$h_{k,a}$: $z \mapsto kz^k(1 - az^{-k}\ln z)^{-1}$	$\tilde{h}_{k,a}$: $\zeta \mapsto k\zeta - \ln k - \ln(1 - a\zeta\exp(-k\zeta))$
5	An almost regular mapping with asymptotic Dulac series at zero $z \mapsto Cz^{\nu} + \sum P_j(z) z^{\nu_j}$, where $C > 0$, $\nu > 0$, $0 < \nu_j \nearrow \infty$, and and the P_j are real polynomials	An almost regular mapping with asymptotic Dulac exponential series at infinity $\zeta \mapsto \nu\zeta - \ln C + \sum Q_j(\zeta) \cdot \exp(-\mu_j\zeta)$, where $C > 0$, $\nu > 0$, $0 < \mu_j \nearrow \infty$, and the Q_j are real polynomials

The most important example is a normalizing cochain (see the example in §1.1) written in the logarithmic chart. Upon transition to the logarithmic chart the cochain F_{norm} becomes a map-cochain denoted by $\tilde{F}_{\text{norm}}$ and defined in the half-plane $\mathbf{C}_a^+$: $\xi \geq a$; a depends on the cochain. The k-partition of a punctured disk by sectors becomes the partition of $\mathbf{C}_a^+$ into half-strips by the rays $\eta = \pi m/k$, $m \in \mathbf{Z}$. The mappings making up $\tilde{F}_{\text{norm}}$ extend analytically to the ε-neighborhoods of the corresponding half-strips in the partition for arbitrary $\varepsilon \in (0, \pi/2)$ (a depends also on ε), and can be expanded in a common asymptotic Dulac exponential series; see row 3 of the table. The modulus of the correction of the coboundary has the upper estimate $\exp(-C\exp k\xi)$ for some $C > 0$ depending on the cochain. The cochain $\tilde{F}_{\text{norm}}$ is periodic: it is preserved under a shift by $2\pi i$. The set of all such cochains corresponding to different values of C and the same value of k is denoted by $\mathscr{N}\mathscr{C}_k$. The set of almost regular germs with affine principal part the identity is denoted by $\mathscr{R}^0$.

We now describe the mappings of the class **TO** in the logarithmic chart. Their appearance sharply complicates the investigation of the monodromy transformations of polycycles; this investigation is relatively simple without them [19]. In a natural chart a mapping Δ of class **TO** is described by the Theorem 2 in §0.4: after a suitable scale change in the inverse image it has the form $\Delta = g \circ \Delta_{\text{st}} \circ F_{\text{norm}}$ where $\Delta_{\text{st}} = C \circ f_0 \circ h_{k,a}$, $C \in \mathbf{R}$, $f_0(z) = \exp(-1/z)$, $h_{k,a} = kz^k/(1 - az^k \ln z)$, and $g\colon (\mathbf{C}, 0) \to (\mathbf{C}, 0)$ is the germ of a holomorphic mapping with linear part the identity. Taking into account the examples in the table given, we get that the mapping Δ, written in the logarithmic chart, has the form

$$\tilde{\Delta} = \tilde{g} \circ (\text{id} - \ln C) \circ \exp \circ \tilde{h}_{k,a} \circ \tilde{F}_{\text{norm}}.$$

The corrections of the mappings $\tilde{g}$ and $\tilde{F}_{\text{norm}}$ decrease exponentially; the mappings $\text{id} - \ln C$ and $\tilde{h}_{k,a}$ have "affine principal part" that does not in general coincide with the identity mapping. This coincidence holds only when $C = 1$ and $k = 1$, respectively. In what follows it is convenient to group together all the affine factors in the composition, and for the remaining factors (not considering the exponential, of course) to make the affine principal part the identity. We mention that

$$(\text{id} - \ln C) \circ \exp = \exp \circ h_0,$$
$$h_0 = \text{id} + \ln(1 - (\exp(-\zeta)\ln C)) \in \mathscr{R}^0;$$

the mapping h_0 is almost regular and has decreasing correction. Next, denote by a_k the affine mapping $\zeta \mapsto k\zeta - \ln k$. Then

$$\tilde{h}_{k,a} = h_1 \circ a_k,$$
$$h_1 = \zeta - \ln\left(1 + \frac{a}{k^2}(\zeta + \ln k)\exp(-\zeta)\right).$$

The exact expression for h_1 is only needed for seeing that $h_1 \in \mathscr{R}^0$. Finally, the composition

$$F = a_k \circ \tilde{F}_{\text{norm}} \circ a_k^{-1}$$

is a map-cochain corresponding to the partition into half-strips of width π by the rays $\eta = \pi l$, $l \in \mathbf{Z}$. Indeed, if the function f_l in $\tilde{F}_{\text{norm}}$ is holomorphic in the half-strip $\eta \in [\pi l/k, \pi(l+1)/k]$, then the function $f_l \circ ((\zeta + \ln k)/k)$ in the tuple $\tilde{F}_{\text{norm}} \circ a_k^{-1}$ is holomorphic in the half-strip $\eta \in [\pi l, \pi(l+1)]$. As before, the correction of the map-cochain F decreases exponentially with rate of order $\exp(-\xi)$. This construction motivates the following definition. Let:

$$\mathcal{N}\mathcal{C} = \bigcup_k a_k \circ \mathcal{N}\mathcal{C}_k \circ a_k^{-1}.$$

As indicated above, all the cochains in this set correspond to one and the same partition of the half-plane $\mathbf{C}_a^+$ (a depends on the cochain $F \in \mathcal{N}\mathcal{C}$) by the rays $\eta = \pi l$, $\xi \geq a$. The partition by these rays of an arbitrary domain in the right half-plane is denoted by Ξ_{st} and is called the standard partition. The half-strip adjacent to $(\mathbf{R}^+, \infty)$ from above in the standard partition of the right half-plane is called the main half-strip of the standard partition. For what follows we need to extend that mapping in a normalizing cochain corresponding to the main half-strip of the standard partition, to a curvilinear half-strip close to the right half-strip $|\eta| \leq \pi/2$. The possibility of such an extension follows from the supplement to Theorem 1 in §0.4A and is given by the next theorem.

THEOREM. *For an arbitrary normalizing cochain* $\tilde{F}_{\mathrm{norm}}$ *and any* $C > 0$ *there exists an* $a > 0$ *such that the mapping in* $\tilde{F}_{\mathrm{norm}}$ *corresponding to the half-strip* $\eta \in [0, \pi]$, $\xi \geq a$ *can be extended analytically to the half-strip* $\Phi_C \Pi(k)$, *where* $\Pi(k) = \{\xi \geq a, |\eta| \leq \pi/2k\}$ *and* $\Phi_C = \zeta + C\zeta^{-2}$, $C > 0$. *As before, the modulus of the correction of the extended mapping has as an upper estimate the decreasing exponential* $\exp(-\mu\xi)$ *for some* $\mu > 0$.

Note that $(\Phi_C \Pi(k), \infty) \subset (\Pi(k), \infty)$ for $C > 0$.

This implies that the cochains in the set $\mathcal{N}\mathcal{C}$ possess the following properties.

1. Each cochain $F \in \mathcal{N}\mathcal{C}$ corresponds to the standard partition of some right half-plane $\mathbf{C}_a^+$, where a depends on F.

2. All the mappings in the cochain extend to the ε-neighborhoods of the corresponding half-strips for some $\varepsilon > 0$.

3. The mapping in the cochain corresponding to the main half-strip of the partition extends holomorphically to the half-strip

$$\Phi_C \Pi, \quad \Pi = \{\xi \geq a, |\eta| \leq \pi/2\}, \quad \Phi_C = \zeta + C\zeta^{-2},$$

for arbitrary $C > 0$ (a depends also on C), and the correction of the extended mapping can be estimated from above by a decreasing exponential.

4. The corrections of all mappings of the cochain in the ε-neighborhoods of the corresponding half-strips can be estimated in modulus from above by the decreasing exponential $\exp(-\mu\xi)$ for some $\mu > 0$ common for all the mappings in F.

5. The correction of the coboundary δF in the ε-neighborhoods of all the rays of the partition can be estimated from above by an iterated exponential:

$$|\delta F - \mathrm{id}| < \exp(-C \exp \xi)$$

for some $C > 0$.

The properties listed above for normalizing cochains of class $\mathscr{N}\mathscr{C}_k$ become these properties under conjugation by the affine mapping $a_k: \zeta \mapsto k\zeta - \ln k$.

Denote by $\mathscr{H}$ the set of germs of mappings $(\mathbf{C}^+, \infty) \to (\mathbf{C}^+, \infty)$ obtained from germs of holomorphic mappings $(\mathbf{C}, 0) \to (\mathbf{C}, 0)$ with linear part the identity by passing to the logarithmic chart (see line 3 of the table):

$$\mathscr{H} = \{-\ln \circ g \circ \exp(-\zeta) | g \in O_0, \ g(0) = 0, \ g'(0) = 1\}.$$

Finally, denote by $\mathscr{A}\!f\!f$ the set of germs of affine mappings $(\mathbf{C}^+, \infty) \to (\mathbf{C}^+, \infty)$ with real coefficients and positive multiplier, and by $\mathscr{M}_{\mathbf{R}}$ ($\mathbf{R}$ for real mappings) the set of germs of mappings, below called real, whose restrictions to $(\mathbf{R}^+, \infty)$ act as $(\mathbf{R}^+, \infty) \to \mathbf{R}$. Then we get finally that

$$\mathbf{TO} \subset (\mathscr{H} \circ \exp \circ \mathscr{R}^0 \circ \mathscr{N}\mathscr{C} \circ \mathscr{A}\!f\!f) \cap \mathscr{M}_{\mathbf{R}} \stackrel{\text{def}}{=} \underline{\mathbf{TO}}.$$

This is all that we need to know about the mappings of class **TO** in what follows. Denote by $\underline{\mathbf{FROM}}$ the set of germs of mappings inverse to the germs in the class $\underline{\mathbf{TO}}$.

THEOREM. *An arbitrary finite composition of restrictions to* $(\mathbf{R}^+, \infty)$ *of germs in the classes* $\underline{\mathbf{TO}}$, $\underline{\mathbf{FROM}}$, *and* $\mathscr{R}$ *either is the identity or does not have fixed points on* $(\mathbf{R}^+, \infty)$.

This theorem is proved below. We proceed to an investigation of the compositions described in it.

§1.3. The composition characteristic, proper choice of semitransversal, and the first step in decomposition of a monodromy transformation

Thus, the monodromy transformation of an elementary polycycle can be decomposed into a composition of germs in the classes $\underline{\mathbf{TO}}$, $\mathscr{R}$, $\underline{\mathbf{FROM}}$. Such a composition is said to be balanced if the numbers of germs in the classes $\underline{\mathbf{TO}}$ and $\underline{\mathbf{FROM}}$ are the same, and unbalanced otherwise.

REMARK. In a natural chart an unbalanced composition is a flat germ if in it there are more mappings of class $\underline{\mathbf{TO}}$ than there are of class $\underline{\mathbf{FROM}}$, and a vertical germ, that is, the inverse of a flat germ, otherwise. An unbalanced composition always has the isolated fixed point zero. Limit cycles cannot accumulate on a polycycle with unbalanced monodromy transformation [21], [4]. Below we consider only balanced compositions of germs in the classes $\underline{\mathbf{TO}}$, $\mathscr{R}$, and $\underline{\mathbf{FROM}}$.

Associated with a composition of N germs in these classes are a function χ and its graph, called the characteristic of the composition. The function χ is defined on the closed interval $[-N, 0]$ and is linear between adjacent integers. If the nth term of the composition is of class $\underline{\mathbf{TO}}$ or $\underline{\mathbf{FROM}}$, then $\chi(-n) = \chi(-n+1) - 1$ or $\chi(-n) = \chi(-n+1) + 1$, respectively. In the remaining cases $\chi(-n) = \chi(-n+1)$, and $\chi(0) = 0$. For a balanced

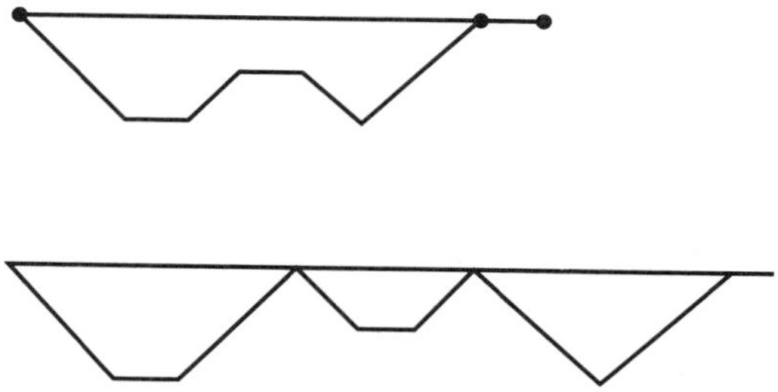

FIGURE 14

composition, $\chi(0) = \chi(-N)$. We remark that the mappings exp and ln belong to **TO** and **FROM**, respectively.

A semitransversal to a polycycle is said to be properly chosen if the decomposition of the monodromy transformation corresponding to it into the factors described above has nonpositive characteristic. A proper choice of semitransversal is always possible, because a change of semitransversal leads to a cyclic permutation of the factors in the composition. Below we consider only balanced compositions with nonpositive characteristic. The oscillation of the characteristic is called the class of the corresponding composition and cycle. The group of all balanced compositions of germs in the classes **TO**, $\mathscr{R}$, and **FROM** with nonpositive characteristic of class n is denoted by G_n. Obviously, $G_0 = \mathscr{R}$. Let $G^{-1} = \mathscr{A\!f\!f}$.

We recall some notation and introduce some. If $\mathscr{A}_1, \dots, \mathscr{A}_m$ are subsets of some group, then $\mathrm{Gr}(\mathscr{A}_1, \dots, \mathscr{A}_m)$ and $\mathrm{Gr}_+(\mathscr{A}_1, \dots, \mathscr{A}_m)$ denote the group and semigroup, respectively, generated by them. The set of all products $a_m \circ \dots \circ a_1$, $a_j \in \mathscr{A}_j$, is denoted by $\mathscr{A}_m \circ \dots \circ \mathscr{A}_1$. Let:

$$\mathrm{Ad}(f)g = f^{-1} \circ g \circ f, \qquad Ag = \mathrm{Ad}(\exp)g,$$
$$A^{-1}g = \mathrm{Ad}(\ln)g.$$

If $\mathscr{A}$ and $\mathscr{B}$ are subsets of some group, then

$$\mathrm{Ad}(\mathscr{A})\mathscr{B} = \mathrm{Gr}(\mathrm{Ad}(a)b \mid a \in \mathscr{A}, b \in \mathscr{B}).$$

If $\mathscr{B}$ is a normal subgroup of the group $G = \mathrm{Gr}(\mathscr{A}, \mathscr{B})$, then any element $g \in G$ can be represented in the form $g = ba$ with $a \in \mathscr{A}$ and $b \in \mathscr{B}$. In this case $G = \mathscr{B} \circ \mathscr{A}$. Let the classes $\mathscr{R}$, $\mathscr{R}^0$, and $\mathscr{A\!f\!f}$ be the same as in §1.2. Then $\mathscr{R} = \mathscr{R}^0 \circ \mathscr{A\!f\!f}$.

We now investigate the simplest properties of compositions of class n.

PROPOSITION 1. $G_n = \mathrm{Gr}(A^k(\mathbf{FROM} \circ \mathscr{R} \circ \mathbf{TO}), \mathscr{R} \mid 1 \le k \le n-1)$.

PROOF. Let us modify the characteristic of a mapping $g \in G_n$ by adding the compositions $\exp^{[k]} \circ \ln^{[k]}$ as shown in Figure 14. Such a modification is possible because $\exp \in$ **TO**, $\ln \in$ **FROM**, and the composition is balanced. We get the composition on the right-hand side of the equality in Proposition 1. ►

PROPOSITION 2. $G_n = \mathrm{Gr}(A^{n-1}(\underline{\mathbf{FROM}} \circ \mathscr{R} \circ \underline{\mathbf{TO}}), G_{n-1})$.

PROOF. The relation $A^k(\underline{\mathbf{FROM}} \circ \mathscr{R} \circ \underline{\mathbf{TO}}) \in G_{n-1}$ is true for $k < n-2$ because $\exp \in \underline{\mathbf{TO}}$ and $\ln \in \underline{\mathbf{FROM}}$. Proposition 2 now follows from Proposition 1.

We now investigate compositions in the class $\underline{\mathbf{FROM}} \circ \mathscr{R} \circ \underline{\mathbf{TO}}$. By the definition at the end of §1.2,

$$\underline{\mathbf{FROM}} \circ \mathscr{R} \circ \underline{\mathbf{TO}} = \mathscr{A\!f\!f} \circ \mathscr{NC}^{-1} \circ \mathscr{R}^0 \circ \ln \circ \mathscr{H} \circ \mathscr{R} \circ \mathscr{H} \circ \exp \circ \mathscr{R}^0 \circ \mathscr{NC} \circ \mathscr{A\!f\!f}.$$

The class $\mathscr{H}$ is not in the class $\mathscr{R}$, because the germs of class $\mathscr{R}$ are always real on $(\mathbf{R}^+, \infty)$, while those of class $\mathscr{H}$ are not always. This consideration, together with the preceding formula, motivates the definition

$$H^0 = \mathrm{Gr}(\mathscr{H}, \mathscr{R}^0).$$

We remark that $\mathscr{R} = \mathscr{R}^0 \circ \mathscr{A\!f\!f}$ and $\mathscr{H} = \mathrm{Ad}(\mathscr{A\!f\!f})\mathscr{H}$. Consequently,

$$\mathscr{H} \circ \mathscr{R} \circ \mathscr{H} \subset H^0 \circ \mathscr{A\!f\!f}.$$

DEFINITION 1. $\mathscr{A}^0 = \mathrm{Gr}(f \in \mathscr{R}^0 \circ \mathscr{N}\mathscr{C} \mid$ there exists a $\tilde{g} \in \mathscr{H} : A\tilde{g} \circ f$ is real). (The set $\mathscr{A}^0$ was defined somewhat differently in the first part: it was generated by the group $\mathscr{A}^0$ defined here and the group $\mathscr{A\!f\!f}$.)

Note that, by definition,

$$\underline{\mathbf{TO}} \subset \mathscr{H} \circ \exp \circ \mathscr{A}^0 \circ \mathscr{A\!f\!f}.$$

Indeed, if $\Delta \in \underline{\mathbf{TO}}$, then the germ of Δ is real on $(\mathbf{R}^+, \infty)$, and

$$\Delta = g \circ \exp \circ h \circ F \circ a, \qquad g \in \mathscr{H},\ h \in \mathscr{R}^0,\ F \in \mathscr{N}\mathscr{C},\ a \in \mathscr{A\!f\!f}.$$

But the composition $\ln \circ \Delta \circ a^{-1}$ is real, and equal to $(Ag) \circ f$, $f = h \circ F$. Consequently, $f \in \mathscr{A}^0$.

This implies that

$$\underline{\mathbf{FROM}} \circ \mathscr{R} \circ \underline{\mathbf{TO}} \subset (\mathscr{A\!f\!f} \circ \mathscr{A}^0 \circ AH^0 \circ \mathscr{A}^0 \circ \mathscr{A\!f\!f}) \cap \mathscr{M}_R.$$

PROPOSITION 3. $G_n \subset \mathrm{Gr}(A^{n-1}[\mathscr{A}^0 \circ AH^0 \circ \mathscr{A}^0 \cap \mathscr{M}_R], G_{n-1})$.

PROOF. In informal language Proposition 3 means that in the expression for a composition in the class G_n of factors of depth n it is possible to exclude the affine germs—to lift them to the depth $n-1$ and include them in the group G_{n-1}. Let us proceed to the formal proof. By Proposition 2 it suffices to prove that

$$A^{n-1}(\underline{\mathbf{FROM}} \circ \mathscr{R} \circ \underline{\mathbf{TO}}) \subset \mathrm{Gr}(A^{n-1}[\mathscr{A}^0 \circ AH^0 \circ \mathscr{A}^0 \cap \mathscr{M}_R], G_{n-1}).$$

For this it suffices to prove that

$$A^n \mathscr{A\!f\!f} \subset G_{n-1}.$$

Indeed,

$$A(\alpha\zeta+\beta) = \ln\circ(\alpha\exp\zeta+\beta)$$
$$= \zeta + \ln\alpha + \ln(1+\beta\alpha^{-1}\exp(-\zeta)) \in \mathscr{R}.$$

From this,

$$A^n\mathscr{Aff} \subset A^{n-1}\mathscr{R} \subset G_{n-1}. \blacktriangleright$$

§1.4. Multiplicatively Archimedean classes. Heuristic arguments

In this section we describe a grading of functions according to rate of decrease that arises naturally in the study of compositions of correspondence mappings. At the end we give brief motivations for the definitions in the next section.

A. Classes of Archimedean equivalence.

DEFINITION 1. A subset of the set of all germs of functions $(\mathbf{R}^+, \infty) \to \mathbf{R}$ is ordered according to growth if the difference of any two germs in this set is a germ of constant sign:

$$f \succ g \Leftrightarrow f - g \succ 0.$$

The sign $\succ$ is used as the "greater than" sign for germs: $f - g \succ 0$ if and only if there exists a representative of the germ $f - g$ that is positive on the whole domain of definition.

DEFINITION 2. Two germs of functions f and g carrying $(\mathbf{R}^+, \infty)$ into $\mathbf{R}$ are said to be multiplicatively Archimedean-equivalent if the ratio of the logarithms of the moduli of these germs is bounded and bounded away from zero. In the language of formulas, $f \sim g$ if and only if there exist c and C such that

$$0 < c < |\ln|f|/\ln|g|| < C.$$

This is clearly an equivalence relation. A class of multiplicatively Archimedean-equivalent germs is called an Archimedean equivalence class.

Everywhere below, a superscript in square brackets denotes the corresponding composition power of the germ of a diffeomorphism:

$$f^{[k]} = f\circ f\circ\cdots\circ f \qquad (k \text{ times}).$$

An exception is the notation for the germ of an inverse mapping: we write f^{-1} instead of $f^{[-1]}$.

EXAMPLES. The germs ξ, 2, $\exp\xi$, $\exp(-\exp\mu\xi)$, $\exp(-\exp^{[2]}(\xi+c))$, and $\exp(-\exp^{[k]}\xi)$ are pairwise multiplicatively Archimedean-nonequivalent for different values of $\mu > 0$, C, and k. The germs $\exp\mu\xi$ and $\exp\nu\xi$, as well as $\exp(-\exp(\xi+\alpha))$ and $\exp(-\exp(\xi+\beta))$, are multiplicatively Archimedean-equivalent for arbitrary real α and β and arbitrary real nonzero μ and ν.

B. Proper groups and the Archimedean classes corresponding to them.

DEFINITION 3. A group of germs of diffeomorphisms $(\mathbf{R}^+, \infty) \to (\mathbf{R}^+, \infty)$ is said to be ordered if it is ordered in the sense of Definition 1.

DEFINITION 4. A group of germs of diffeomorphisms $(\mathbf{R}^+, \infty) \to (\mathbf{R}^+, \infty)$ is said to be k-proper if:

1°. the germs of the group differ from linear germs by a bounded correction;

2°. the group contains the germ $A^k(\mu\xi)$ for an arbitrary $\mu > 0$;

3°. the group is ordered.

EXAMPLES. 1. The group $\mathscr{A}\!\mathit{ff}$ is 0-proper, but not 1-proper.

It is clear that $\mathscr{A}\!\mathit{ff}$ is 0-proper. If it were 1-proper, then $A(\mu + \xi) = \xi + \ln(\xi + \mu \exp(-\xi)) \in \mathscr{A}\!\mathit{ff}$, which cannot be for $\mu \neq 0$.

2. The group of almost regular germs of diffeomorphisms (see the definition in §0.3B) is 1- and 2-proper, but not 3-proper. Let us prove this.

The requirement 1° follows from the definition. We investigate the requirement 2° for $k = 1$, 2, and 3.

$k = 1$. Let us verify that $A(\mu\xi) \in \mathscr{R}$. We have that

$$A(\mu\xi) = \xi + C, \qquad C = \ln \mu;$$

the germ of a translation is almost regular.

$k = 2$. Let us verify that $A^2(\mu\xi) \in \mathscr{R}$. We have that

$$A^2(\mu\xi) = A(\xi + C) = \ln(\exp \xi + C) = \xi + \ln(1 + C\exp(-\xi)).$$

The germ of the right-hand side can be expanded in an asymptotic and even convergent Dulac series in the right half-plane, and hence is almost regular.

$k = 3$. Let us prove that $A^3(\mu\xi) \notin \mathscr{R}$. We have that

$$\begin{aligned} A^3(\mu\xi) &= A(\xi + \ln(1 + C\exp(-\xi)) \\ &= \xi + \ln[1 + \exp(-\xi)\ln(1 + C\exp(-\exp\xi))]. \end{aligned}$$

The function on the right-hand side extends holomorphically to some quadratic standard domain, but cannot be expanded in an asymptotic Dulac series there: on the ray $(\mathbf{R}^+, \infty)$ the second term decreases more rapidly than any exponential, and on the ray $\operatorname{Im} \zeta = \pi$ it is equal to

$$-\xi + O(1) \quad \text{as } \xi \to \infty, \quad C \neq 0, 1.$$

In §§1.5–1.7 we give definitions that enable us to decompose the correction of a monodromy transformation into terms whose Archimedean classes can be obtained with the help of the following general construction.

Let G be an arbitrary k-proper group. For any $g \in G$ we denote by $\mathscr{A}_g^k$ the Archimedean class of the germ $\exp(-\exp^{[k]} \circ g)$. Let:

$$\mathscr{A}_G^k = \{\mathscr{A}_g^k | g \in G\}.$$

3. The set $\mathscr{A}^0_{\mathscr{A}\!\mathit{ff}}$ consists of the unique Archimedean class with representative $\exp \xi$.

4. The set $\mathscr{A}^1_{\mathscr{R}}$ consists of the Archimedean classes of germs

$$f_\mu = \exp(-\exp \mu\xi), \qquad \mu > 0,$$

and only of them.

Indeed, let $g \in \mathscr{R}$ be an arbitrary almost regular germ. Then there exist positive constants μ and C such that

$$\mu(\xi - C) \prec g \prec \mu(\xi + C).$$

Consequently, setting $a = \exp(-\mu C)$ and $b = \exp(\mu C)$, we get that

$$(f_\mu)^a = f_\mu \circ (\xi - C) \prec \exp(-\exp g) \prec f_\mu \circ (\xi + C) = (f_\mu)^b.$$

This means that the germs f_μ and $\exp(-\exp g)$ are multiplicatively Archimedean-equivalent.

The properties of the Archimedean classes constructed are used repeatedly in what follows.

C. Archimedean classes corresponding to a proper group. The main role in comparison of Archimedean classes in the set $\mathscr{A}^k_G$ (the group G is k-proper) is played by the compositions $A^{-k}g$, $g \in G$; see Proposition 2 below. We begin with a study of these compositions.

PROPOSITION 1. *Suppose that G is a k-proper group. Then for any germ $g \in G$ there exists a limit at infinity—a generalized multiplier of the germ $A^{-k}g$:*

$$\lambda_k(g) = \lim_{(\mathbf{R}^+,\infty)} A^{-k}g/\xi,$$

which is zero, positive, or infinite.

PROOF. Let f and g be arbitrary germs in the group G. Since f and g are germs of diffeomorphisms, the following relations are equivalent:

$$f \prec g \quad \text{and} \quad A^{-k}f \prec A^{-k}g.$$

Consequently, for any $\lambda > 0$ one of the following three relations holds:

$$A^k(\lambda\xi) \prec g, \quad A^k(\lambda\xi) = g, \quad A^k(\lambda\xi) \succ g,$$

or, what is equivalent,

$$\lambda\xi \prec A^{-k}g, \quad \lambda\xi = A^{-k}g, \quad \lambda\xi \succ A^{-k}g. \blacktriangleright$$

Therefore,

$$\sup\{\lambda \geq 0 | \lambda\xi \prec A^{-k}g\} = \lim_{(\mathbf{R}^+,\infty)} A^{-k}g/\xi = \lambda_k(g).$$

REMARK 1. Thus, the following mapping is defined:

$$\lambda_k \colon G \to 0 \cup \mathbf{R}^+ \cup \infty, \qquad g \mapsto \lambda_k(g).$$

Let

$$\lambda_k^{-1}(0) = G_{\text{slow}}^-, \quad \lambda_k^{-1}(\mathbf{R}^+) = G_{\text{rap}}, \quad \lambda_k^{-1}(\infty) = G_{\text{slow}}^+,$$
$$G_{\text{slow}} = G \backslash G_{\text{rap}} = G_{\text{slow}}^- \cup G_{\text{slow}}^+.$$

In other words, G_{slow}^-, G_{rap}, and G_{slow}^+ are the subsets of G consisting of those germs g such that the composition $A^{-k} g$ increases more slowly than any linear germ, like a linear germ, and more rapidly than any linear germ, respectively.

REMARK 2. Obviously, G_{rap} is a group, while G_{slow}^- and G_{slow}^+ are semigroups. The designations rap (rapid) and slow indicate the rapidity and slowness of decrease of the corrections. Moreover, $G_{\text{slow}}^{\pm} \circ G_{\text{rap}} = G_{\text{slow}}^{\pm}$.

EXAMPLES. 1. Let $G = \mathscr{A}\!f\!f$, a 0-proper group. Then $G_{\text{rap}} = G$ and $G_{\text{slow}} = \varnothing$.

2. Let $G = \mathscr{R}$, a 1-proper group; recall that $\mathscr{R}$ is the set of all almost regular germs; see §0.3. Let $\mu_1(g) = \lim_{(\mathbf{R}^+, \infty)} g/\xi$. Then $g = \xi(\mu + o(1))$ on $(\mathbf{R}^+, \infty)$, $\mu = \mu_1(g)$, and

$$A^{-1} g = \exp[(\mu + o(1)) \ln \xi] = \xi^{\mu + o(1)}.$$

Consequently, for $G = \mathscr{R}$ and $k = 1$,

$$G_{\text{rap}} = \{g | \mu_1(g) = 1\}, \qquad G_{\text{slow}}^{\pm} = \{g | \mu_1(g) \gtrless 1\}.$$

The following three propositions enable us to compare germs in Archimedean classes belonging to the sets $\mathscr{A}_G^k$ for the same or different k and G.

PROPOSITION 2. *Suppose that G is a k-proper group, and $f, g \in G$. Then the Archimedean classes $\mathscr{A}_f^k$ and $\mathscr{A}_g^k$ coincide if and only if $h = f \circ g^{-1} \in G_{\text{rap}}$. In the language of formulas,*

$$\mathscr{A}_f^k \equiv \mathscr{A}_g^k \Leftrightarrow f \circ g^{-1} \in G_{\text{rap}}.$$

PROOF. The germs $\varphi = \exp(-\exp^{[k]} \circ f)$ and $\psi = \exp(-\exp^{[k]} \circ g)$ are multiplicatively Archimedean equivalent if and only if the analogous equivalence holds for the germs

$$\exp(-A^{-k} h) \text{ and } \exp(-\xi), \quad \text{where } h = f \circ g^{-1}. \tag{$*$}$$

Then by the definition of the group G_{rap}, there exists a λ such that $A^{-k} h = (\lambda + o(1))\xi$.

This implies the multiplicatively Archimedean equivalence of the germs in $(*)$.

Conversely, if the germs in $(*)$ are equivalent, then the germ $A^{-k} h$ does not increase more rapidly nor more slowly than a linear germ, that is, $h \in G_{\text{rap}}$. ►

DEFINITION 5. Let G be a k-proper group, and let $f, g \in G$. We say that $f \prec\!\prec g$ in G if $f \circ g^{-1} \in G_{\text{slow}}^-$.

PROPOSITION 3. *Suppose that* G *is a* k*-proper group,* $f, g \in G$, $f \prec\prec g$, $\varphi \in \mathscr{A}_f^k$, $\psi \in \mathscr{A}_g^k$, *and* $\varphi \to 0$ *and* $\psi \to 0$ *on* $(\mathbf{R}^+, \infty)$. *Then* $|\psi| \prec |\varphi|$ *on* $(\mathbf{R}^+, \infty)$.

PROOF. By definition, the germs $|\varphi|$ and $|\psi|$ belong to the Archimedean classes of the germs $\tilde{\varphi} = \exp(-\exp^{[k]} \circ f)$ and $\tilde{\psi} = \exp(-\exp^{[k]} \circ g)$, respectively, and can be estimated from above and from below by positive powers of them. Therefore, it suffices to prove that for any $\lambda > 0$

$$\tilde{\psi} \prec \tilde{\varphi}^{\lambda}.$$

This is equivalent to the inequality $\exp(-\xi) \prec \exp(-\lambda A^{-k} h)$, where $h = f \circ g^{-1}$.

The last inequality follows from the fact that the germ h increases at infinity more slowly than any linear germ, by the definition of the semigroup G^-_{slow}.

PROPOSITION 4. *Suppose that* G *and* $\tilde{G}$ *are* k*- and* m*-proper groups, respectively,* $k < m$, $f \in G$, $g \in \tilde{G}$, $\varphi \in \mathscr{A}_f^k$, $\psi \in \mathscr{A}_g^m$, *and* $\varphi \to 0$ *and* $\psi \to 0$ *on* $(\mathbf{R}^+, \infty)$. *Then* $|\psi| \prec |\varphi|$ *on* $(\mathbf{R}^+, \infty)$.

PROOF. Let

$$\tilde{\varphi} = \exp(-\exp^{[k]} \circ f), \qquad \tilde{\psi} = \exp(-\exp^{[m]} \circ g).$$

As above, the germs φ and $\tilde{\varphi}$, as well as ψ and $\tilde{\psi}$, are multiplicatively Archimedean-equivalent. Therefore, as above, it suffices to prove that $\tilde{\psi} \prec \tilde{\varphi}^{\lambda}$ for any $\lambda > 0$.

This is equivalent to the inequality

$$\exp(-\sigma) \prec \exp(-\lambda\xi) \tag{$**$}$$

for any $\lambda > 0$, where

$$\sigma = \exp^{[m-k]} \circ A^{-k} h, \qquad h = g \circ f^{-1}.$$

PROPOSITION 5. *The germ* σ *defined above increases on* $(\mathbf{R}^+, \infty)$ *more rapidly than any linear germ.*

PROOF. It suffices to prove the proposition for $m - k = 1$; further compositions with an exponential only increase the growth. Accordingly, let $\sigma = \exp \circ A^{-k} h$. We prove that for any $C > 0$,

$$A^{-k} h \succ C \ln \xi \tag{$***$}$$

This inequality can be proved by induction on k. Requirement 1 in the definition of a proper group is used here; it implies that the germs f and g differ from a linear germ by a bounded correction. Consequently, the germ h has the same property. Therefore, $h \succ \varepsilon\xi$ for some $\varepsilon > 0$. This gives the induction base: the inequality $(***)$ for $k = 0$.

INDUCTION STEP. Suppose that the inequality $(***)$ has been proved for some k. Let us prove it for $k+1$. We have that

$$A^{-(k+1)}h \succ A^{-1}(2\ln\xi) = \exp(2\circ\ln\circ\ln\xi) = (\ln\xi)^2 \succ C\ln\xi$$

for arbitrary $C > 0$. The inequality $(***)$ is proved.

Consequently, for arbitrary $C > 0$,

$$\sigma \succ \exp\circ C\ln\xi = \xi^C. \blacktriangleright$$

The inequality $(**)$, and with it Proposition 4, follows immediately from Proposition 5. ►

D. Motivations for the basic definitions. Normalizing cochains can be expanded in Dulac series with real coefficients, and are uniquely determined by these series. Therefore, any germ of a composition $\operatorname{Re} F\circ\exp^{[k]}\circ f$, $F \in \mathscr{N}\mathscr{C} - \mathrm{id}$, $f \in G^{k-1}$, where G^{k-1} is a k-proper group (the notation G^{k-1} is used below for the group defined in §1.7), belongs to the Archimedean class $\mathscr{A}_f^k$. Suppose that $k < m$, $g \in G^{m-1}$. Then it follows from Proposition 4 that

$$\mathscr{A}_f^k \cdot \mathscr{A}_g^m \subset \mathscr{A}_g^m \qquad ({}^*{}_*{}^*)$$

On the other hand, the class

$$\mathscr{N}_{G^{m-1}}^m \overset{\text{def}}{=} (\mathscr{N}\mathscr{C} - \mathrm{id})\circ\exp^{[m]}\circ G^{m-1}$$
$$= \{F\circ\exp^{[m]}\circ g \mid F \in \mathscr{N}\mathscr{C},\ g \in G^{m-1}\}$$

is not invariant under the analogous multiplication:

$$\mathscr{N}_{G^{k-1}}^k \cdot \mathscr{N}_{G^{m-1}}^m \not\subset \mathscr{N}_{G^{m-1}}^m; \qquad ({}_*{}^*{}_*)$$

the reason for this is indicated below. For what follows it is necessary to include the sets $\mathscr{N}_{G^{k-1}}^k$ in broader sets that belong to the classes $\mathscr{A}_{G^{k-1}}^k$ and satisfy the analogue of the formulas $({}^*{}_*{}^*)$. We explain the formula $({}_*{}^*{}_*)$. Suppose that $F_1, F_2 \in \mathscr{N}\mathscr{C} - \mathrm{id}$, $f \in G^{k-1}$, $g \in G^{m-1}$. The relations

$$F_1\circ\exp^{[k]}\circ f,\ F_2\circ\exp^{[m]}\circ g \in \mathscr{N}_{G^{m-1}}^m$$

would mean that $F \overset{\text{def}}{=} F_1\circ\rho\cdot F_2 \in \mathscr{N}\mathscr{C} - \mathrm{id}$, where $\rho = \exp^{[k]}\circ f\circ g^{-1}\circ\ln^{[m]}$. The composition $F_1\circ\rho$ is defined, roughly speaking, as the tuple $\{f_j\circ\rho\}$ of functions on the domains $\sigma\Xi_j$, where $\sigma = \rho^{-1}$, the f_j are the functions of the tuple F_1, and the Ξ_j are the corresponding domains of the partition Ξ^{F_1}. See §2.3 for the precise definition. If the Ξ_j are rectilinear half-strips, then the $\sigma\Xi_j$ are not necessarily such half-strips. The composition $F_1\circ\rho$ corresponds to the partition obtained from the standard partition by rays $\eta = \pi j$, $j \in \mathbf{Z}$, under the action of the diffeomorphism σ. On the other hand, the germs of class $\mathscr{N}\mathscr{C} - \mathrm{id}$ correspond to the standard partition. Thus, $F \notin \mathscr{N}\mathscr{C} - \mathrm{id}$ in general.

The diffeomorphisms defined in §1.6 and then used to construct regular cochains of class n are obtained mostly with the help of this construction.

We proceed to the detailed definitions.

§1.5. Standard domains and admissible germs of diffeomorphisms

Two half-strips are used repeatedly in the constructions to follow: right and standard. A right half-strip is defined by the formula

$$\Pi = \{\zeta | \xi \geq a,\ |\eta| < \pi/2\}, \qquad a \geq 0.$$

A standard half-strip is defined by

$$\Pi_* = \Phi\Pi, \qquad \Phi = \zeta + \zeta^{-2}.$$

DEFINITION 1. A standard domain is a domain that is symmetric with respect to the real axis, belongs to the right half-plane, and admits a real conformal mapping onto the right half-plane that has derivative equal to $1+o(1)$ and extends to the δ-neighborhood of the part of the domain outside a compact set for some $\delta > 0$.

REMARK. The correction of the conformal mapping in the previous definition increases more slowly than $\varepsilon|\xi|$ at infinity for each $\varepsilon > 0$.

An important example of a standard domain is given by

PROPOSITION 1. *The exponential of a standard half-strip is a standard domain.*

PROOF. Indeed,

$$\begin{aligned}\exp \Pi_* &= \exp \circ \Phi\Pi = \exp \circ \Phi \circ \ln(\mathbf{C}^+ \backslash K) \\ &= A^{-1}\Phi(\mathbf{C}^+ \backslash K),\end{aligned}$$

where K is the disk $|\zeta| \leq \exp a$, and a is the same as in the definition of the right half-strip Π. Further,

$$\begin{aligned}[A^{-1}(\zeta + \zeta^{-2})]' &= \exp\left(\ln\zeta + \frac{1}{\ln^2\zeta}\right) \cdot \left(1 - \frac{2}{\ln^3\zeta}\right) \cdot \zeta^{-1} \\ &= \left(\exp\left(\frac{1}{\ln^2\zeta}\right)\right) \cdot (1 + o(1)) \\ &= 1 + o(1) \quad \text{in } (\mathbf{C}^+, \infty).\end{aligned}$$

Consequently, the real conformal mapping

$$A^{-1}\Phi\colon\ \mathbf{C}^+ \backslash K \to \exp \Pi_*$$

on $(\mathbf{C}^+, \infty)$ has derivative of the form $1 + o(1)$. The inverse mapping

$$\psi\colon\ \exp\Pi_* \to \mathbf{C}^+ \backslash K$$

can be extended to the δ-neighborhood of the part of $\exp\Pi_*$ outside some compact set and also has derivative of the form $1 + o(1)$. Further, there exists a conformal mapping $\psi_0\colon \mathbf{C}^+\backslash K \to \mathbf{C}^+$ with correction tending to zero as $\zeta \to \infty$ (it is given by the Zhukovskiĭ function if K is the unit disk). Therefore, the mapping $\psi_0 \circ \psi$ is real, can be extended to the δ-neighborhood of the part of $\exp \Pi_*$ outside some compact set, and has derivative of the form $1 + o(1)$, that is, it satisfies the requirements imposed in the definition of a standard domain. ►

DEFINITION 2. Let $\mathbf{\Omega}$ be some set of standard domains. The germ of the diffeomorphism $\sigma_{\mathbf{R}}\colon (\mathbf{R}^+, \infty) \to (\mathbf{R}^+, \infty)$ is said to be admissible of class $\mathbf{\Omega}$, or $\mathbf{\Omega}$-admissible, if:

1°. the inverse germ ρ admits a biholomorphic extension to some standard domain, and for each standard domain $\Omega \in \mathbf{\Omega}$ there exists a standard domain $\tilde{\Omega} \in \mathbf{\Omega}$ such that ρ maps $\tilde{\Omega}$ biholomorphically into Ω, and, moreover,

2°. the derivative ρ' is bounded in $\tilde{\Omega}$,

3°. there exists a $\mu > 0$ such that $\operatorname{Re} \rho < \mu\xi$ in $\tilde{\Omega}$,

4°. for each $\nu > 0$,

$$\exp \operatorname{Re} \rho \succ \nu\xi \quad \text{in } \tilde{\Omega}.$$

The extension of the germ $\sigma_{\mathbf{R}}$ to the domain $\rho\tilde{\Omega}$ is denoted by σ and also called an $\mathbf{\Omega}$-admissible germ. To speak of an admissible and not an $\mathbf{\Omega}$-admissible germ means by definition that $\mathbf{\Omega}$ is understood to be the class of all standard domains.

DEFINITION 3. The germ of a diffeomorphism $\sigma_{\mathbf{R}}\colon (\mathbf{R}^+, \infty) \to (\mathbf{R}^+, \infty)$ is said to be nonessential of class $\mathbf{\Omega}$ if it admits a biholomorphic extension to a standard half-strip Π_* for some a and there exists a standard domain of class $\mathbf{\Omega}$ that belongs to $\sigma\Pi_*$.

EXAMPLES. 1. The germ $\sigma = \exp \circ \mu$ with $\mu > 1$ is nonessential. Indeed, the image $\sigma\Pi_*$ contains the part of the right half-plane $\mathbf{C}^+$ outside some compact set.

2. The germ $\sigma = \exp$ is nonessential by Proposition 1.

3. The germ $\sigma = \exp$ is not admissible: requirement 4° of Definition 2 fails.

4. The germs of $\sigma = \exp \circ \mu$ with $0 < \mu < 1$, of $\sigma = \zeta^\mu$ with $\mu \geq 1$, and of $\sigma \in \mathscr{Aff}$ are admissible.

REMARK. A standard half-strip has two opposing properties: it is not too broad and not too narrow. On the one hand, the main function of each map-cochain of class $\mathscr{NC}$ extends to a standard half-strip for sufficiently large a dependent on the cochain; as a rule, it is impossible to implement such an extension to a right half-strip Π. On the other hand, the exponential of a standard half-strip is a standard domain.

§1.6. Germs of regular map-cochains, RROK

Two map-cochains or two functional cochains are said to be equivalent if there exists a standard domain in which they are defined and coincide. An equivalence class of map-cochains or functional cochains is called the germ of a map-cochain or of a functional cochain. The representatives of a germ are considered in standard domains, by definition.

A. Regular partitions. Let $\mathbf{\Omega}$ be a set (class) of standard domains.

DEFINITION 1. Suppose that a partition Ξ is given in a standard domain $\Omega \in \mathbf{\Omega}$. The image of the partition Ξ under the action of an admissible germ of a diffeomorphism σ of class $\mathbf{\Omega}$ is the partition $\sigma_*\Xi$ of a standard domain $\tilde{\Omega} \in \mathbf{\Omega}$ in which a representative, carrying $\tilde{\Omega}$ into Ω, of the germ $\rho = \sigma^{-1}$ is defined and biholomorphic. The domains of the partition $\sigma_*\Xi$ are defined by the equalities

$$(\sigma_*\Xi)_j = \sigma(\Xi_j \cap \rho\tilde{\Omega}),$$

where the Ξ_j are the domains of the partition Ξ. The domain $(\sigma_*\Xi)_j$, by definition, corresponds to the domain Ξ_j.

EXAMPLES. Pictured in Figure 15 are the images of the standard partition under the action of the diffeomorphisms $\exp\circ\mu$, $\mu \in (0, 1)$; ζ^μ, $\mu > 1$; $\mu\zeta$, $\mu > 0$. In the first case the domains of the image partition are ordinary sectors, in the second they are "parabolic sectors," and in the third they are horizontal half-strips.

DEFINITION 2A. The standard partition Ξ_{st} is the partition of a domain in $\mathbf{C}$ by the rays $\eta = \pi j$, $j \in \mathbf{Z}$. The strip $\eta \in [\pi(j-1), \pi j]$ is denoted by Π_j.

We modify the standard partition. Recall that

$$\Pi_*^{(\varepsilon)} = \Phi_{1-\varepsilon}\Pi, \quad \Pi = \{\xi \ge a, |\eta| \le \tfrac{\pi}{2}\}, \quad \Phi_{1-\varepsilon} = \zeta + (1-\xi)\zeta^{-2}.$$

Let $\Pi_* = \Pi_*^{(0)}$ and $\Pi_{\text{main}} = \Pi_1 \cup \Pi_*$.

DEFINITION 2B. The modified standard partition Ξ_M of class $\mathbf{\Omega}$ (M for modified) is a partition of a standard domain Ω of class $\mathbf{\Omega}$ into the half-strips $\Pi_j \cap \Omega$, $j \in \mathbf{Z}\backslash\{0, 1\}$, and the half-strips Π_{main}, $\Pi_0\backslash\Pi_{\text{main}}$ (see

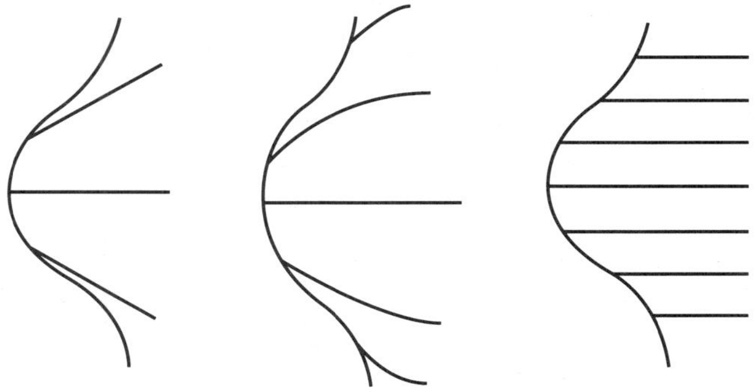

FIGURE 15

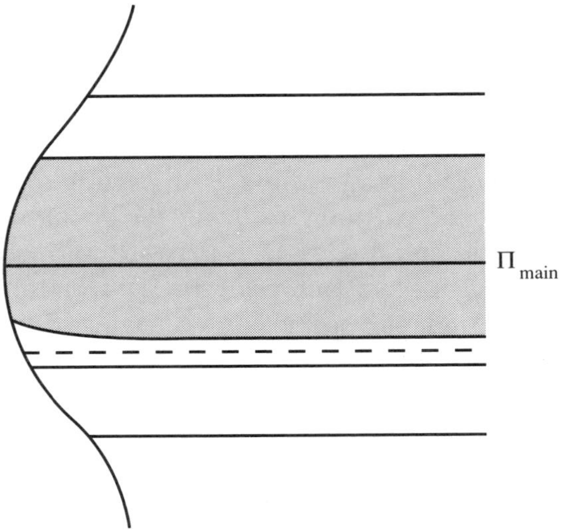

FIGURE 16

Figure 16). The set Π_{main} is called the main domain of the modified standard partition. Its lower boundary line is denoted by $\mathscr{L}_0$ and called special. The remaining boundary lines of the partition Ξ_M, the rays $\eta = \pi j$, $j \neq 0$, are said to be standard.

The union of all boundary lines of an arbitrary partition Ξ is denoted by $\partial\Xi$.

DEFINITION 3. An **R**-regular partition of a standard domain is defined to be a product of finitely many images of the standard partition under the action of admissible diffeomorphisms:

$$\Xi = \prod_1^N \sigma_{j*}\Xi_{\text{st}}.$$

By definition, this partition has type $\boldsymbol{\sigma}_{\mathbf{R}}$, where $\boldsymbol{\sigma} = (\sigma_1, \dots, \sigma_N)$, and numerical type N.

DEFINITION 4. Let $\boldsymbol{\sigma} = (\sigma_1, \dots, \sigma_N)$ be a tuple of admissible diffeomorphisms ordered by the rate of their growth at infinity: $\sigma_j \succ \sigma_{j+1}$ on $(\mathbf{R}^+, \infty)$. The modified regular partition, or partition of type $(\boldsymbol{\sigma}, k)$ and numerical type N is defined to be the product

$$\Xi = \prod_1^k \sigma_{j*}\Xi_M \cdot \prod_{k+1}^N \sigma_{j*}\Xi_{\text{st}}. \qquad (*)$$

DEFINITION 5. A tuple of $N+1$ partitions of type $\boldsymbol{\sigma}_{\mathbf{R}}$, $(\boldsymbol{\sigma}, 1), \dots, (\boldsymbol{\sigma}, N)$, where $\boldsymbol{\sigma} = (\sigma_1, \dots, \sigma_N)$, is called a regular partition of type $\boldsymbol{\sigma}$. The partitions of types $\boldsymbol{\sigma}_{\mathbf{R}}$ and $(\boldsymbol{\sigma}, k)$ are called the **R**-realization and the k-realization, respectively, of a partition of type $\boldsymbol{\sigma}$, and also an R-regular partition and a k-regular partition.

DEFINITION 6. The generalized ε-neighborhood of the main domain of a modified standard partition is defined to be the set

$$\Pi_{\text{main}}^{(\varepsilon)} = \Pi_*^{(\varepsilon)} \cup \Pi_1^{\varepsilon},$$

where Π_1^{ε} is the ε-neighborhood of the half-strip Π_1, and $\Pi_*^{(\varepsilon)} = \Phi_{1-\varepsilon}\Pi$ (see the reminder at the beginning of §1.6).

Denote by $\mathscr{D}^{\varepsilon}$ the ε-neighborhood of a set $\mathscr{D}$. For uniformity, all the ε-neighborhoods of the domains of the standard partition and of the domains of the modified standard partition aside from the main domain, as well as the generalized ε-neighborhood of the main domain will be called generalized ε-neighborhoods of the domains of the standard or modified standard partition. Further, the $(\boldsymbol{\sigma}, \varepsilon)$-neighborhood or generalized ε-neighborhood of a domain of an $\mathbf{R}$-regular (k-regular) partition of type $\boldsymbol{\sigma}_{\mathbf{R}}$ (respectively, $(\boldsymbol{\sigma}, k)$) given by the intersection of the domains $\sigma_j \Xi_j$, where the Ξ_j are the domains of the standard (for $j \geq k+1$) or modified standard (for $j \leq k$) partition, is defined to be the intersection of the images of the generalized ε-neighborhoods of these domains under the action of the corresponding diffeomorphisms σ_j. We proceed to the precise definition.

NOTATION. Let Ξ_j be a domain of a standard or modified standard partition. Its generalized ε-neighborhood is denoted by $\Xi_j^{(\varepsilon)}$.

DEFINITION 7. Let $\mathscr{U} = \bigcap_1^N \sigma_j(\Xi_j \cap \rho_j \tilde{\Omega})$, a domain of an $\mathbf{R}$-regular or k-regular partition of type $\boldsymbol{\sigma}_{\mathbf{R}}$ or $(\boldsymbol{\sigma}, k)$, $\boldsymbol{\sigma} = (\sigma_1, \dots, \sigma_N)$. Then the generalized ε-neighborhood $\mathscr{U}^{(\varepsilon)}$ of the domain $\mathscr{U}$ has the form

$$\mathscr{U}^{(\varepsilon)} = \bigcap_1^N \sigma_j(\Xi_j^{(\varepsilon)} \cap \rho_j \tilde{\Omega}).$$

Such a neighborhood is also called a $(\boldsymbol{\sigma}_{\mathbf{R}}, \varepsilon)$- or $(\boldsymbol{\sigma}, k, \varepsilon)$-neighborhood of the domain $\mathscr{U}$ of the partition (depending on the type of partition).

The generalized ε-neighborhood of a common boundary line $\mathscr{L}$ of two domains of an $\mathbf{R}$- or k-regular partition is defined to be the intersection of the generalized ε-neighborhoods of the corresponding domains; denote it by $\mathscr{L}^{(\varepsilon)}$.

DEFINITION 8. All the boundary lines of an $\mathbf{R}$-regular partition are said to be standard. A boundary line $\mathscr{L}$ of a modified regular partition $(*)$ of type $(\boldsymbol{\sigma}, k)$ is said to be standard if it belongs to the image of a boundary ray of a modified standard partition under the action of one of the diffeomorphisms σ_j, $1 \leq j \leq k$, or to the image of a boundary ray of the standard partition under the action of one of the diffeomorphisms σ_j, $k+1 \leq j \leq N$. The remaining boundary lines of the partition $(*)$ are said to be special. The union of the generalized ε-neighborhoods of all the standard boundary lines of a regular partition is called the generalized ε-neighborhood of the boundary of the partition and denoted by $\partial\Xi^{(\varepsilon)}$.

REMARKS. 1. A special boundary line of a modified regular partition of a domain Ω is the image of a special boundary line of a modified standard partition under the action of one of the diffeomorphisms in the tuple $\boldsymbol{\sigma}$ and is at the same time not the image of a boundary ray of the standard or modified standard partition under the action of another diffeomorphism in the tuple $\boldsymbol{\sigma}$.

2. Special boundary lines of a modified regular partition do not in general belong to the generalized ε-neighborhood of the boundary of the partition. An $\mathbf{R}$-regular partition does not have special boundary lines.

DEFINITION 9. A regular partition of type $\boldsymbol{\sigma}_{\mathbf{R}}$ or $(\boldsymbol{\sigma}, k)$ is said to be rigged if for all sufficiently small $\varepsilon > 0$ and all $C > 0$ the real-valued function "rigging cochain" defined as follows is considered in the generalized ε-neighborhood of the boundary of the partition. The functions in the tuple forming the rigging cochain are defined in generalized ε-neighborhoods of the standard boundary lines of the partition. The tuple function given in a neighborhood of the boundary line $\mathscr{L}$ is denoted by $m_{\boldsymbol{\sigma}, C, \varepsilon, \mathscr{L}}$ and is equal to

$$m_{\boldsymbol{\sigma}, C, \varepsilon, \mathscr{L}} = \sum \exp(-C \exp \operatorname{Re} \rho_j), \qquad \rho_j = \sigma_j^{-1}.$$

The summation is over all j such that $\mathscr{L}$ is the image of a boundary line of the partition in the case of an $\mathbf{R}$-regular partition, and over those j such that

$$\begin{aligned} &\mathscr{L} \subset \sigma_j(\partial \Xi_M \backslash \mathscr{L}_0) && \text{for } 1 \leq j \leq k, \\ &\mathscr{L} \subset \sigma_j \partial \Xi_{\mathrm{st}} && \text{for } k+1 \leq j \leq N \end{aligned}$$

in the case of a regular partition of type $(\boldsymbol{\sigma}, k)$. The so-defined cochain is denoted by $m_{\boldsymbol{\sigma}, \varepsilon, C}$ or simply m_C if $\boldsymbol{\sigma}$ and ε are fixed beforehand.

REMARK. The ray $(\mathbf{R}^+, \infty)$ is always a line of an $\mathbf{R}$-regular partition. This is also true for a modified partition of type $(\boldsymbol{\sigma}, k)$ whose numerical type is greater than k. If the numerical type of a partition of type $(\boldsymbol{\sigma}, k)$ is equal to k, then the ray $(\mathbf{R}^+, \infty)$ is not a boundary line of the partition.

B. Regular cochains.

The following definition is fundamental in this section.

DEFINITION 10. An ε-extendable germ of an $\mathbf{R}$-regular (k-regular) cochain of class $\boldsymbol{\Omega}$ is a germ with a representative (called an $\mathbf{R}$-regular or k-regular functional cochain, and also a cochain of class $(\boldsymbol{\sigma}, \mathbf{R})$ or $(\boldsymbol{\sigma}, k)$) defined in a standard domain of class $\boldsymbol{\Omega}$ depending on the germ such that:

1°. The corresponding partition is an $\mathbf{R}$-regular partition of type $\boldsymbol{\sigma}_{\mathbf{R}}$ (k-regular of type $(\boldsymbol{\sigma}, k)$), where $\boldsymbol{\sigma} = (\sigma_1, \dots, \sigma_N)$, the germs σ_j are admissible germs of class $\boldsymbol{\Omega}$, and the partition is rigged;

2°. The functions forming the cochain extend holomorphically to generalized ε-neighborhoods of the domains of the partition that correspond to them;

3°. The modulus of the cochain can be estimated from above by the function $\exp \nu\xi$ for some $\nu \in \mathbf{R}$ depending on the cochain, with the cochain called a rapidly decreasing cochain if ν can be taken < 0 in this estimate, and a weakly decreasing cochain if the modulus of the cochain can be estimated from above by the function $C|\zeta|^{-5}$ for some $C > 0$;

4°. The functions forming the coboundary of the cochain admit analytic extension to the generalized ε-neighborhoods of the standard boundary lines of the partition and can be estimated in modulus there from above by the corresponding functions of the rigging cochain $\mathscr{M}_C$ for some $C > 0$ depending on the cochain.

REMARK. Suppose that $\boldsymbol{\sigma} = (\sigma_1, \dots, \sigma_N)$, and Ω is a standard domain in which the diffeomorphisms $\rho_j = \sigma_j^{-1}$ are defined. Regular partitions of type $\boldsymbol{\sigma}_{\mathbf{R}}$ and $(\boldsymbol{\sigma}, k)$ and rigging cochains of them coincide in the upper half-plane. This motivates the following definition.

DEFINITION 11. A regular cochain F of type $\boldsymbol{\sigma} = (\sigma_1, \dots, \sigma_N)$ is a tuple of $N+1$ cochains of type $\boldsymbol{\sigma}_R$ and (σ, k), $1 \le k \le N$, that coincide in the intersection of a standard domain with the upper half-plane. The cochains of the tuple are called the $\mathbf{R}$-realization and the k-realization of the cochain F, respectively, and denoted by $F_{\mathbf{R}}$ and $F_{(k)}$. The parameter N is called the numerical type of the cochain. If all the cochains in the tuple are rapidly (weakly) decreasing, then the tuple itself is said to be a rapidly (weakly) decreasing cochain.

The set of all regular functional cochains is denoted by $\mathscr{F}\mathscr{C}_{\text{reg}}$, and the rapidly decreasing regular functional cochains by $\mathscr{F}\mathscr{C}_{\text{reg}}^{+}$ ($\mathscr{F}\mathscr{C}$ for functional cochains).

DEFINITION 12. An ε-extendable germ of a regular map-cochain is a germ whose correction is the germ of an ε-extendable rapidly decreasing regular functional cochain. The germ of a weakly regular ε-extendable map-cochain is defined similarly: in the preceding definition the correction must decrease not rapidly, but weakly. The sets of all regular and weakly regular map-cochains are denoted by $\mathscr{M}\mathscr{C}_{\text{reg}}$ and $\mathscr{M}\mathscr{C}_{\text{wr}}$ ($\mathscr{M}\mathscr{C}$ for mapping cochains, and wr for weakly regular).

DEFINITION 13. Let $\mathscr{D}$ be an arbitrary set consisting of germs of admissible diffeomorphisms. The germ of a functional cochain or map-cochain is regular of type $\mathscr{D}$ if the corresponding partition is of type $\boldsymbol{\sigma} = (\sigma_1, \dots, \sigma_N)$, with $\sigma_j \in \mathscr{D}$. Notation:

$$\mathscr{F}\mathscr{C}_{\text{reg}}(\mathscr{D}), \qquad \mathscr{M}\mathscr{C}_{\text{reg}}(\mathscr{D}).$$

The sets of germs of regular rapidly or weakly decreasing functional cochains of class $\mathscr{D}$ and the set of weakly regular map-cochains of class $\mathscr{D}$ are denoted by

$$\mathscr{F}\mathscr{C}_{\text{reg}}^{+}(\mathscr{D}), \qquad \mathscr{F}\mathscr{C}_{\text{wr}}(\mathscr{D}), \qquad \mathscr{M}\mathscr{C}_{\text{wr}}(\mathscr{D}),$$

respectively.

EXAMPLES OF REGULAR MAP-COCHAINS. 1. The mappings of class $\mathscr{N}\mathscr{C}$. They satisfy the requirements of Definition 10, as mentioned in §1.2.

2. Simple map-cochains. These are the regular map-cochains corresponding to the class of quadratic standard domains for which the germs of the admissible diffeomorphisms σ_j are linear.

3. Sectorial map-cochains are analogous cochains, the only difference being that the corresponding germs of admissible diffeomorphisms are exponentials: $\sigma_j = \exp \circ \mu_j \zeta$, $\mu_j \in (0, 1)$.

COMMENTS. Map-cochain of class $\mathcal{N}\mathcal{C}$ are simple. In Chapter II we prove general theorems implying that map-cochains of class $\mathcal{R} \circ \mathcal{N}\mathcal{C}$ are also simple. The germs of class $\mathrm{Ad}(\mathcal{A}\!f\!f)\mathcal{N}\mathcal{C}$ possess the same property. Finally, germs of simple map-cochains form a group with the "composition" operation; namely, for compositions there arise intersections of images of the standard partition under the action of different dilations with positive coefficient. We write an explicit formula for the rigging cochain corresponding to a regular partition of type $\boldsymbol{\sigma}_{\mathbf{R}}$, $\sigma = (\mu_1, \dots, \mu_N)$; it follows from the general Definition 5, of course. The rigging cochain function defined in a neighborhood of the boundary (horizontal) ray $\mathcal{L}$ is equal to

$$m_{\boldsymbol{\sigma}, \varepsilon, C, \mathcal{L}} = \sum \exp(-C \exp \mu_j^{-1} \xi).$$

The summation is over all j such that $\mathcal{L} \subset \mu_j \partial \Xi_{\mathrm{st}}$.

The germs of the sectorial map-cochains also form a group with the "composition" operation. The function of the corresponding rigging cochain defined in the generalized ε-neighborhood of the ray $\mathcal{L}$ of the partition (lying on a line through 0; the neighborhood itself is the part of a sector outside a compact set) is equal to

$$m_{\boldsymbol{\sigma}, \varepsilon, C, \mathcal{L}} = \sum \exp(-C|\zeta|^{\mu_j^{-1}}).$$

The summation is over those j for which

$$\mathcal{L} \subset \exp \circ \mu_j (\partial \Xi_{\mathrm{st}} \cap \mu_j^{-1} \circ \ln \tilde{\Omega}).$$

C. Special sets of admissible germs.

The main goal of this and the next section is to introduce the groups G^n, which contain the monodromy transformations of class n and are $(n+1)$-proper in the sense of Definition 4 in B in §1.4. The elements of G^n are defined directly, not in terms of a product of generators, like monodromy transformations of class n. Their corrections do not oscillate—this is a basic fact proved below. The elements of G^n are defined with the use of regular functional cochains. The regular functional cochains in the definition of the group G^n correspond to partitions constructed with the help of G^{n-1}. Thus, the definition is inductive: in defining G^n we assume that G^{n-1} has already been defined. The induction base: $G^{-1} = \mathcal{A}\!f\!f$ and $G^0 = \mathcal{R}$. The subsequent definitions represent the induction step.

REMARK. The fact that the group $G^0 = \mathcal{R}$ is ordered follows from the Phragmén-Lindelöf theorem. The proof that G^n is proper is our basic goal.

For an arbitrary ordered group G denote by G^- (G^+) the set $\{g \in G | g \preceq \mathrm{id}\}$ ($g \succeq \mathrm{id}$), where id is the identity of the group. Recall that the subgroup G_{rap} and the semigroups G^-_{slow} and G^+_{slow} are defined for an arbitrary proper group G; see C in §1.4. Thus, by the induction hypothesis, the sets G^m_{rap} and $G^{m\pm}_{\mathrm{slow}}$ are defined for $m \le n-1$.

DEFINITION 14. Let $\sigma: (\mathbf{R}^+, \infty) \to (\mathbf{R}^+, \infty)$ be the germ of a diffeomorphism. The generalized exponent of the germ σ is defined to be the limit $\mu(\sigma) = \lim_{(\mathbf{R}^+, \infty)} (A\sigma)'$, if it exists.

The following definitions are especially important. They introduce the sets of admissible germs to be used in constructing all subsequent partitions. All germs in the group G^m, $m \le n-1$, are real. Definitions 15 and 16 below introduce germs of diffeomorphisms defined on $(\mathbf{R}^+, \infty)$. Lemma 5.1 below says something about the possibility of extending them biholomorphically.

DEFINITION 15. Let $\mu_m(g) = \mu(A^{-m} g)$,

$$L^m = \mathrm{Gr}(\mathrm{Ad}(A^{1-m} g \circ \ln) f | f \in \mathscr{A}^0, \ g \in G^{m-1}, \ \mu_m(g) = 0),$$

$$\mathscr{L}^m = \mathrm{Gr}(\mathrm{Ad}(A^{1-m} g \circ \ln) \tilde{f} | \tilde{f} = A(g_1) f, \ f \in \mathscr{A}^0,$$

$$g_1 \in \mathscr{H}^0, \ \tilde{f} \in \mathscr{M}_{\mathbf{R}}, \ g \in G^{m-1}, \ \mu_m(g) = 0).$$

REMARK. The germs of the set $\mathscr{L}^m$ are the germs of the set L^m, "corrected up to real germs." For $m = 0$ and $g \in G^{-1} = \mathscr{A}\!f\!f$ we have $\mu_m(g) = 1$. For $m = 1$ and $g \in G^0 = \mathscr{R}$ we have $\mu_m(g) \in (0, \infty)$. Therefore, $\mathscr{L}^m = \varnothing$ for $m \le 1$. Take $g \circ \varnothing = g$.

DEFINITION 16. a. $\mathscr{D}^n_0 = A^{-n} G^{n-1^-}_{\mathrm{slow}} \circ \exp$.

b. $\mathscr{D}^m_1 = \mathscr{D}^m_0 \cup \mathscr{D}^m_* \circ \mathscr{L}^m$, where

$$\mathscr{D}^m_* = A^{-m}(G^{m-1^+}_{\mathrm{slow}} \cup G^{m-1}_{\mathrm{rap}}), \qquad m \le n-1,$$

$$\mathscr{D}^m_{\mathrm{slow}} = A^{-m} G^{m-1^+}_{\mathrm{slow}}, \qquad \mathscr{D}^m_{\mathrm{rap}} = A^{-m} G^{m-1}_{\mathrm{rap}}.$$

LEMMA 5.1_n. a. *All the germs in Definitions* 15 *and* 16 *admit a biholomorphic extension to* $(S^\forall, \infty)$, *where* $S^\forall$ *is an arbitrary sector of the form* $|\arg \zeta| < \alpha$, $\pi/2 < \alpha < \pi$.

The lemmas whose numbers begin with 5 are proved in Chapter V for all $n > 1$; they are proved earlier for $n = 1$.

D. Standard domains of class n.

DEFINITION 17. A standard domain of class 0 is a quadratic standard domain. A standard domain of class $n \ge 1$ and type 1 is a domain that can be specified in the form $\Omega = A^{-n} g(\mathbf{C}^+ \backslash K)$, where $g \in G^{n-1}_{\mathrm{rap}}$, and K is a disk such that the mapping $A^{-n} g$ is biholomorphic in the domain $\mathbf{C}^+ \backslash K$ (as shown below, such a disk always exists).

EXAMPLE 1. A quadratic standard domain is of class 1 and type 1.

Indeed, by definition, a quadratic standard domain has the form

$$\Omega = \varphi(\mathbf{C}^+), \qquad \varphi = \zeta + C\sqrt{\zeta + 1}.$$

We prove that $A\varphi \in G^0$. Indeed,

$$\begin{aligned} A(\zeta + C\sqrt{\zeta+1}) &= \ln(\exp\zeta + C\sqrt{\exp\zeta + 1}) \\ &= \zeta + \ln(1 - \exp(-\zeta/2)\sqrt{1 + \exp(-\zeta)}). \end{aligned}$$

EXAMPLE 2. Standard domains of class 1 have the following property.

PROPOSITION 1. (³) *For any positive number* $\nu < 1$ *there exists a standard domain of class* 1 *such that in it*

$$|\zeta|^\nu \prec \varepsilon\xi$$

for all $\varepsilon > 0$.

PROOF. The desired domain is taken in the form

$$\Omega_{C,\mu} = \psi_{C,\mu}\mathbf{C}^+, \quad \psi_{C,\mu} = \mathrm{id} + C\zeta^\mu, \qquad \mu \in (\nu, 1).$$

We prove that it is standard of class 1. Indeed,

$$\begin{aligned} A\psi_{C,\mu} &= \ln\circ(\exp + C\exp\circ\mu) \\ &= \mathrm{id} + \ln(1 + C\exp\circ(\mu - 1)) \in \mathscr{R} = G^0. \end{aligned}$$

It is proved that $\psi_{C,\mu} \in G^0_{\mathrm{rap}}$.

Next, the boundary of the domain $\Omega_{C,\mu}$ has the asymptotics

$$\xi = (C_1 + o(1))|\eta|^\mu, \qquad C_1 > 0.$$

Consequently, in $\Omega_{C,\mu}$

$$|\zeta| \prec O(1)\xi^{1/\mu}.$$

From this,

$$|\zeta|^\nu = O(1)\xi^{\nu/\mu} = o(1)\xi. \blacktriangleright$$

EXAMPLE 3.

PROPOSITION 2. *The logarithmic standard domain*

$$\Omega_{L,C} = \left(\mathrm{id} + C\frac{\zeta}{\ln\zeta}\right)(\mathbf{C}^+\backslash K)$$

is a standard domain of class 2. *In it*

$$|\zeta| \prec \varepsilon\xi\ln^3|\zeta|$$

for all $\varepsilon > 0$.

PROOF. We prove the first assertion. It must be verified that

$$A^2\left(\zeta + C\frac{\zeta}{\ln\zeta}\right) = g, \qquad g \in G^1.$$

(³) The relation $f \prec g$ for functions on the domain Ω with boundary including infinity means, by definition, that there is a neighborhood of infinity in Ω in which $f < g$ pointwise.

Indeed,

$$A^2\left(\zeta + C\frac{\zeta}{\ln\zeta}\right) = A\left(\zeta + \ln\left(1+\frac{C}{\zeta}\right)\right)$$
$$= \zeta + \ln[1 + (\exp(-\zeta))\ln(1 + C\exp(-\zeta))] \in G^0 \subset G^1.$$

This proves the first assertion. Let us prove the second. The boundary of the logarithmic domain has the asymptotics

$$\xi = \frac{C + o(1)}{\ln^2|\eta|}\eta$$

as $|\eta| \to \infty$. Consequently, in $\Omega_{L,C}$,

$$|\zeta| \le \xi \ln^2|\zeta| O(1) \prec \varepsilon\xi \ln^3|\zeta|$$

for all $\varepsilon > 0$. ▶

Below we define standard domains of class n and type 2 for $n > 1$. Standard domains of class n and types 1 and 2 together are called standard domains of class n.

The set of all standard domains of class n is denoted by $\mathbf{\Omega}_n$. The admissible germs of class $\mathbf{\Omega}_n$ (see Definition 2 in §1.5) are called admissible germs of class n for brevity. The definition of G^n given below in §1.7 uses the sets $\mathscr{D}_0^n$ and $\mathscr{D}_1^{n-1}$.

LEMMA 5.1_n.b. *The sets* $\mathscr{D}_0^n$ *and* $\mathscr{D}_1^{n-1}$ *consist of admissible germs of class* n.

LEMMA 5.2_n.a. *A standard domain of class* n *is standard in the sense of the definition in* §1.4.

Lemma 5.2_n.b is formulated below. Lemma 5.2_n.a is proved for $n = 0$ in §1.5, Proposition 1.

E. Equivalent and negligible diffeomorphisms.

EXAMPLE 4. Description of the subsets G^0_{rap}, G^{0+}_{slow}, and G^{0-}_{slow}.

By definition, G^0 is the group of almost regular germs. For an arbitrary such germ g the limit $\mu(g) = \lim_{(\mathbf{R}^+,\infty)} g'$ exists. We prove that the germs of the group G^0_{rap} are distinguished by the requirement that $\mu(g) = 1$. Indeed,

$$A^{-1}g = A^{-1}((\mu + o(1))\zeta) = \exp\circ(\mu + o(1))\ln\zeta$$
$$= \exp\circ\ln\zeta^{\mu+o(1)} = \zeta^{\mu+o(1)}, \qquad \mu = \mu(g).$$

This implies that $g \in G^{0+}_{\text{slow}}$ for $\mu > 1$, and $g \in G^{0-}_{\text{slow}}$ for $\mu < 1$. Further, for $\mu = 1$,

$$g(\zeta) = \zeta + O(1)$$

because g can be expanded in an asymptotic Dulac series.

Consequently, $A^{-1}g = \zeta \cdot O(1)$, that is, $g \in G^0_{\text{rap}}$.

EXAMPLE 5. The set $\mathscr{D}_1^0$ is the set $\mathscr{A}\!f\!f$ by definition. The germs in this set are admissible for an arbitrary class of standard domains.

To investigate the set $\mathscr{D}_0^1$ it is useful to "simplify" the germs of this set. To do this we need

DEFINITION 18. A diffeomorphism of a domain containing the point at infinity on the boundary is said to be negligible in this domain if its correction decreases in this domain more rapidly than $|\zeta|^{-5}$. Two admissible germs σ and $\tilde{\sigma}$ of class $\mathbf{\Omega}$ are said to be equivalent if there exists a standard domain Ω of class $\mathbf{\Omega}$ such that the composition quotient $\sigma^{-1}\circ\tilde{\sigma}$ is negligible in the domain $\tilde{\sigma}^{-1}\Omega$.

Definition 18 is motivated by the following circumstance. Suppose that the admissible germs σ and $\tilde{\sigma}$ of class $\mathbf{\Omega}$ are equivalent, and the domain Ω is the same as in Definition 13. Let $\mathscr{D}$ be a domain of the standard partition or a modified standard partition, and let $\mathscr{D}^\varepsilon$ and $\mathscr{D}^{\varepsilon/2}$ be its generalized ε- and $\varepsilon/2$-neighborhoods. Then there exist two disks $K\subset K'$ such that

$$\tilde{\sigma}(\mathscr{D}^{\varepsilon/2}\cap\Omega\backslash K')\subset\sigma(\mathscr{D}^{\varepsilon}\cap\Omega\backslash K).$$

Indeed, this inclusion is equivalent to the following:

$$\sigma^{-1}\circ\tilde{\sigma}(\mathscr{D}^{\varepsilon/2}\cap\Omega\backslash K')\subset\mathscr{D}^{\varepsilon}\cap\Omega\backslash K.$$

The margin between $\mathscr{D}^{\varepsilon/2}$ and $\mathscr{D}^{\varepsilon}$, that is, the function $\operatorname{dist}(\zeta,\partial\mathscr{D}^{\varepsilon})$ for $\zeta\in\mathscr{D}^{\varepsilon/2}$, does not decrease more rapidly than $|\zeta|^{-3}$ at infinity (4), while the correction of the germ $\sigma^{-1}\tilde{\sigma}$, extended to $\tilde{\sigma}^{-1}(\Omega\backslash K)$, decreases more rapidly than $|\zeta|^{-5}$. The disk K can be chosen so that the correction of this germ is less than the margin everywhere in $\tilde{\sigma}^{-1}(\Omega\backslash K)$. This yields the necessary inclusion.

This means that in the definition of a regular functional cochain the admissible germs of diffeomorphisms can be replaced by equivalent germs. Namely, if the germ σ_j is equivalent to the germ $\tilde{\sigma}_j$, $j=1,\dots,N$, then the germ of a (σ,ε)-extendable regular functional cochain f of type $\sigma=(\sigma_1,\dots,\sigma_N)$ is simultaneously a $(\tilde{\sigma},\varepsilon/2)$-extendable germ of type $\tilde{\sigma}=(\tilde{\sigma}_1,\dots,\tilde{\sigma}_N)$. Indeed, the functions in the tuple f, which are defined in (σ,ε)-neighborhoods of the domains of a regular partition of type σ, are simultaneously defined in $(\tilde{\sigma},\varepsilon/2)$-neighborhoods of the domains of a regular partition of type $\tilde{\sigma}$. The main function of the tuple f extends to the (σ,ε)-neighborhood of the extended main domain of the partition of type σ, and hence to the $(\tilde{\sigma},\varepsilon/2)$-neighborhood of the extended main domain of the partition of type $\tilde{\sigma}$. The growth estimates for the functions in the tuple f in the $(\tilde{\sigma},\varepsilon/2)$-neighborhoods of the domains of the partition of type $\tilde{\sigma}$ are the same as in the (σ,ε)-neighborhoods of the domains of the partition of type σ. The rigging cochains for the partitions of types σ and $\tilde{\sigma}$ are comparable, that is,

---

(4) This is verified in §2.1 below.

for each rigging cochain of one partition there exists a rigging cochain of the other partition that is defined possibly in more narrow neighborhoods of the boundary lines of the partition and majorizes the first rigging cochain. This follows from the equality

$$\operatorname{Re}\tilde{\rho}_j = \operatorname{Re}(\tilde{\rho}_j \circ \rho_j^{-1}) \circ \rho_j = \operatorname{Re}\rho_j + o(1); \qquad \rho_j = \sigma_j^{-1}, \ \tilde{\rho}_j = \tilde{\sigma}_j^{-1}.$$

EXAMPLE 6. In suitable quadratic standard domains the germs of the class $\mathscr{D}_0^1$ are equivalent to germs of the form $\exp(\mu\zeta + C)$, $\mu \in (0, 1)$.

Indeed, the group $\mathscr{R}$ decomposes into the product $\mathscr{R} = \mathscr{A}\!f\!f \circ \mathscr{R}^0$ (recall that $\mathscr{R}^0$ is the group of almost regular germs with exponentially decreasing correction). Suppose that $g \in \mathscr{R}$ is an arbitrary germ, $g = ag_0$, $a \in \mathscr{A}\!f\!f$, $g_0 \in \mathscr{R}^0$. Then the germs $\sigma = \exp g$ and $\tilde{\sigma} = \exp a$ are equivalent in the domain $\tilde{\sigma}^{-1}\Omega$ for some quadratic standard domain Ω, and even for the domain $\mathbf{C}^+ \backslash K$, where K is a sufficiently large disk; it is necessary only that the half-strip $\tilde{\sigma}^{-1}(\mathbf{C}^+ \backslash K)$ lie in the domain of definition of the diffeomorphism $\sigma^{-1} \circ \tilde{\sigma}$. Indeed, $\sigma^{-1} \circ \tilde{\sigma} = g_0^{-1}$; the correction of the germ g_0^{-1} decreases exponentially in some quadratic standard domain, and hence in the domain $a^{-1} \circ \ln(\mathbf{C}^+ \backslash K)$.

REMARK. The examples after Definition 13 now admit a new interpretation. Namely, simple map-cochains are regular map-cochains of class $\mathscr{D}_1^0$, while sectorial map-cochains are regular map-cochains of class $\mathscr{D}_0^1$. We now change the definition of the class $\mathscr{D}_0^1$: the diffeomorphisms of class $\mathscr{D}_0^1$ will be taken to be all the mappings of the form $\exp \mu\zeta$, $\mu \in (0, 1)$, and only they. With this change in the definition of the class, the set of regular cochains of class $\mathscr{D}_0^1$ itself remains unchanged. It is such cochains that are considered in the first part.

§1.7. Main definitions: standard domains, superexact asymptotic series, and regular functional cochains of class n

The series and cochains defined in this subsection have a class, a rank, and a type taking the values 0 or 1. Pairs of nonnegative integers, the first of which denotes the class and the second the rank, are lexicographically ordered. The definitions are given with the help of transfinite induction on the class and rank. Moreover, the groups G^n, which include G_n, are defined; the identity theorem is proved below just for their elements. As before, let $G^{-1} = \mathscr{A}\!f\!f$.

INDUCTION BASE. $n = 0$.

DEFINITION 1°. A standard domain of class 0 is a quadratic standard domain (see Definition 1 in §0.3 A).

The rank is not defined for superexact asymptotic series and regular cochains of class 0.

DEFINITION 2°. A STAR-0 is a Dulac series; a real STAR-0 is a Dulac series with real terms. The set of all STAR-0 is denoted by Σ^0, and the set of all real STAR-0 by Σ^0_R.

The regular functional cochains of class n defined below have two properties: regularity and expandability. Expandability of the germ of a functional cochain in a STAR-0 means by definition that there exists a representative of this germ defined in a standard domain of class 0 depending on the germ and such that for every $\nu > 0$ there is a partial sum of the STAR-0, depending only on the germ and not on ν, that approximates the germ with accuracy $o(\exp(-\nu\xi))$ in the chosen domain.

DEFINITION 3°. The germ of a regular functional cochain of class 0 and type l is defined when $l = 1$ to be the germ of a regular functional cochain of type $\mathscr{D}_1^0 = \mathscr{A}\!f\!f$, expandable in a STAR-0; when $l = 0$ it is an almost regular germ, that is, a germ of class $\mathscr{R}$. The set of all germs of regular functional cochains of class 0 and type l is denoted by $\mathscr{F}\mathscr{C}_l^0$, and the set of rapidly decreasing germs of regular functional cochains of class 0 and type l by $\mathscr{F}\mathscr{C}_{l+}^0$; the subsets of the preceding sets consisting of weakly real germs (see Definition 5 below) are denoted by $\mathscr{F}\mathscr{C}_{\mathbf{R}l}^0$ and $\mathscr{F}\mathscr{C}_{\mathbf{R}l+}^0$, respectively.

DEFINITION 4°. Let $J^0 = \mathrm{Ad}(\mathscr{A}\!f\!f)\mathscr{A}^0$.

DEFINITION 5. A cochain φ is said to be weakly real if it is defined in a domain symmetric with respect to $\mathbf{R}$ and $\varphi(\overline{\zeta}) = \overline{\varphi(\zeta)}$ on $(\mathbf{R}^+, \infty)$. An equivalent definition: $R\varphi = \varphi$ and $I\varphi = 0$ on $(\mathbf{R}^+, \infty)$, where $R\varphi(\zeta) = (\varphi(\zeta) + \overline{\varphi(\overline{\zeta})})/2$ and $I\varphi(\zeta) = (\varphi(\zeta) - \overline{\varphi(\overline{\zeta})})/2i$.

REMARK. As a rule, the cochains considered below have a jump on the real axis; more precisely, the real axis is a boundary line for the corresponding partition. Therefore, the functions in the tuple φ that correspond to the domains adjacent to the real axis do not coincide on the real axis. For a weakly real cochain these functions are conjugate on the real axis. In particular, the coboundary of such a cochain, restricted to $(\mathbf{R}^+, \infty)$, is equal to twice the imaginary part of the cochain on $(\mathbf{R}^+, \infty)$.

INDUCTION STEP. Suppose that $n \geq 1$ and $r \geq 0$. Assume that for all $(m, q) < (n, r)$ we have already defined: the sets $\Sigma_0^{m,q}$ and $\Sigma_1^{m-1,q}$ of STAR of class m, rank q, and type 0, as well as those of class $m-1$, rank q, and type 1; the sets $\mathscr{F}\mathscr{C}_0^{m,q}$ and $\mathscr{F}\mathscr{C}_1^{m-1,q}$ of germs of regular functional cochains of class m, rank q, and type 0 and of class $m-1$, rank q, and type 1; the group $G^m (m \leq n-1)$; the set of standard domains of class m $(m \leq n)$.

The unions $\bigcup_q \sum_0^{m,q}$, $\bigcup_q \sum_1^{m-1,q}$, $\bigcup_q \mathscr{F}\mathscr{C}_0^{m,q}$, and $\bigcup_q \mathscr{F}\mathscr{C}_1^{m-1,q}$ are denoted by $\sum_0^m$, $\sum_1^{m-1}$, $\mathscr{F}\mathscr{C}_0^m$, and $\mathscr{F}\mathscr{C}_1^{m-1}$, respectively. The subsets of the sets $\mathscr{F}\mathscr{C}_0^m$ and $\mathscr{F}\mathscr{C}_1^{m-1}$ consisting of rapidly decreasing germs are denoted by $\mathscr{F}\mathscr{C}_{0+}^m$ and $\mathscr{F}\mathscr{C}_{1+}^{m-1}$, respectively.

We now define standard domains and germs of admissible diffeomorphisms of class n. Below we encounter both assertions such as "there exists a standard domain of class $n \ldots$" (having some property) and assertions such as "for each standard domain of class $n \ldots$" (some assertion holds). Therefore, the set of standard domains must be sufficiently rich on the one hand, and not too broad on the other hand.

Denote by $\mathscr{L}(\mathscr{A})$ the **R**-linear hull of a set $\mathscr{A}$.

DEFINITION 1^n. A standard domain of class n and type 1 is defined to be a domain of the form $A^{-n}g(\mathbf{C}^+\backslash K)$, where $g \in G_{\mathrm{rap}}^{n-1}$, and K is a disk such that the mapping $A^{-n}g$ is biholomorphic in $\mathbf{C}^+\backslash K$; it is proved in Chapter V that such a disk always exists.

A standard domain of class n and type 2 $(n \geq 2)$ is defined to be a domain of the form

$$A^{-1}(\mathrm{id} + \psi)(\mathbf{C}^+\backslash K),$$

where K is a disk such that the previous mapping is biholomorphic in $\mathbf{C}^+\backslash K$, $\mathrm{id} + \psi \in \mathscr{M}_{\mathbf{R}}$, and ψ has the following form:

$$\psi = \exp\varphi, \qquad \varphi \in [\mathscr{L}(\mathscr{F}_{\mathbf{R},\mathrm{id}}^{n-2}, \mathscr{F}_{++}^{n-2}, \mathscr{F}_{+}^{n-1})\backslash\mathscr{L}(\mathscr{F}_{++}^{n-2}, \mathscr{F}_{+}^{n-1})] \circ \ln^{[n-1]},$$

$$\mathscr{F}_{++}^{n-2} = \{\tilde{\varphi} \in \mathscr{F}_{+g}^{n-2} | g \in G_{\mathrm{slow}}^{n-3^+}\}, \qquad \mathrm{Re}\,\varphi \to -\infty, \ \varphi' \to 0$$

in $(\Pi^\forall, \infty)$, where $\Pi^\forall$ is a rectilinear half-strip of the form $\xi \geq 0$, $|\eta| \leq a$ with $a > 0$ arbitrary.

A standard domain of class n is a domain of one of the types defined above.

An admissible germ of class n is an admissible germ of class $\mathbf{\Omega}_n$, where $\mathbf{\Omega}_n$ is the set of all standard domains of class n; see Definition 1^n.

Everywhere below in Chapter I, $l \in \{0, 1\}$.

The next lemma strengthens the analogous assertion in §1.6.

LEMMA 5.2.a. *An arbitrary standard domain of class* n *is standard in the sense of Definition* 1 *in* §1.5.

For brevity let

$$\mathscr{F}_{l,g}^m = \mathscr{F}\mathscr{C}_l^m \circ \exp^{[m]} \circ g, \qquad \mathscr{F}_l^m = \bigcup_{g\in G^{m-1}} \mathscr{F}_{l,g}^m,$$

$$\mathscr{F}_{l+,g}^m = \mathscr{F}\mathscr{C}_{l+}^m \circ \exp^{[m]} \circ g, \qquad \mathscr{F}_{l+}^m = \bigcup_{g\in G^{m-1}} \mathscr{F}_{l+,g}^m.$$

We proceed to the definition of STAR-$(n-l, r)_l$. The definition consists of three parts. In the first part we define a set E^n of exponents of STAR-(n, r) that is independent of r, in the second we define a set $\mathscr{K}_l^{n-l,r}$ of coefficients that depends on r, and in the third we define the series themselves.

DEFINITION 2^n. I. The set E^n of exponents of STAR-n is the set of partial sums of STAR-$(n-1)_1$ of type 1, which has the following additional properties. Let $\mathbf{e}$ be in E^n. Then:

1°. the sum $\mathbf{e}$ is weakly real with nonnegative principal exponents;

2°. the limit

$$\nu(\mathbf{e}) = \lim_{(\mathbf{R}^+, \infty)} \mathbf{e}/\exp^{[n]}$$

exists, called the principal exponent of the term with exponent $\mathbf{e}$;

3°. there exists a $\mu > 0$ and a standard domain of class n in which

$$|\mathrm{Re}\,\mathbf{e} \circ \ln^{[n]}| < \mu\xi, \qquad |\mathbf{e} \circ \ln^{[n]}| < \mu|\zeta|;$$

4°. $\mathrm{Im}\,\mathbf{e} \to 0$ on $(\mathbf{R}^+, \infty)$.

The mapping $\nu: E^n \to \mathbf{R}$, $\mathbf{e} \mapsto \nu(\mathbf{e})$, is called the principal exponents mapping.

II. For $r = 0$ the set $\mathscr{K}_l^{n-l,r}$ is the set $\mathscr{L}(\mathscr{F}_1^{n-1-l})$. For $r \geq 1$,

$$\mathscr{K}_l^{n-l,r} = \mathscr{L}(\mathscr{F}_1^{n-1-l}, \mathscr{F}_{l+g}^{n-l,r-1} | g \in G_{\mathrm{slow}}^{n-1-l^-}).$$

III. A STAR-$(n-l, r)_l$ is a formal series of the form

$$\Sigma = \sum_{j=1}^{\infty} a_j \exp \mathbf{e}_j, \qquad \mathbf{e}_j \in E^n, \ a_j \in \mathscr{K}_l^{n-l,r}, \ \nu(\mathbf{e}_j) \to -\infty.$$

The term $a \exp \mathbf{e}$ is called the term of the STAR-n with exponent $\mathbf{e}$, coefficient a, and principal exponent $\nu(\mathbf{e})$.

DEFINITION 3^n. I. A functional cochain φ is said to be expandable in a STAR-$(n-l)_l$, denoted by Σ, if there is a standard domain Ω of class n such that there exists for every $\nu > 0$ a partial sum of the series Σ approximating φ with accuracy $o(\exp(-\nu \,\mathrm{Re} \exp^{[n-l]}))$ in the domain $\ln^{[n-l]}\Omega$.

II. A functional cochain F is called a regular cochain of class $n-l$, rank r, and type l if it has the following two properties: regularity and expandability.

Regularity: F is a regular of class $\mathscr{F}\mathscr{C}_{\mathrm{reg}}(\mathscr{D}_l^{n-l})$ in the sense of the definitions in §1.6 in some standard domain of class n.

Expandability: the composition $F \circ \exp^{[n-l]} = \varphi$ can be expanded in a STAR-$(n-l, r)$.

The set of all germs of regular functional cochains of class n, rank r, and type 0 (respectively, class $n-1$, rank r, and type 1) is denoted by $\mathscr{F}\mathscr{C}_0^{n,r}$ ($\mathscr{F}\mathscr{C}_1^{n-1,r}$, respectively). The subsets of rapidly decreasing and (or) weakly real cochains are denoted by $\mathscr{F}\mathscr{C}_+^{n,r}$, $\mathscr{F}\mathscr{C}_{\mathbf{R}0+}^{n,r}$, $\mathscr{F}\mathscr{C}_{\mathbf{R}0}^{n,r}$, $\mathscr{F}\mathscr{C}_{1+}^{n-1,r}$, etc. The union with respect to rank of a collection of all sets in this list differing only in rank is denoted like the corresponding sets, but without indicating the rank. For example, $\bigcup_r \mathscr{F}\mathscr{C}_{\mathbf{R}1+}^{n-1,r} = \mathscr{F}\mathscr{C}_{\mathbf{R}1+}^{n-1}$.

DEFINITION 4^n. Let:

$$\begin{aligned} H^n &= \mathrm{Gr}(\mathrm{id} + \mathscr{F}_{0+}^n), \\ J^{n-1} &= \mathrm{Ad}(G^{n-1})A^{n-1}\mathscr{A}^0, \\ G^n &= G^{n-1} \circ J^{n-1} \circ H^n \cap \mathscr{M}_{\mathbf{R}}. \end{aligned}$$

This concludes the basic definitions of sets of germs and series for what follows.

EXAMPLES. 1. Let $n = 1$. Then E^n is the set of sums of real quasipolynomials with exponents in $[0, 1]$, where a quasipolynomial with exponent 1 has degree zero or is equal to zero. A STAR-$(1, 0)$ has the form

$$\Sigma = \sum a_j \exp \mathbf{e}_j, \qquad a_j \in \mathscr{F}\mathscr{C}^0, \ \mathbf{e}_j \in E^1, \ \nu(\mathbf{e}_j) \to -\infty.$$

REMARK. A representation of a term in a STAR-$(1, 0)$ in the form

$$a \exp \mathbf{e}, \qquad a \in \mathscr{K}^{1,0}, \ \mathbf{e} \in E^1,$$

is not unique. For example,

$$(\exp \zeta)\exp(-\exp \zeta) = \exp \circ [(-1 + \exp(-\zeta)) \cdot \exp \zeta].$$

The exponents on the right-hand side and left-hand side are equal to $-\exp \zeta$ and $-\exp \zeta + 1$, respectively, while the coefficients are $\exp \zeta$ and 1. Examples of substantially more complicated nonuniqueness arise in the expressions for STAR-(m, r) when $(m, r) > (1, 0)$. Because of this, equality to a STAR-m is not defined algebraically ("after a suitable renumbering, the terms with the same indices coincide"), but analytically. Namely, we use

DEFINITION 6. A STAR-m is equal to zero if for every $\nu > 0$ all its partial sums from some point on (depending on ν) equal $o(\exp(-\nu \exp^{[m]}))$ on the real axis as $\xi \to \infty$.

Let us return to the examples.

2. For $h \in \mathscr{R}^0$ the composition Ah can be expanded in the domain $\ln \Omega$ in a STAR-$(1, 0)$ of the form

$$\Sigma = \zeta + \sum \frac{P_j(\exp \zeta)}{\exp k_j \zeta} \exp(-\nu_j \exp \zeta), \qquad 0 < \nu_j \to \infty, \ k_j \in \mathbf{Z}_+,$$

where the P_j are real polynomials, and $\Omega \in \mathbf{\Omega}_0$.

3. A STAR-$(1, 1)_1$ has the form

$$\Sigma^{1,1} = \sum (a_j + \sum b_{jk} \circ \exp \circ \mu_{jk} \zeta) \exp(\mathbf{e}_j),$$

where $a_j \in \mathscr{F}\mathscr{C}_1^0$, $b_{jk} \in \mathscr{F}\mathscr{C}_{1+}^{1,0}$, $\mathbf{e}_j \in E^1$, and $\mu_{jk} \in (0, 1)$. For example,

$$(\exp \zeta + \exp(-\exp \zeta/2))\exp(\exp \zeta)$$

is a term of an STAR-$(1, 1)_1$.

In this example the coefficient in parentheses is given by an explicit formula. In the general case the coefficient can be the sum of a simple functional cochain and decreasing cochains of class n and rank r; this sum cannot be expressed with the help of finitely many operations in terms of elementary functions.

4. Suppose that $h \in H^0$ and $n \geq 1$. Then $A^n h \in H^n$.

PROOF. We prove more, namely:

$$A^n h - \mathrm{id} \in \mathscr{F}_{0+\mathrm{id}}^{n,0}.$$

Regularity. By definition, it suffices to prove that

$$(A^n h - \mathrm{id}) \circ \ln^{[n]} \in \mathscr{F}\mathscr{C}^+_{\mathrm{reg}}(\mathscr{D}^n_0).$$

We prove that the expression on the left-hand side gives a holomorphic function; then we get an upper estimate for the correction of this function. By definition,

$$(A^n h - \mathrm{id}) \circ \ln^{[n]} = \ln^{[n]} \circ h - \ln^{[n]} \overset{\mathrm{def}}{=} \tilde{\varphi}.$$

The function on the right-hand side is holomorphic in a quadratic standard domain far enough to the right, because both h and $\ln^{[n]}$ have this property, and the correction $h - \mathrm{id}$ is exponentially small. Further, the function $|\tilde{\varphi}|$ decreases like some exponential $\exp(-\nu\xi)$, because the function $h - \mathrm{id}$ has this property, while the derivative of the function $\ln^{[n]}$ is bounded. This proves that the composition $A^n h$, regarded as a functional cochain, is regular.

Expandability is proved by induction on n. For $n = 1$ this is an assertion of Example 2. Suppose that expandability has been proved for $m = n - 1$: for every $\nu > 0$ there exists an expansion

$$A^{n-1}h = \mathrm{id} + \varphi, \qquad \varphi = \sum_1^N a_j \exp(-\nu_j \exp^{[n-1]}) + R,$$

where $\nu_j > 0$, $a_j \in \mathscr{K}^{n-1,0}$, and $R = o(|\exp(-\nu \exp^{[n-1]}|)$, in the domain $\ln^{[n-1]}\Omega$, Ω being a quadratic standard domain. Then

$$A^n h = \ln \circ (\exp + \varphi \circ \exp) = \mathrm{id} + \ln\left(1 + \frac{\varphi \circ \exp}{\exp}\right)$$

The expandability of the last term can now be proved with the help of the Taylor formula for the logarithm. ▶

5. Suppose that $h \in J^0$ and $n \geq 1$. Then $A^{n-1}h \in J^{n-1}$. This is an immediate consequence of the definition of the set J^{n-1}.

§1.8. The multiplicative and additive decomposition theorems

Here and in §1.10 we formulate assertions to be proved below in Chapters II and IV. Each assertion has an index, used in referring to it. The formulations contain an integer parameter n used as the index, with respect to which induction will be carried out.

We precede the formulation of the theorems by

DEFINITION 1. a. If f and g belong to G^k, then $f \prec\!\prec g$ in G^k if and only if $f \circ g^{-1} \in G^{k^-}_{\mathrm{slow}}$.

b. $(k, f) \prec (m, g)$ if and only if $f \in G^{k-1}$, $g \in G^{m-1}$, and either $k < m$, or $k = m$ and $f \prec\!\prec g$ in G^{m-1}.

REMARK. Definition 1a is a particular case of Definition 5 in §1.4, because G^k is $(k+1)$-proper, as will be shown below.

MULTIPLICATIVE DECOMPOSITION THEOREM, MDT_n. 1°. *G^n is a group.* 2°. $G_n \subset G^n$.

ADDITIVE DECOMPOSITION THEOREM, ADT_n. *For an arbitrary germ $g \in G^n$ the following expansion is valid:*

$$g = a + \sum \varphi_j + \sum \psi_j, \tag{$*$}$$

$$a \in \mathscr{A}\!f\!f, \quad \varphi_j \in \mathscr{F}^{k_j}_{1+f_j}, \quad k_j \leq n-1,$$

$$\psi_j \in \mathscr{F}^{n}_{0+g_j}, \quad f_j \in G^{k_j - 1}, \quad g_j \in G^{n-1},$$

$$(k_j, f_j) \prec (k_{j+1}, f_{j+1}), \quad g_j \prec\!\prec g_{j+1} \quad \text{in } G^{n-1};$$

the first nonzero term given by the formula $(*)$ in the expansion for $g - \mathrm{id}$ is weakly real.

REMARKS. 1. The theorem MDT_n enables us to represent an arbitrary monodromy transformation $\Delta \in G^n$ of class n as a composition

$$\Delta \in \mathscr{A}\!f\!f \circ J^0 \circ (H^1 \circ J^1) \circ \cdots \circ (H^{n-1} \circ J^{n-1}) \circ H^n.$$

The corrections of germs of class $H^k \circ J^k$ decrease no more slowly than $\exp(-\exp^{[k]} \mu\xi)$ on $(\mathbf{R}^+, \infty)$, where $\mu > 0$ depends on the germ.

2. The main theorem is the additive decomposition theorem; the multiplicative theorem is needed mainly in order to derive from it the additive theorem.

3. The assertion of the ADT_n about the first (lowest, principal) term in the expansion of the correction $g - \mathrm{id}$ being weakly real enables us to get a lower estimate of it, and at the same time to get a lower estimate for the whole correction. This is done in the next section.

§1.9. Reduction of the finiteness theorem to auxiliary results

We prove Theorem V in a formulation more general than in §0.1.

IDENTITY THEOREM. *Let $g \in G^n \cap \mathrm{Fix}_\infty$. Then $g = \mathrm{id}$.*

PROOF. In the expansion for g given by the theorem ADT_n three cases are relevant for the formula $(*)$ in §1.8.

Case 1: $a \neq \mathrm{id}$. In this case the correction $g - \mathrm{id}$ is nonzero, because the difference $a - \mathrm{id}$ has this property, and the remaining term in the expansion in $(*)$ in §1.8 tend to zero on $(\mathbf{R}^+, \infty)$. Consequently, $g \notin \mathrm{Fix}_\infty$ in Case 1.

Case 2: $a = \mathrm{id}$ in the expansion in $(*)$ in §1.8, but $g \neq \mathrm{id}$. We prove that in this case $g \notin \mathrm{Fix}_\infty$. Let φ be the first term in $(*)$ in §1.8 after id: $\varphi \in \mathscr{F}^k_{+f}$ for some k such that $0 \leq k \leq n$ and $f \in G^{k-1}$.

The next two theorems (the proofs of which occupy an essential place in the subsequent text) show that $|\varphi| \in \mathscr{A}^k_f$.

THE PHRAGMÉN-LINDELÖF THEOREM FOR COCHAINS, PL_n. *If a regular functional cochain of class* $\mathcal{F}\mathcal{C}_0^n$ *or* $\mathcal{F}\mathcal{C}_1^m$ *decreases more rapidly than any exponential* $\exp(-\nu\xi)$ $(\nu > 0)$ *on* $(\mathbf{R}^+, \infty)$ *for* $m \leq n-1$, *then it is identically equal to zero on* $(\mathbf{R}^+, \infty)$.

This theorem is proved in Chapter III. It can be used as follows. By the definition of the set $\mathcal{F}_{+f}^k$, there exists a rapidly decreasing cochain $F \in \mathcal{F}\mathcal{C}_+^k$ such that

$$\varphi = F \circ \exp^{[k]} \circ f, \qquad F \not\equiv 0.$$

By the Phragmén-Lindelöf theorem, the STAR-k for the composition $F \circ \exp^{[k]}$ is not identically equal to zero. Otherwise,

$$F \circ \exp^{[k]} = o(\exp(-\nu \exp^{[k]}))$$

for arbitrary $\nu > 0$ on $(\mathbf{R}^+, \infty)$, and hence $F \equiv 0$ on $(\mathbf{R}^+, \infty)$, by the Phragmén-Lindelöf theorem.

After this, the germ $|\varphi|$ can be estimated from below with the help of the theorems to follow.

LEMMA ON A LOWER ESTIMATE FOR THE PARTIAL SUMS OF A STAR-k, LEPS_k. *Let* Σ *be a partial sum of a weakly real STAR-*k *whose principal exponents are greater than* ν *for all terms. Then on* $(\mathbf{R}^+, \infty)$,

$$|\Sigma| \succ \exp(\nu \exp^{[k]}).$$

COROLLARY. *Let* F *be a nonzero weakly real regular cochain of class* k. *Then for some* $\nu > 0$,

$$|F| \succ \exp(-\nu\xi) \quad \text{on } (\mathbf{R}^+, \infty).$$

This lemma and its corollary are proved in §4.10. The reduction "lemma ⇒ corollary" is based on the fact that, of two terms with different principal exponents in a STAR-k, the one with larger principal exponent is larger in modulus on $(\mathbf{R}^+, \infty)$.

The corollary immediately yields the inequality

$$|\varphi| \succ \exp(-\nu \exp^{[k]} \circ f).$$

On the other hand, the cochain F is rapidly decreasing; therefore, for some $\varepsilon > 0$,

$$|\varphi| \prec \exp(-\varepsilon \exp^{[k]} \circ f) \quad \text{on } (\mathbf{R}^+, \infty).$$

This implies that the germ $|\varphi|$ is multiplicatively Archimedean-equivalent to the germ $\tilde{\varphi} = \exp(-\exp^{[k]} \circ f)$, that is, it belongs to the class $\mathcal{A}_f^k$. We now prove that

$$|g - \mathrm{id}| \succ \tilde{\varphi}^\nu (1 + o(1)). \tag{$*$}$$

To do this, note that an arbitrary term of the expansion $(*)$ in §1.8 coming after φ has the form

$$\psi \in F_1 \circ \exp^{[m]} \circ g, \qquad F_1 \in \mathcal{F}\mathcal{C}_+^m, \ g \in G^{m-1}, \ (k, f) \prec (m, g).$$

Since F_1 is a rapidly decreasing cochain, $|\psi| \prec \exp(-\varepsilon \exp^{[m]} \circ g)$ for some $\varepsilon > 0$. Setting $\tilde{\psi} = \exp(-\exp^{[m]} \circ g)$, we get that $|\psi| \prec \tilde{\psi}^{\varepsilon}$. But the germ $\tilde{\psi}$ belongs to the Archimedean class $\mathscr{A}_g^m$. Since $(k, f) \prec (m, g)$, Propositions 3 and 4 in §1.4 are applicable. They imply that $\psi \in |\varphi| \cdot o(1)$. The relation in $(*)$ is obtained from this. Consequently, $g \notin \mathrm{Fix}_{\infty}$ also in this case.

Case 3: $g = \mathrm{id}$. In this case and, as shown above, only in this case the condition of the theorem holds: $g \in G^n \cap \mathrm{Fix}_{\infty}$.

In the next two sections we prove the decomposition theorems modulo the lemmas in §1.10.

§1.10. Group properties of regular map-cochains

Here we formulate lemmas used in §1.11 to derive the decomposition theorems. The proofs of these lemmas occupy the main part of the text. The lemmas are marked by an index used to refer to them. They are formulated for cochains of classes n and $n-1$ and are proved by induction on n; the index indicates the class.

THE FIRST SHIFT LEMMA, SL 1_n.

a. $\mathscr{F}\mathscr{C}^n_{0(+)} \circ \exp^{[n]} \circ G^{n-1}_{\mathrm{rap}} \subset \mathscr{F}\mathscr{C}^n_{0(+)} \circ \exp^{[n]}$.

b. $\mathscr{F}\mathscr{C}^{n-1}_{1(+)} \circ \exp^{[n-1]} \circ G^{n-2}_{\mathrm{rap}} \subset \mathscr{F}\mathscr{C}^{n-1}_{1(+)} \circ \exp^{[n-1]}$.

Here and below, the plus in parentheses means that the assertion is true both with the plus without the parentheses, and without it.

In order not to repeat similar assertions in what follows, we adopt the

CONVENTION. Let n be fixed. Then $\mathscr{D}^m$, $\mathscr{F}^m_{(+)}$, $\mathscr{F}^m_{(+)g}$, $\mathscr{F}^{m(+)}_{\mathrm{reg}}$, etc., stand for $\mathscr{D}^m_1$, $\mathscr{F}^m_{1(+)}$, $\mathscr{F}^m_{1(+)g}$, $\mathscr{F}^{m(+)}_{1,\mathrm{reg}}$, etc., if $m \leq n-1$, and they stand for $\mathscr{D}^n_0$, $\mathscr{F}^n_{0(+)}$, $\mathscr{F}^n_{0(+)g}$, $\mathscr{F}^{n(+)}_{0,\mathrm{reg}}$, etc., if $m = n$.

The preceding convention is based on the fact that the value of n is fixed everywhere in Chapters I–IV. All assertions whose designation contains a nonnegative integer parameter (denoted by m, as a rule) are proved by induction on that parameter. They are trivial for $m = 0$ and are proved for $m \leq n-1$ by the induction hypothesis. The convention singles out the n for which these assertions are proved.

With this convention Lemma $\mathrm{SL1}_n$ takes the following form.

LEMMA $\mathrm{SL1}_n$. *Let $m = n-1$ or $m = n$. Then*

$$\mathscr{F}\mathscr{C}^m_{(+)} \circ \exp^{[m]} \circ G^{m-1}_{\mathrm{rap}} \subset \mathscr{F}\mathscr{C}^m_{(+)} \circ \exp^{[m]}.$$

To formulate Lemma $\mathrm{SL1}_m$ and the following assertions of this section for $m \leq n-1$ it is necessary to decipher $\mathrm{SL1}_n$ and the other assertions by using the preceding convention, and then to replace n by m in the resulting formulation.

THE SECOND SHIFT LEMMA, $\mathrm{SL2}_n$. *Let $m = n-1$ or $m = n$, and suppose that $(k, f) \prec (m, g)$ and $\varphi \in \mathscr{F}_f^k$. Then*

$$\varphi \circ (\mathrm{id} + \mathscr{F}_{+g}^m) \subset \varphi + \mathscr{F}_{+g}^m.$$

THE THIRD SHIFT LEMMA, $\mathrm{SL3}_n$.a. *Let $m = n-1$ or $m = n$, and suppose that $f \succ\succ g$ in G^{m-1} or $f \circ g^{-1} \in G_{\mathrm{rap}}^{m-1}$. Then*

$$\mathscr{F}_{(+)f}^m \circ (\mathrm{id} + \mathscr{F}_{+g}^m) \subset \mathscr{F}_{(+)f}^m.$$

b. $(\mathrm{id} + \mathscr{F}_{+g}^m)^{-1} = \mathrm{id} + \mathscr{F}_{+g}^m$ *for an arbitrary* $g \in G^{m-1}$.

THE FOURTH SHIFT LEMMA, $\mathrm{SL4}_n$.a. $J^{n-1} \subset \mathrm{Gr}(\mathrm{id} + \mathscr{F}_{1+}^{n-1})$. *Moreover, generating elements in the group J^{n-1} can be chosen in the set* $\mathrm{id} + \mathscr{F}_{1+}^{n-1}$.

b. $\mathscr{F}_{0(+)g}^n \circ J^{n-1} \subset \mathscr{F}_{0(+)g}^n$.

We emphasize once more that, by the induction hypothesis, all these lemmas are assumed to be proved for $m \leq n-1$ (n a positive integer), as are Theorems MDT_m and ADT_m. The induction base—proofs of the lemmas for $m = 0$—is contained in §4.1.

§1.11. Proofs of the decomposition theorems

A. Proof of the multiplicative decomposition theorem, MDT_n**.** The second assertion of the theorem follows immediately from the first and the relation

$$G_n \subset \mathrm{Gr}(A^n H^0, J^{n-1}, G^{n-1});$$

see §1.3. Let us prove the first assertion. For this it suffices to prove:

$$\text{I. } G^n \circ G^n = G^n; \qquad \text{II. } (G^n)^{-1} = G^n.$$

The second assertion follows again from the first and the fact that all three factors in the definition of G^n are groups. The first assertion is a consequence of the following three assertions:

$$\begin{aligned} &1^\circ.\ \mathrm{Ad}(G^{n-1})J^{n-1} = J^{n-1}; \\ &2^\circ.\ \mathrm{Ad}(G^{n-1})H^n = H^n; \\ &3^\circ.\ \mathrm{Ad}(J^{n-1})H^n = H^n. \end{aligned} \qquad (*)$$

The assertion 1° follows immediately from the definition of the set J^{n-1}. The second is a direct consequence from the following new definition of the group H^n:

$$H^n = \mathrm{Ad}(G^{n-1})(\mathrm{id} + \mathscr{F}_{0+\mathrm{id}}^n). \qquad (**)$$

PROPOSITION 1. *The new definition $(**)$ of the group H^n is equivalent to the original one; see §1.7.*

The proof of the proposition is given below in B.

We prove the assertion 3°. Let $h \in J^{n-1}$ and $f \in H^n$. It must be proved that $\mathrm{Ad}(h)f \in H^n$. It suffices to consider the case when f is a generating element of H^n in the sense of the old definition, that is,

$$f = \mathrm{id} + \psi, \qquad \psi \in \mathscr{F}^n_{0+g}, \ g \in G^{n-1}.$$

By part a in Lemma $\mathrm{SL4}_n$, the group J^{n-1} is generated by the elements in $\mathrm{id} + \mathscr{F}^{n-1}_{1+}$. It suffices to prove the inclusion $\mathrm{Ad}(h)f \in H^n$ for generating elements h of J^{n-1}. Accordingly, let $h \in \mathrm{id} + \mathscr{F}^{n-1}_{1+}$. Then, by part b of Lemma $\mathrm{SL3}_n$,

$$h^{-1} = \mathrm{id} + \varphi, \qquad \varphi \in \mathscr{F}^{n-1}_{+}.$$

Further, by Lemma $\mathrm{SL2}_n$,

$$\begin{aligned} h^{-1} \circ f &= (\mathrm{id} + \varphi) \circ (\mathrm{id} + \psi) = \mathrm{id} + \psi + \varphi \circ (\mathrm{id} + \psi) \\ &= \mathrm{id} + \varphi + (\psi + \tilde{\psi}) = h^{-1} + (\psi + \tilde{\psi}), \qquad \tilde{\psi} \in \mathscr{F}^n_{0+g}. \end{aligned}$$

Consequently, $\psi + \tilde{\psi} = \psi_1 \in \mathscr{F}^n_{0+g}$, since the set $\mathscr{F}^n_{0+g}$ is closed under addition. Therefore,

$$\mathrm{Ad}(h)f = \mathrm{id} + \psi_1 \circ h, \qquad \psi_1 \circ h \in \mathscr{F}^n_{0+g},$$

by b in Lemma $\mathrm{SL4}_n$. Assertion 3° is proved.

B. The proof of Proposition 1 is similar. It suffices to prove that for all $g \in G^{n-1}$

$$\mathrm{Ad}(g)(\mathrm{id} + \mathscr{F}^n_{0+\mathrm{id}}) = \mathrm{id} + \mathscr{F}^n_{0+g}. \qquad (***)$$

To do this we prove the more general assertion that for all $f, g \in G^{n-1}$

$$\mathrm{Ad}(g)(\mathrm{id} + \mathscr{F}^n_{0+f}) \subset \mathrm{id} + \mathscr{F}^n_{0+fg}. \qquad ({}_{*}{}^{*}{}_{*})$$

For $f = \mathrm{id}$ the latter assertion proves than the left-hand side of the equality in $(***)$ belongs to the right-hand side. Replacing g and f in the formula $({}_{*}{}^{*}{}_{*})$ by g^{-1} and g, respectively, we get the reverse inclusion.

By Theorem ADT_{n-1}, which falls under the induction hypothesis, g^{-1} admits the expansion $(*)$ in §1.8 for $m = n-1$. According to Lemma $\mathrm{SL2}_n$, for an arbitrary germ $\varphi \in \mathscr{F}^n_{0+f}$,

$$g^{-1} \circ (\mathrm{id} + \varphi) = g^{-1} + \psi, \qquad \psi \in \mathscr{F}^n_{0+f}.$$

Consequently,

$$g^{-1} \circ (\mathrm{id} + \varphi) \circ g = \mathrm{id} + \psi \circ g, \qquad \psi \circ g \in \mathscr{F}^n_{0+fg},$$

which proves the assertion $({}_{*}{}^{*}{}_{*})$. ▶

C. The proof of the additive decomposition theorem ADT_n is by induction on n. For $n = 0$ the assertion is trivial: by definition, $G^0 = \mathscr{R}$, $\mathscr{R} = \mathscr{A}\!f\!f + \mathscr{R}^0$, and $\mathscr{R}^0 \subset \mathscr{F}\mathscr{C}_+^0$.

According to the induction hypothesis, Theorem ADT_{n-1} is proved for the group G^{n-1}. Using the assertion in part a of Lemma $\mathrm{SL4}_n$ and the definition of the group H^n, we get from this that if $g \in G^n$, then

$$g \in (\mathscr{A}\!f\!f + \mathscr{L}(\mathscr{F}_{1+}^k | 0 \le k \le n-2,\ \mathscr{F}_{0+}^{n-1})) \circ \mathrm{Gr}(\mathrm{id} + \mathscr{F}_{1+}^{n-1}) \circ \mathrm{Gr}(\mathrm{id} + \mathscr{F}_{0+}^n).$$

From Lemmas $\mathrm{SL2}_n$ and $\mathrm{SL3}_n$,

$$g \in \mathscr{A}\!f\!f + \mathscr{L}(\mathscr{F}_{1+}^k | 0 \le k \le n-1,\ \mathscr{F}_{0+}^n).$$

This is the expansion $(*)$ in §1.8. We prove that in this expansion it can be assumed that

$$(k_j, g_j) \prec (k_{j+1}, g_{j+1})$$

if the expansion itself, with the convention in §1.9 taken into account, is written in the form

$$g = a + \sum \varphi_j, \qquad a \in \mathscr{A}\!f\!f,\ \varphi_j \in \mathscr{F}_{+g_j}^{k_j}. \tag{${}^{*}_{*}{}^{*}$}$$

We order the terms on the left-hand side of the last expansion in such a way that either $k_j < k_{j+1}$, or $k_j = k_{j+1}$ and $g_j \preceq g_{j+1}$. If, furthermore, $g_j \prec\!\!\prec g_{j+1}$ in $G^{k_j - 1}$ for all $k_j = k_{j+1}$, then our assertion is proved. Suppose that the last inequality is violated, that is, $g_j \circ g_{j+1}^{-1} = h \in G_{\mathrm{rap}}^{k-1}$ for some j: $k_j = k_{j+1} = k$. Then

$$\varphi_j \in \mathscr{F}_{+g_j}^k = \mathscr{F}_{+h\circ g_{j+1}}^k = \mathscr{F}_{+h}^k \circ g_{j+1} \overset{①}{\subset} \mathscr{F}_{+\mathrm{id}}^k \circ g_{j+1} = \mathscr{F}_{+g_{j+1}}^k.$$

The inclusion ① follows from Lemma $\mathrm{SL1}_n$. Formula $(*)$ in §1.8 is proved under the additional condition that $g_j \prec\!\!\prec g_{j+1}$ for $k_j = k_{j+1}$.

It remains to prove the assertion that the first nonzero term φ in the expansion for $g - \mathrm{id}$ given by formula $({}^{*}_{*}{}^{*})$ is weakly real. The main role in the proof is played by the assertion in the next subsection.

D. A criterion for being weakly real. *The* **R**-*realization of a regular functional cochain of class* $\mathscr{F}\mathscr{C}_0^n$ *or* $\mathscr{F}\mathscr{C}_1^m$ *for* $m \le n-1$ *is weakly real if and only if the imaginary part of the cochain decreases on* $(\mathbf{R}^+, \infty)$ *more rapidly than any exponential* $\exp(-\lambda\xi)$, $\lambda > 0$.

PROOF. *Necessity.* Suppose that the **R**-realization of a regular functional cochain F is weakly real, and let F^u and F^l be the functions in the tuple F corresponding to the domains of the partition that are adjacent to $(\mathbf{R}^+, \infty)$ from above and from below. Since the realization is weakly real,

$$\delta F = 2 \operatorname{Im} F^u$$

on $(\mathbf{R}^+, \infty)$. On the other hand,

$$|\delta F| < \sum \exp(-C \exp \rho_j), \qquad \binom{*}{*}$$

where $\rho_j^{-1} = \sigma_j$ are the admissible germs used to construct the partition corresponding to F. By property 4° of admissible germs (see Definition 2 in §1.5),

$$|\delta F| \prec \exp(-\nu\xi) \qquad \binom{**}{**}$$

for all $\nu > 0$. This proves the necessity.

Sufficiency. With the previous notation, suppose that $|\mathrm{Im}\, F^u| \prec \exp(-\nu\xi)$ on $(\mathbf{R}^+, \infty)$ for all $\nu > 0$. Then on $(\mathbf{R}^+, \infty)$

$$|(IF)^l| = \tfrac{1}{2i}(F^l - \overline{F^u})| \le |\tfrac{1}{2}\delta F| + |\mathrm{Im}\, F^u|.$$

By inequality $\binom{**}{**}$, the first term decreases more rapidly than any exponential; the second term has the same property by assumption; the cochain IF is regular of class, mentioned above. Consequently, by the Phragmén-Lindelöf theorem, PL_n, $IF|_{(\mathbf{R}^+, \infty)} \equiv 0$, which is equivalent to the cochain F being weakly real.

E. Conclusion of the proof of the additive decomposition theorem. This subsection serves as a continuation of **C**. We prove that the first nonzero term given by $({}^*{}_*{}^*)$ for $g - \mathrm{id}$ (denote it by φ) is weakly real. If $\varphi = a - \mathrm{id} \neq 0$, then φ is real, since a is real. Suppose that $a = \mathrm{id}$. Then $\varphi = F \circ \exp^{[k]} \circ f$ for some k, F, f: $0 \le k \le n$, $F \in \mathscr{F}\mathscr{C}_+^k$, and $f \in G^{k-1}$. As shown in §1.9, for any term ψ following after φ in the expansion $(*)$ in §1.8 we have the estimate

$$|\psi \circ f^{-1} \circ \ln^{[k]}| \prec \exp(-\nu\xi)$$

on $(\mathbf{R}^+, \infty)$ for arbitrary $\nu > 0$. Further,

$$0 = \mathrm{Im}(g - \mathrm{id}) \circ f^{-1} \circ \ln^{[k]}|_{(\mathbf{R}^+, \infty)} = (\mathrm{Im}\, F + o(\exp(-\nu\xi)))|_{(\mathbf{R}^+, \infty)}$$

for arbitrary $\nu > 0$. This and the criterion for being weakly real imply that the $\mathbf{R}$-realization of F is weakly real. Theorem ADT_n is proved.

* * *

To prove the identity theorem it remains to prove the lemmas in §1.10, the Phragmén-Lindelöf theorem, the theorem on a lower estimate, and its corollary. The proofs of the lemmas make up the main point, and are reminiscent of proofs of identities: it is required to establish that certain precisely described sets belong to other precisely described sets. However, it is these proofs that involve the most technical difficulties. They are carried out in Chapters II and IV and are based on results in Chapter V. The Phragmén-Lindelöf theorem is proved in Chapter III, modulo some lemmas, which are also proved in Chapter V. Finally, the theorem on a lower estimate and its corollary are proved in §4.10. This section concludes the investigation of the first four chapters and simultaneously serves as the foundation on which the investigations in Chapter V are constructed.

CHAPTER II

Function-Theoretic Properties of Regular Functional Cochains

Here we prove the "function-theoretic" part of the lemmas in §1.10. The corresponding formulations are obtained if cochains of class $\mathscr{F}\mathscr{C}^m$ are replaced everywhere in these lemmas by cochains of class $\mathscr{F}\mathscr{C}_{\text{reg}}(\mathscr{D}^m)$, keeping thereby the regularity requirement and waiving the decomposability requirement. The lemmas obtained are denoted by $\text{SL}1\text{–}4_{n,\text{reg}}$; their detailed formulations are given below, directly before the proofs.

The simplest properties of regular cochains are discussed at the beginning of the chapter.

§2.1. Differential algebras of cochains

Recall that for each class $\mathbf{\Omega}$ of standard domains the regular functional cochains of class $\mathbf{\Omega}$ were defined in §1.6, and the $\mathbf{\Omega}$-admissible germs of diffeomorphisms were defined in §1.5. We assume below that the class $\mathbf{\Omega}$ is fixed and contains a domain in the intersection of any two domains in it; for arbitrary n the standard domains of class n have this property (Lemma 5.2_n in §2.4 below). It is also assumed that $|\zeta| \prec \xi^3$ in the domains of class $\mathbf{\Omega}$; for example, quadratic standard domains have this property (see §0.3), as well as all domains belonging to them. Explicit reminders about the class $\mathbf{\Omega}$ are not given below.

DEFINITION 1. Two regular function cochains are said to be equivalent if they coincide in the intersection of some standard domain with the upper half-plane. An equivalence class of regular functional cochains is called the germ (at infinity—this refinement is often omitted) of a regular cochain.

Since regular cochains admit several realizations, operations on them require special definition. Namely, all the realizations of a single cochain coincide in the upper half-plane. In view of this circumstance all operations are carried out on the restrictions of regular cochains to the intersection of standard domains with the upper half-plane.

LEMMA 1. *Let $\mathscr{D}$ be an arbitrary set of admissible germs. Then the germs of class $\mathscr{F}\mathscr{C}_{\text{reg}}(\mathscr{D})$ form a differential algebra. This means that sums, differences, products, and derivatives of such cochains in the intersection of some* *(depending on the germs) standard domain with the upper half-plane can be extended to the whole standard domain as regular cochains.*

PROOF. The proof of Lemma 1 follows almost immediately from the definitions. Let F_1 and F_2 be two regular cochains of type $\boldsymbol{\sigma}_1$ and $\boldsymbol{\sigma}_2$, and let $F = F_1 + F_2$ be their sum in the upper half-plane (the difference and product are investigated similarly). We prove that F extends to a regular cochain of type $\boldsymbol{\sigma} = \boldsymbol{\sigma}_1 + \boldsymbol{\sigma}_2$. Let

$$\boldsymbol{\sigma}_1 = (\sigma_1, \dots, \sigma_N), \quad \boldsymbol{\sigma}_2 = (\tilde{\sigma}_1, \dots, \tilde{\sigma}_M), \quad \boldsymbol{\sigma} = (\tilde{\tilde{\sigma}}_1, \dots, \tilde{\tilde{\sigma}}_K),$$

where in each sequence the next germ increases on $(\mathbf{R}^+, \infty)$ more slowly than the preceding one in the sense of Definition 1 in §1.4A. For arbitrary k with $1 \le k \le K$ we construct the k-realization $F_{(k)}$ of the cochain F; the $\mathbf{R}$-realization of it is constructed similarly. Take k_1 and k_2 such that $\sigma_{k_1} \succeq \tilde{\tilde{\sigma}}_k \succ \sigma_{k_1+1}$, and $\tilde{\sigma}_{k_2} \succeq \tilde{\tilde{\sigma}}_k \succ \tilde{\sigma}_{k_2+1}$. Let $\widetilde{F}_1 = F_1(k_1)$ and $\widetilde{F}_2 = F_2(k_2)$ be the k_1- and k_2-realizations of the cochains F_1 and F_2. We prove that $\widetilde{F} = \widetilde{F}_1 + \widetilde{F}_2$ is a regular cochain of type $(\boldsymbol{\sigma}, k)$ (see Definition 10 in §1.6); it will be the desired realization of F.

We verify all the requirements of Definition 10 in §1.6. Let Ξ_1 and Ξ_2 be modified regular partitions of types $(\boldsymbol{\sigma}_1, k_1)$ and $(\boldsymbol{\sigma}_2, k_2)$ corresponding to $\widetilde{F}_1$ and $\widetilde{F}_2$. Then their product $(\boldsymbol{\sigma}_2, k_2)$ is a modified regular partition of type $(\boldsymbol{\sigma}, k)$.

By the definition of a sum of cochains, the cochain $\widetilde{F}$ corresponds to the partition Ξ. The ε-extendibility of the sum and its exponential estimate follow from the analogous properties of the terms. This verifies requirements 1–3 of the definition of regular cochains (Definition 10 in §1.6).

It remains to verify the estimate of the coboundary. Corresponding to the product of two partitions that are $\mathbf{R}$-regular or modified regular is the sum of the rigging cochains. Further,

$$\delta(\widetilde{F}_1 + \widetilde{F}_2) = \delta\widetilde{F}_1 + \delta\widetilde{F}_2.$$

This implies at once that the coboundary of the sum can be estimated from above by a corresponding rigging cochain. This proves Lemma 1 for a sum and difference of cochains.

We estimate the coboundary of the product:

$$|\delta(\widetilde{F}_1\widetilde{F}_2)| < |\widetilde{F}_1|\,|\delta\widetilde{F}_2| + |\widetilde{F}_2|\,|\delta\widetilde{F}_1| \le (\exp\nu\xi)(m^1 + m^2).$$

Here m^1 and m^2 are rigging cochains of the partitions Ξ_1 and Ξ_2 that majorize the coboundaries $\delta\widetilde{F}_1$ and $\delta\widetilde{F}_2$.

But the product of a rigging cochain of a regular partition by an exponential is majorized by some other rigging cochain corresponding to the same partition. Indeed, by requirement 4° in the definition of admissible germs (Definition 2 in §1.5), for an arbitrary admissible germ there exists a standard domain Ω (this domain is also of class n for germs of class n) such that for arbitrary $\nu > 0$,

$$\operatorname{Re}\exp\rho \succ \nu\xi \quad \text{in } \Omega.$$

Consequently,

$$|\exp(-C\exp\rho + \nu\xi)| \prec \left|\exp\left(-\tfrac{C}{2}\exp\rho\right)\right| \quad \text{in } \Omega.$$

This ends the proof of Lemma 1 for sums, differences, and products.

The proof that derivatives are regular is based on the Cauchy estimate and uses

PROPOSITION 1. *Suppose that σ is an admissible germ, $0 < \delta < \varepsilon$, Π_k is the half-strip $\xi \geq a$, $\eta \in [\pi k, \pi(k+1)]$, Π_k^ε and Π_k^δ are its ε- and δ-neighborhoods, Π_{main} is the main domain of the modified standard partition, $\Pi_{\text{main}}^{(\varepsilon)}$ and $\Pi_{\text{main}}^{(\delta)}$ are its generalized ε- and δ-neighborhoods (see Definition 6 in §1.6), and Ω is a domain belonging to a standard domain of the class $\mathbf{\Omega}$ described at the beginning of the section. Then the margin between the domains imbedded one in the other (the distance from points of the smaller domain to the boundary of the larger domain) for the pair of domains $\sigma\Pi_k^\varepsilon \cap \Omega^\varepsilon$, $\sigma\Pi_k^\delta \cap \Omega^\delta$ exceeds a constant, while for the pair of domains $\sigma\Pi_{\text{main}}^{(\varepsilon)} \cap \Omega^\varepsilon$, $\sigma\Pi_{\text{main}}^{(\delta)} \cap \Omega^\delta$ it exceeds the function $C \cdot |\varepsilon - \delta|\,|\zeta|^{-3}$; here Ω^ε and Ω^δ are the ε-neighborhood and δ-neighborhood of Ω.*

Recall that the inequality $|\zeta| \prec \xi^3$ holds in a standard domain of class $\mathbf{\Omega}$.

The regularity of the derivative of a regular cochain now follows from the fact that the product of a rigging cochain of a regular partition by a power is majorized by some other rigging cochain of the same partition; this was proved above for a product by an exponential. This easily yields a proof of the lemma modulo a proof of the proposition, and we present it in detail.

PROOF OF PROPOSITION 1. We begin with a remark. Let σ be a biholomorphic mapping of a planar domain, and assume that the inverse mapping has bounded derivative: $\sigma^{-1} = \rho$, $|\rho'| < C$. Then σ decreases the distance between points no more than by a factor of C. Indeed, if ζ_1 and ζ_2 are arbitrary points in the domain of σ and $\omega_j = \sigma(\zeta_j)$, for $j = 1, 2$, then $\zeta_j = \rho(\omega_j)$ and $|\zeta_1 - \zeta_2| < C|\omega_1 - \omega_2|$. Therefore,

$$|\sigma(\zeta_1) - \sigma(\zeta_2)| = |\omega_1 - \omega_2| > |\zeta_1 - \zeta_2|/C.$$

This immediately implies the first assertion of the proposition, since the derivative of the germ inverse to an admissible germ is bounded according to Definition 2 in §1.5.

We prove the second assertion. It follows from the definition of the generalized ε-neighborhood of the half-strip Π_{main} (Definition 6 in §1.6) that there exists a constant $C_1 > 0$ such that $\operatorname{dist}(\zeta, \partial\Pi_{\text{main}}^{(\varepsilon)}) \geq C_1(\varepsilon - \delta)|\zeta|^{-3}$ for $\zeta \in \Pi_{\text{main}}^{(\delta)}$. Consequently, by the remark at the beginning of the proof,

$$\operatorname{dist}(\sigma(\zeta), \partial\sigma\Pi_{\text{main}}^{(\varepsilon)}) \geq C_1(\varepsilon - \delta)/|\rho(\zeta)|^3 \sup|\rho'|.$$

But since the derivative $|\rho'|$ is bounded, the function $|\rho(\zeta)|$ increases no more rapidly than $|\zeta|$; consequently, there exists a constant $C_2 > 0$ such that $|\rho(\zeta)| < C_2|\zeta|$. This implies the proposition, and with it Lemma 1. ►►

§2.2. Completeness

We fix a standard domain Ω of class $\mathbf{\Omega}$. In this domain we consider the space $\mathscr{F}$ of regular functional cochains having the following properties:

All the cochains correspond to a single partition Ξ that is **R**-regular or modified regular, are ε-extendible with a common value of ε, and can be estimated from above in modulus by the exponential $a \cdot \exp \nu\xi$ with a common exponent ν and coefficient a depending on the cochain;

The coboundaries of the cochains can be estimated from above in modulus by one and the same rigging cochain m for the partition Ξ, multiplied by a constant depending on the cochain, $|\delta F| \leq C_F m$.

Let $\partial\Xi^{(\varepsilon)}$ be the domain of the rigging cochain m. We introduce in the space $\mathscr{F}$ the norm

$$\|F\| = \sup_{\Omega} |F \exp \nu\xi| + \sup_{\partial\Xi^{(\varepsilon)}} |\delta F/m|.$$

LEMMA 2. *The space $\mathscr{F}$ is complete in the norm introduced.*

PROOF. This is an immediate consequence of the Weierstrass theorem on completeness of the space of holomorphic functions with the C-norm. ►

COROLLARY 1. *Let $\varphi: (\mathbf{C}, 0) \to (\mathbf{C}, 0)$ be a holomorphic function. Then $\varphi \circ \mathscr{F}\mathscr{C}^+_{\text{reg}}(\mathscr{D}) \subset \mathscr{F}\mathscr{C}^+_{\text{reg}}(\mathscr{D})$.*

PROOF. Let $F \in \mathscr{F}\mathscr{C}^+_{\text{reg}}(\mathscr{D})$. Then $\varphi \circ F = \sum \varphi^{(j)}(0)F^j/j!$. By Lemma 1, the partial sums of this series are cochains in $\mathscr{F}\mathscr{C}_{\text{reg}}(\mathscr{D})$. Since the Taylor series for φ converges, the partial sums form a Cauchy sequence in $\mathscr{F}$. ►

COROLLARY 2.

$$\exp \mathscr{F}\mathscr{C}^+_{\text{reg}}(\mathscr{D}) \subset 1 + \mathscr{F}\mathscr{C}^+_{\text{reg}}(\mathscr{D}).$$

PROOF. The function $\varphi: \zeta \mapsto \exp \zeta - 1$ satisfies the condition of Corollary 1. ►

COROLLARY 3.

$$\ln(1 + \mathscr{F}\mathscr{C}^+_{\text{reg}}(\mathscr{D}) \subset \mathscr{F}\mathscr{C}^+_{\text{reg}}(\mathscr{D}),$$

$$1/(1 + \mathscr{F}\mathscr{C}^+_{\text{reg}}(\mathscr{D})) \subset \mathscr{F}\mathscr{C}^+_{\text{reg}}(\mathscr{D}).$$

PROOF. The proof is analogous to the preceding proof. ►

§2.3. Shifts of functional cochains by slow germs

In this section we deal with compositions of cochains with germs growing slowly at infinity in a certain sense (slow shift of the independent variable of the cochain).

DEFINITION 1. Let F be a functional cochain defined in a domain Ω of class $\mathbf{\Omega}$, $F=\{F_j\}$, and let ρ be the germ of a diffeomorphism $(\mathbf{R}^+, \infty) \to (\mathbf{R}^+, \infty)$ that is extendible to a domain $\tilde{\Omega} \in \mathbf{\Omega}$ such that $\rho\tilde{\Omega} \subset \Omega$. Then the functional cochain $F \circ \rho$ is the tuple $\{F_j \circ \rho\}$ of functions, taken on the domain Ω.

Suppose that in the preceding definition F is a regular functional cochain, Ω and $\tilde{\Omega}$ are standard domains of class $\mathbf{\Omega}$ such that $\rho\tilde{\Omega} \subset \Omega$, Ω^+ and $\tilde{\Omega}^+$ are their intersections with the upper half-plane, F^+ is the restriction of F to Ω^+, and $F^+ \circ \rho$ is the cochain defined above on $\tilde{\Omega}^+$. We say that $F \circ \rho$ is a regular cochain of class $\mathscr{D}$ and numerical type N if $F^+ \circ \rho$ is the restriction to $\tilde{\Omega}^+$ of each of the $N+1$ realizations of some regular cochain of class $\mathscr{D}$ and numerical type N.

LEMMA 3. 1. *Let* $\mathscr{D}_1$ *and* $\mathscr{D}_2$ *be arbitrary sets of germs of admissible diffeomorphisms of class* $\mathbf{\Omega}$, *and let* σ *be the germ of a diffeomorphism such that*

$$\sigma\mathscr{D}_1 \subset \mathscr{D}_2 \cup \{\textit{the set of nonessential germs of class } \mathbf{\Omega}\}.(^5) \qquad (*)$$

2. *Suppose that* $\operatorname{Re}\rho < \mu\xi$ *in* Ω *for some* $\mu > 0$ *and* $\Omega \in \mathbf{\Omega}$.

3. *Let* F *be a regular functional cochain of type* $\boldsymbol{\sigma} = (\sigma_1, \dots, \sigma_M)$, $\sigma_j \in \mathscr{D}_1$, *and let the nonessential germs in the sequence* $\sigma \circ \boldsymbol{\sigma} = (\sigma\circ\sigma_1, \dots, \sigma\circ\sigma_M)$ *(if there are any) go first and in succession.*

Then $F \circ \rho$ *is a regular functional cochain of type* $\mathscr{D}_2$ *or a holomorphic function that increases more slowly than an exponential. If in addition* F *is a rapidly decreasing cochain, and the germ* σ *has the property that there exist a positive constant* ε *and a standard domain* $\tilde{\Omega}$ *of class* $\mathbf{\Omega}$ *such that*

$$\operatorname{Re}\rho \succ \varepsilon\xi$$

in $\tilde{\Omega}$, *then* $F \circ \rho$ *is again a rapidly decreasing cochain.*

PROOF. Let F be a cochain of type $\boldsymbol{\sigma} = (\sigma_1, \dots, \sigma_M)$. We consider three cases; in each of them we indicate the realizations of F that give realizations of the cochain $F \circ \rho$.

CASE 1. All the germs $\tilde{\sigma}_j = \sigma \circ \sigma_j$ are nonessential. In this case let $\tilde{F}$ be an M-realization of F. By the definition of $\mathbf{\Omega}$-nonessential germs, the domain $\tilde{\Omega}$ of class $\mathbf{\Omega}$ can be chosen such that

$$\tilde{\rho}_j\tilde{\Omega} \subset \Pi_{\text{main}}, \qquad \tilde{\rho}_j = \tilde{\sigma}_j^{-1}.$$

The main domain of a partition of type $(\boldsymbol{\sigma}, M)$ has the form

$$\Omega_M = \bigcap_1^M \sigma_j\Pi_{\text{main}}.$$

Let F^u be the corresponding function in the tuple $\tilde{F}$.

(^{5}) Below we write "nonessential germs" instead of "nonessential germs of class $\mathbf{\Omega}$" when it is clear what class $\mathbf{\Omega}$ is meant.

In view of the preceding inclusion,

$$\rho\widetilde{\Omega} \subset \Omega_M.$$

Therefore, in Case 1 the composition $F \circ \rho$ extends from $(\mathbf{R}^+, \infty)$ to the holomorphic function $F^u \circ \rho$ on $\widetilde{\Omega}$; $F^u \circ \rho$ is estimated below.

CASE 2. Among the germs $\sigma \circ \sigma_j$ there are essential and nonessential germs. We construct the $\mathbf{R}$-realization of the cochain $F \circ \rho$; the l-realization for $l \geq 1$ is constructed analogously. By assumption, the nonessential germs in the sequence $\{\sigma \circ \sigma_j\}$ go first and in succession; let $\sigma \circ \sigma_k$ be the last of them. Take the k-realization $F_{(k)}$ of the cochain F. Let $\Omega_k = \bigcap_1^k \sigma_j \Pi_{\text{main}}$. Arguing as in Case 1, we choose a standard domain $\widetilde{\Omega}$ of class $\mathbf{\Omega}$ such that $\rho\widetilde{\Omega} \subset \Omega_k$. The partition of type $(\boldsymbol{\sigma}, k)$ in the domain Ω_k coincides with the $\mathbf{R}$-regular partition Ξ of type $(\sigma_{k+1}, \dots, \sigma_M)$, and the cochain $\widetilde{F} = F_{(k)} |_{\Omega_k}$ is a tuple $\{F_j\}$ of functions that are holomorphic in the domains of this partition. Let $\Xi_j \subset \Omega_k$ be a domain in the partition Ξ, F_j the corresponding function in the tuple $\widetilde{F}$, and $\sigma_* \Xi_j = \sigma(\Xi_j \cap \rho\widetilde{\Omega})$ (this domain can also be empty). The cochain $\widetilde{F} \circ \rho$ is the tuple of functions $F_j \circ \rho$, defined on the domains $\sigma_* \Xi_j$. The corresponding partition of the domain $\widetilde{\Omega}$ is $\mathbf{R}$-regular of type

$$\tilde{\boldsymbol{\sigma}} = (\sigma \circ \sigma_{k+1}, \cdots, \sigma \circ \sigma_M).$$

The $\mathbf{R}$-realization of the cochain $F \circ \rho$ is constructed.

The l-realization of the cochain $F \circ \rho$ has the form $F_{(k+l)} \circ \rho$, where $F_{(k+l)}$ is the $(k+l)$-realization of the cochain F. It can be verified similarly that the composition $F_{(k+l)} \circ \rho$ corresponds on the domain $\widetilde{\Omega}$ to a partition of type $(\tilde{\boldsymbol{\sigma}}, l)$, where $\tilde{\boldsymbol{\sigma}}$ is the same as above.

The cochain and coboundary estimates are obtained below.

CASE 3. All the germs $\sigma \circ \sigma_j$ are essential. In this case the k-realization of the cochain F corresponds to the k-realization of the cochain $F \circ \rho$; more precisely, if $\Xi_{(k)}$ is the k-realization of a partition of type $\boldsymbol{\sigma}$ of the domain Ω, then $\sigma_* \Xi_{(k)}$ is the k-realization of a partition of type $\sigma \circ \boldsymbol{\sigma}$ of the domain $\widetilde{\Omega}$.

We now estimate the cochain $F \circ \rho$ and the coboundary $\delta(F \circ \rho) = (\delta F) \circ \rho$; in the first case the coboundary is empty.

By assumption, $|F| \prec \exp \nu\xi$ in Ω for some $\nu > 0$. Consequently, $|F \circ \rho| \prec \exp \nu \operatorname{Re} \rho$ in $\widetilde{\Omega}$. But by assumption, the domain $\widetilde{\Omega}$ can be chosen so that for some $\mu > 0$

$$\operatorname{Re} \rho < \mu\xi \quad \text{in } \widetilde{\Omega}.$$

Consequently,

$$|F \circ \rho| \prec \exp \nu\mu\xi \quad \text{in } \widetilde{\Omega}.$$

If $\nu < 0$, and $\operatorname{Re} \rho \succ \varepsilon\xi$ in $\widetilde{\Omega}$, then $|F \circ \rho| \prec \exp(-|\nu\varepsilon|\xi)$ in $\widetilde{\Omega}$: the cochain $F \circ \rho$ is rapidly decreasing.

We estimate the coboundary $\delta F \circ \rho$. The formula for a rigging cochain is precisely suited to accommodate a shift by a germ ρ such that the condition $(*)$ holds with $\sigma = \rho^{-1}$. Suppose that in the $(\boldsymbol{\sigma}, \varepsilon)$-neighborhood of a standard boundary line $\mathscr{L}$ of the partition Ξ the corresponding rigging function is equal to

$$m_{\mathscr{L}} = \sum \exp(-C \exp \operatorname{Re} \rho_j).$$

Then in the $(\tilde{\boldsymbol{\sigma}}, \varepsilon)$-neighborhood of the boundary line $\sigma(\mathscr{L} \cap \rho\tilde{\Omega})$ of the partition $\sigma_* \Xi$ the function

$$m_{\mathscr{L}} \circ \rho = \sum \exp(-C \exp \operatorname{Re} \rho_j \circ \rho)$$

is, by definition, a rigging cochain function corresponding to the partition $\sigma_* \Xi$. Therefore, the estimate $|\delta F| < m$ implies that $|\delta F \circ \rho| < m \circ \rho$, and $m \circ \rho$ is a rigging cochain of the partition $\sigma_* \Xi$. ▶

Lemma 3 admits a generalization that is used in the proof of the fourth shift lemma.

LEMMA 3 BIS. *Assume all the conditions of Lemma* 3 *except one: instead of* $(*)$ *we suppose that the set* $\sigma\mathscr{D}_1$ *consists of nonessential germs and germs equivalent to germs of the set* $\mathscr{D}_2$ (*see Definition* 18 *in* §1.6E). *Then the conclusions of Lemma* 3 *hold.*

PROOF. Suppose that F is a cochain of type $\boldsymbol{\sigma} = (\sigma_1, \dots, \sigma_M)$, the germs $\sigma \circ \sigma_1, \dots, \sigma \circ \sigma_k$ are nonessential, and the germs $\sigma \circ \sigma_j$ with $k+1 \le j \le M$ are equivalent to germs $\tilde{\sigma}_j \in \mathscr{D}_2$. Then $F \circ \rho$ is a regular functional cochain of type $\tilde{\boldsymbol{\sigma}} = (\tilde{\sigma}_{k+1}, \dots, \tilde{\sigma}_M)$.

It follows from Lemma 3 that $F \circ \rho$ is a cochain of type $\boldsymbol{\sigma}_k = (\sigma \circ \sigma_{k+1}, \dots, \sigma \circ \sigma_M)$ that is ε-extendible for some $\varepsilon > 0$. The definition of equivalent germs and the equivalence of $\tilde{\sigma}_j$ and $\sigma \circ \sigma_j$ imply that $F \circ \rho$ is an $(\varepsilon/2)$-extendible cochain of type $\tilde{\boldsymbol{\sigma}}$ in some standard domain Ω of class $\tilde{\Omega}$.

The rigging cochains of realizations of the partitions of type $\boldsymbol{\sigma}_k$ and $\tilde{\boldsymbol{\sigma}}$ differ from each other at most by a factor of 2 in the exponential, as follows from the inequality

$$\exp \operatorname{Re}(\mathrm{id} + o(1)) \circ \tilde{\rho}_j \prec 2 \exp \tilde{\rho}_j.$$

This proves that the coboundary of the cochain $F \circ \rho$ can be estimated from above by the rigging cochain of the partition of type $\tilde{\boldsymbol{\sigma}}$, and concludes the proof of the lemma. ▶

§2.4. The first shift lemma, regularity: $\mathrm{SL1}_{n,\mathrm{reg}}$

A. Formulation of the basic lemma and the auxiliary lemma.

Recall that $\mathscr{D}^m = \mathscr{D}_0^n$ for $m = n$, and $\mathscr{D}^m = \mathscr{D}_1^m$ for $m \le n-1$.

LEMMA $\mathrm{SL1}_{n,\mathrm{reg}}$. *Suppose that* $m = n$ *or* $m = n-1$. *Then*

$$\mathscr{F}\mathscr{C}^{(+)}_{\mathrm{reg}}(\mathscr{D}^m) \circ A^{-m} G^{m-1}_{\mathrm{rap}} \subset \mathscr{F}\mathscr{C}^{(+)}_{\mathrm{reg}}(\mathscr{D}^m).$$

The plus in parentheses means that the assertion is true both with the plus (without the parentheses) and without it.

PROOF. Recall that the set of germs $A^{-m}G^{m-1}_{\mathrm{rap}}$ is denoted by $\mathscr{D}^m_{\mathrm{rap}}$. Lemma 5.1_m asserts that it consists of admissible germs. We need the following additional assertions, which we precede by

DEFINITION 1. A class $\mathbf{\Omega}$ of standard domains is said to be proper if:

(1°) for any $C > 0$ an arbitrary domain of class $\mathbf{\Omega}$ contains a domain of the same class whose distance from the boundary of the first is not less than C;

(2°) the intersection of any two domains of class $\mathbf{\Omega}$ contains a domain of the same class.

LEMMA 5.2_n. a. *Standard domains of class* n *are standard in the sense of Definition* 1 *in* §1.5. (*The set of these domains is denoted by* $\mathbf{\Omega}_n$.)

b. *Germs of the classes* $\mathscr{D}^{n-1}_{\mathrm{rap}}$ *and* $\mathscr{D}^{n-1}_{\mathrm{rap}} \circ \mathscr{L}^{n-1}$ *are admissible of class* $\mathbf{\Omega}_n$.

c. *Suppose that* $g \in G^{n-1}_{\mathrm{rap}}$, $\sigma = A^{-n}g$, *and* $\rho = \sigma^{-1}$. *Then there exist an* $\varepsilon > 0$, *a* $\mu > 0$, *and a standard domain* Ω *of class* $\mathbf{\Omega}_n$ *such that in* Ω

$$\operatorname{Re}\rho \succ \varepsilon\xi, \qquad |\sigma| \prec \mu|\zeta|.$$

d. *The class* $\mathbf{\Omega}_n$ *is proper in the sense of Definition* 1 *above.*

This lemma is proved in §5.4 for $n > 1$ and in subsection C of this section for $n = 1$.

B. Conclusion of the proof of the first shift lemma (regularity).

We verify that the set $\mathscr{D}^m$ and the germ σ satisfy the condition of Lemma 3 if in it $\mathscr{D}_1 = \mathscr{D}_2 = \mathscr{D}^m$. Namely, we prove

PROPOSITION 1. 1°. $\mathscr{D}^n_{\mathrm{rap}} \circ \mathscr{D}^n_0 \subset \mathscr{D}^n_0$.

2°. $\mathscr{D}^{n-1}_{\mathrm{rap}} \circ \mathscr{D}^{n-1}_1 \subset \mathscr{D}^{n-1}_1$.

PROOF. By Definition 16 in §1.6,

$$\mathscr{D}^n_{\mathrm{rap}} \circ \mathscr{D}^n_0 = A^{-n}(G^{n-1}_{\mathrm{rap}} \circ G^{n-1^-}_{\mathrm{slow}}) \circ \exp.$$

But it follows from the definitions of G^{n-1}_{rap} and G^{n-1}_{slow} that

$$G^{n-1}_{\mathrm{rap}} \circ G^{n-1^-}_{\mathrm{slow}} = G^{n-1^-}_{\mathrm{slow}}. \tag{$*$}$$

Indeed, G^{n-1} is a group; a composition of germs, one of which (after application of the homomorphism A^{-n}) grows more slowly than a linear germ and the other is comparable with a linear germ, again grows more slowly than a linear germ (see also C in §1.4).

2°. It remains to prove the equality

$$\mathscr{D}_{\text{rap}}^{n-1} \circ \mathscr{D}_{*}^{n-1} = \mathscr{D}_{*}^{n-1}.$$

By the definition of $\mathscr{D}_{*}^{n-1}$, the last equality is equivalent to the equality

$$G_{\text{rap}}^{n-2} \circ (G_{\text{slow}}^{n-2^{+}} \cup G_{\text{rap}}^{n-2}) = G_{\text{slow}}^{n-2^{+}} \cup G_{\text{rap}}^{n-2},$$

which, in turn, follows from the equality $(*)$ for $n-2$ instead of $n-1$ if we pass to the complement:

$$G_{\text{slow}}^{n-2^{+}} \cup G_{\text{rap}}^{n-2} = G^{n-2} \backslash G_{\text{slow}}^{n-2^{-}}.$$

The condition, "in the tuple $\{\sigma \circ \sigma_j\}$ the nonessential germs go first and in succession," is satisfied because, as proved above, there are no nonessential germs in this tuple under the conditions of Lemma $\mathrm{SL1}_{n,\text{reg}}$.

The conditions imposed on ρ in Lemma 3 follow from Lemma 5.2_nb and c. Namely, the inequality $(*)$ of §2.3 follows from Lemma 5.2_nb and the requirement 3° in the definition of admissible germs (Definition 2 in §1.5). The inequality $\operatorname{Re}\rho \succ \varepsilon\xi$ is contained in Lemma 5.2_nc.

All the conditions in Lemma 3 have been checked. Lemma $\mathrm{SL1}_{n,\text{reg}}$ follows from Lemma 3. ►

C. Proof of Lemma 5.2_1. The proof below enters in the induction base, a step which makes up the main part of the work.

(a) Standard domains of class 1 are only of type 1 (see Definition 1^n in §1.7). A domain of this class has the form

$$\Omega = A^{-1}(f+c)(\mathbf{C}^{+} \backslash K), \qquad f \in \mathscr{R}^0, \quad c \in \mathbf{R}.$$

On $(\mathbf{R}^{+}, \infty)$ we have that $f = \mathrm{id} + \varphi$, $\varphi = o(1)$, and

$$A^{-1}(f+c) = \zeta \exp(C + \varphi \circ \ln).$$

The mapping

$$\psi = \zeta \cdot \exp \varphi \circ \ln$$

also carries the domain $\mathbf{C}^{+} \backslash K_1$ into the domain Ω, where K_1 is another disk in general. The germ φ is real on $(\mathbf{R}^{+}, \infty)$. Also, $\psi' = 1 + o(1)$ in $(\mathbf{C}^{+}, \infty)$. Consequently, there exists a mapping $\psi^{-1} + o(1) \colon \Omega \to \mathbf{C}^{+}$ with derivative equal to $1 + o(1)$ on (Ω, ∞). This verifies the requirement in the definition of a standard domain (Definition 1 in §1.5) for the domain Ω.

(b) The germs of class $\mathscr{D}_{\text{rap}}^{0}$ are admissible of class Ω_1.

REMARK. The assertion (b) takes precisely this form because $\mathscr{L}^0 = \varnothing$. Recall that $G^{-1} = \mathscr{A}\!f\!f$. Consequently,

$$\mathscr{D}_{\text{rap}}^{0} = G_{\text{rap}}^{-1} = \mathscr{A}\!f\!f.$$

The admissibility of affine real germs is obvious.

(c) Suppose that $g \in G^0_{\text{rap}}$, $\sigma = A^{-1}g$, and $\rho = \sigma^{-1}$. Then there exist an $\varepsilon > 0$, a $\mu > 0$, and a standard domain Ω of class $\mathbf{\Omega}_1$ such that in Ω

$$\operatorname{Re}\rho \succ \varepsilon\xi, \qquad |\sigma| \prec \mu|\zeta|.$$

In the proof of assertion (a) it was already established that

$$A^{-1}g = \mu\zeta(1 + o(1))$$

for some $\mu > 0$. This implies the second of the inequalities proved.

We prove that in some standard domain of class 1

$$\operatorname{Re}\rho = (\mu + o(1))\xi$$

for some $\mu > 0$. This implies that in this domain

$$\varepsilon\xi \prec \operatorname{Re}\rho \prec \tilde{\mu}\xi$$

for some positive ε and $\tilde{\mu}$, and that proves the first inequality in Lemma 5.2_1 c.

Accordingly, let

$$g^{-1} = f + C, \qquad f \in \mathscr{R}^0, \; f = \mathrm{id} + \varphi,$$
$$|\varphi| \prec \exp(-\varepsilon\xi)$$

for some $\varepsilon > 0$. Let $\mu = \exp C$. Then

$$\rho = A^{-1}g^{-1} = \mu\zeta \exp\circ\varphi\circ\ln = \mu\zeta(1 + \varphi_1),$$

where

$$|\varphi_1| = |(\exp(-\varepsilon\ln|\zeta|))(1 + o(1))| \prec |\zeta|^{-\varepsilon/2}.$$

By Proposition 1 in subsection D of §1.6, there exists a standard domain of class 1 such that in it

$$|\zeta|^{1-\varepsilon/2} = o(\xi).$$

In this domain

$$\operatorname{Re}\rho = (\mu + o(1))\xi,$$

which is what was required. Assertion (c) of Lemma 5.2_1 is verified.

(d) To prove the first requirement in the definition of properness it suffices to verify that the shift belongs to the group $A^{-1}G^0_{\text{rap}}$, and then to use the fact that the derivative $(A^{-1}g)'$ is bounded in $(\mathbf{C}^+, \infty)$ for $g \in G^0_{\text{rap}}$. It is obvious (and was already verified above) that $A(\mathrm{id} + C) \in G^0_{\text{rap}}$.

We verify that together with any two domains Ω_1 and Ω_2 the class $\mathbf{\Omega}_1$ also contains a domain belonging to their intersection. Let

$$\Omega_j = A^{-1}g_j(\mathbf{C}^+ \backslash K_j), \qquad g_j \in G^0_{\text{rap}}, \; j = 1, 2.$$

We find a domain

$$\widetilde{\Omega} = A^{-1}\tilde{g}(\mathbf{C}^+ \backslash \widetilde{K}), \qquad \tilde{g} \in G^0_{\text{rap}},$$

such that $\tilde{\Omega} \subset \Omega_1 \cap \Omega_2$. Suppose that

$$g_1^{-1} \circ g_2 = \mathrm{id} + C + \varphi, \qquad \varphi \succ 0, \text{ and } \varphi \to 0 \text{ on } (\mathbf{R}^+, \infty).$$

The inequality $\varphi \succ 0$ can be achieved by changing the numbering. In view of the group property of the set G^0_{rap} and the expandibility of germs in this group in Dulac series, we get that $\varphi \equiv 0$ or

$$\varphi = (P(\zeta)\exp(-\nu\zeta))(1 + o(1)) \quad \text{in } (\Pi, \infty),$$

where P is a real polynomial and $\nu > 0$. We consider two cases: $\nu \in (0, 1)$; $\nu \geq 1$ or $\varphi \equiv 0$.

CASE 1°: $\nu \in (0, 1)$. In this case it is possible to take $\tilde{g} = g_1$, and the disk K_2 can be chosen sufficiently large. We prove that

$$(\Omega_2, \infty) \subset (\Omega_1, \infty),$$

or

$$(\exp \circ g_2 \circ \ln)(\mathbf{C}^+ \backslash \tilde{K}) \subset (\exp \circ g_1 \circ \ln)(\mathbf{C}^+ \backslash K_1),$$

or

$$g_2 \tilde{\Pi} \subset g_1 \Pi,$$

where $\tilde{\Pi} = \ln(\mathbf{C}^+ \backslash \tilde{K})$ and $\Pi = \ln(\mathbf{C}^+ \backslash K_1)$ are right half-strips (see the beginning of §1.5). It suffices to prove the inclusion

$$g_1^{-1} \circ g_2 \tilde{\Pi} \subset \Pi,$$

or the inequality

$$|\operatorname{Im}(g_1^{-1} \circ g_2)| \prec \eta$$

in (Π, ∞), or the equality $\operatorname{sgn} \operatorname{Im} \varphi = -\operatorname{sgn} \eta$ in (Π, ∞). By the Cauchy-Riemann condition and the fact that φ is real, this equality follows from the inequality

$$\operatorname{Re} \varphi' \prec 0$$

in (Π, ∞).

The last inequality is easy to verify. The germ φ' can be expanded in an asymptotic Dulac series; therefore, in (Π, ∞)

$$\varphi' = -(\exp(-\nu\zeta))Q(\zeta)(1 + o(1)),$$

where Q is a real polynomial. The arguments of the last two terms tend to zero, and

$$\arg(-\exp(-\nu\zeta)) \in \left(\frac{\pi}{2} + \varepsilon, \frac{3\pi}{2} - \varepsilon\right) f \quad \text{in } (\Pi, \infty)$$

for some $\varepsilon > 0$, since $\nu \in (0, 1)$. This implies that $\operatorname{Re} \varphi' \prec 0$ in (Π, ∞). This concludes the proof in case 1°.

CASE 2°: $\nu \geq 1$ OR $\varphi \equiv 0$. In this case we take $\tilde{g} \in G^0_{\mathrm{rap}}$ such that

$$g_j^{-1} \circ \tilde{g} = \mathrm{id} + C_j + \varphi_j, \qquad \varphi_j \to 0, \ \varphi_1 = \exp(-\zeta/2).$$

This can be done, since

$$\mathrm{id} + C_1 + \exp(-\zeta/2) \in G^0_{\mathrm{rap}},$$

and G^0_{rap} is a group. Note that $\varphi_2 = (\exp(-\zeta/2))(1 + o(1))$. This follows from the requirements singling out case 2° and implying that the correction of the composition $g_1^{-1} \circ g_2$ is small (modulo the shift). The fact that

$$(g_j^{-1} \circ \tilde{g})(\Pi, \infty) \subset (\Pi, \infty)$$

can now be proved just as in Case 1°. This concludes the proof of assertion 2, and with it that of the whole of Lemma 5.2_1. ►

REMARK. The proof in Case 1 gives us in the simplest variant the ideas that are basic for Chapter V.

§2.5. Shifts of cochains by cochains

In this section we prove for functional cochains general lemmas lying at the basis of the proof of the function-theoretic shift lemmas $\mathrm{SL}\,2_{n,\mathrm{reg}}$ and $\mathrm{SL}\,3_{n,\mathrm{reg}}$ formulated schematically in the introduction to the chapter, and in detail in §§2.7 and 2.8.

LEMMA 4. *Suppose that $\mathbf{\Omega}$ is a proper class of standard domains, $\mathscr{D}_1$ and $\mathscr{D}_2$ are two sets of $\mathbf{\Omega}$-admissible germs, and σ is the germ of a diffeomorphism $(\mathbf{R}^+, \infty) \to (\mathbf{R}^+, \infty)$ satisfying the following conditions*:

1. *σ can be extended to a complex domain in such a way that*

$$\sigma \circ \mathscr{D}_1 \subset \mathscr{D}_2 \cup \{\text{a set of nonessential germs}\}(^6);$$

2. *for any domain Ω of class $\mathbf{\Omega}$ there exists a domain $\tilde{\Omega}$ of the same class such that $\rho\tilde{\Omega} \subset \Omega$, where $\rho = \sigma^{-1}$;*
3. *there exists a standard domain Ω of class $\mathbf{\Omega}$ in which*

$$\operatorname{Re}\rho \prec \alpha\xi \quad \text{for arbitrary } \alpha > 0. \qquad (*)$$

Then for arbitrary $F_1 \in \mathscr{F}\mathscr{C}_{\mathrm{reg}}(\mathscr{D}_1)$ and $F_2 \in \mathscr{F}\mathscr{C}^+_{\mathrm{reg}}(\mathscr{D}_2)$,

$$F \overset{\mathrm{def}}{=} F_1 \circ (\rho + F_2) - F_1 \circ \rho \in \mathscr{F}\mathscr{C}^+_{\mathrm{reg}}(\mathscr{D}_2).$$

PROOF. For the proof it must be verified that the functional cochain F satisfies all the requirements of Definition 10 in §1.6. The first two requirements describe domains to which the functions of the tuple F extend holomorphically. The following elementary consideration is crucial for the investigation of these domains. Suppose that we are given two domains, the first containing the second, and two functions f_1 and f_2, the first defined in the first (larger) domain and the second in the second (smaller) domain. Suppose, moreover, that the modulus of f_2 is less than the distance from the points

---

(6) See the footnote (5) after Lemma 3.

of the smaller domain to the boundary of the larger. Then the composition $f_1 \circ (\mathrm{id} + f_2)$ is holomorphic in the smaller domain.

Below, this consideration is applied to nested pairs of generalized neighborhoods of the domains of a regular partition, and Proposition 1 of §2.1 is used.

Let us proceed directly to the proof of Lemma 4.

1. We construct a standard domain of class $\mathbf{\Omega}$ in which all the requirements of the definition of a regular functional cochain are satisfied (Definition 10 in §1.6) for F. For this we take a standard domain Ω_1 of class $\mathbf{\Omega}$ in which the inequality given by requirement 3° of Definition 10 in §1.6 is satisfied for F_1: there exists a $\nu > 0$ such that

$$|F_1| < \exp \nu\xi \quad \text{in the domain } \Omega_1.$$

Next, for each $\delta > 0$ we construct a domain Ω_2^δ in which

$$|F_2| < \delta, \qquad |F_2| < \delta|\zeta|^{-4}, \tag{$**$}$$

and the image $\rho\Omega_2^\delta$ belongs to Ω_1 together with its 2δ-neighborhood. The inequalities $(**)$ hold because the cochain F_2 is rapidly decreasing.

We prove that such domains can be constructed. First take a domain Ω_2 such that in it

$$|F_2| < \exp(-\mu\xi)$$

for some $\mu > 0$, and $\Omega_2 \subset \mathbf{C}^+$. If $\Omega \subset \Omega_1 \cap \Omega_2$ is chosen so that the distance $\operatorname{dist}(\partial\Omega_2, \Omega)$ is sufficiently large, then the inequalities $(**)$ will be satisfied in this domain. Next, take a domain Ω of class $\mathbf{\Omega}$ such that in it the inequality $(**)$ holds and $\operatorname{dist}(\partial\Omega_1, \Omega) > 2\delta$. By condition 2 of the lemma, there exists a domain $\widetilde{\Omega}$ of class $\mathbf{\Omega}$ such that $(**)$ holds in it and $\rho\widetilde{\Omega} \subset \Omega$. By property 2 of the domains in the proper class $\mathbf{\Omega}$, there exists a domain Ω_2^δ belonging to the class $\mathbf{\Omega}$ and to the intersection $\widetilde{\Omega} \cap \Omega$. This domain is the desired one.

2. We prove that the restriction of the composition F to the intersection of the domain Ω_2^δ with the upper half-plane extends to Ω_2^δ to form cochains corresponding to the $\mathbf{R}$- and k-realizations of a partition Ξ of type $\boldsymbol{\sigma} = \sigma \circ \sigma^{F_1} \cup \sigma^{F_2}$, where $\sigma = \rho^{-1}$, and $\sigma^{F_1} = \boldsymbol{\sigma}_1$ and $\sigma^{F_2} = \boldsymbol{\sigma}_2$ are the types of the cochains F_1 and F_2, respectively. The constructions are analogous to those in the proof of Lemma 3 in §2.3 and Lemma 1 in §2.1. We exclude the nonessential germs of the tuple $\sigma \circ \sigma^{F_1}$, and denote the resulting tuple by $\tilde{\boldsymbol{\sigma}}$. For arbitrary $l > 0$ the l-realization $\widetilde{\Xi}$ of the partition Ξ is obtained as the product of k_1- and k_2-realizations of the partitions of type $\boldsymbol{\sigma}$ and $\boldsymbol{\sigma}_1$ for suitable k_1 and k_2, and the $\mathbf{R}$-realization of Ξ is obtained as the product of $\mathbf{R}$-realizations of the partitions of type $\tilde{\boldsymbol{\sigma}}$ and $\boldsymbol{\sigma}_2$.

We construct the l-realization $\widetilde{F}$ of the cochain F; the $\mathbf{R}$-realization is constructed analogously (only more simply). Let $\widetilde{F}_1$ and $\widetilde{F}_2$ be the k_1- and

k_2-realizations of the cochains F_1 and F_2, respectively, and let

$$\widetilde{F} = \widetilde{F}_1 \circ (\mathrm{id} + \widetilde{F}_2) - \widetilde{F}_1.$$

Suppose that the cochains $\widetilde{F}_1$ and $\widetilde{F}_2$ are ε-extendible, and $\widetilde{\Xi}_1$ and $\widetilde{\Xi}_2$ are the corresponding partitions. By Proposition 1 in §2.1, the margin between the ε- and $(\varepsilon/2)$-neighborhoods of the domains of the partition exceeds $c|\zeta|^{-3}$ for some $c > 0$. Take δ such that $\delta|\zeta|^{-3} < c|\rho|^{-3}$; this is possible because ρ increases no faster than $|\zeta|$. We take the domain Ω_2^δ for such a δ and prove that the cochain $\widetilde{F}$ is $(\varepsilon/2)$-extendible.

Let $\mathscr{U}$ be an arbitrary domain in the partition

$$\Xi = \sigma_* \widetilde{\Xi}_1 \cup \widetilde{\Xi}_2.$$

Then there exist a domain V of the partition $\widetilde{\Xi}_1$ and a domain W of the partition $\widetilde{\Xi}_2$ such that

$$U = \sigma_* V \cap W, \quad \text{where } \sigma_* V = \sigma V \cap \Omega_2^\delta.$$

Let f_1 and f_2 be the functions in $\widetilde{F}_1$ and $\widetilde{F}_2$ corresponding to the domains V and W. The function f_1 extends to the $(\boldsymbol{\sigma}^{F_1}, \varepsilon)$-neighborhood of V; call it $V^{(\varepsilon)}$; the function f_2 extends to the $(\boldsymbol{\sigma}^{F_2}, \varepsilon)$-neighborhood $W^{(\varepsilon)}$ of W. Then the composition $f_1 \circ (\rho + f_2)$ is defined in the $(\boldsymbol{\sigma}, \varepsilon/2)$-neighborhood $W^{(\varepsilon/2)}$ of W. Indeed, the composition $f_1 \circ \rho$ is defined in the $(\tilde{\boldsymbol{\sigma}}, \varepsilon)$-neighborhood $(\sigma_* V)^{(\varepsilon)}$ of the domain $\sigma_* V$; the function f_2 is defined in the domain $W^{(\varepsilon)}$. The function $\rho + f_2$ is defined in $\mathscr{U}^{(\varepsilon)}$ and carries $\mathscr{U}^{(\varepsilon/2)}$ into $V^{(\varepsilon)}$, since

$$\mathscr{U}^{(\varepsilon/2)} \subset (\sigma_* V)^{(\varepsilon/2)}, \qquad \rho\sigma_* V^{(\varepsilon/2)} \subset V^{(\varepsilon/2)}$$

and $|f_2| < \operatorname{dist}(V^{(\varepsilon/2)}, \partial V^{(\varepsilon)}) \circ \rho$. Indeed, the distance on the right-hand side is greater than

$$c|\rho|^{-3} > \delta|\zeta|^{-3} \geq |f_2|.$$

This proves that the cochain $\widetilde{F}$ is $(\varepsilon/2)$-extendible.

The analogous assertion for the $\mathbf{R}$-realization of the cochain F is proved by replacing the estimate $|f_2| < \delta|\zeta|^{-3}$ by $|f_2| < \delta$; the arguments are parallel to those above.

3. We get an upper estimate of the modulus of the cochain $\widetilde{F}$: we show that there exist $c > 0$ and $\varkappa > 0$ such that in Ω_2^δ

$$|\widetilde{F}| < c \exp(-\varkappa\xi).$$

By a condition of the lemma, there exist positive constants c_1, c_2, μ, and ν such that $|F_1| < c_1 \exp \nu\xi$ in Ω_1 and $|F_2| < c_2 \exp(-\mu\xi)$ in Ω_2^δ. In the δ-neighborhood of Ω_2^δ we have the estimate

$$|F| \leq \max_{\theta \subset [0, 1]} |F_1' \circ (\rho + \theta F_2)|\, |F_2|.$$

The domain $\rho\Omega_2^\delta$ belongs to the domain Ω_1 together with its 2δ-neighborhood. Moreover, $|F_2| \le \delta$ and $0 \le \theta \le 1$. Therefore, $(\rho + \theta F_2)\Omega_2^\delta$ belongs to the domain Ω_1 together with its δ-neighborhood. Consequently, by Cauchy's inequality,

$$|F_1' \circ (\rho + \theta F_2)| < \delta^{-1} C_1 \exp(\nu \operatorname{Re}(\rho + \theta F_2)) \le \delta^{-1} C_1' \exp \nu \operatorname{Re} \rho$$

for arbitrary $\theta \in [0, 1]$, where $C_1' = C_1 \exp \nu\delta$; recall that $|F_2| < \delta$ in Ω_2^δ. Next, by a condition of the lemma, for any $\alpha > 0$ the domain Ω_2^δ can be chosen so that in it $\operatorname{Re} \rho < \alpha\xi$. Take α small enough that $\nu\alpha - \mu = -\varkappa < 0$. Then $|F| < C \exp(-\varkappa\xi)$, and requirement 4 is verified.

4. *Estimate of the coboundary.* Let $\mathscr{L}$ be a standard boundary line of the partition Ξ^F. Three cases are possible: $\mathscr{L}$ is a standard boundary line of both partitions of type $\sigma \circ \boldsymbol{\sigma}^{F_1}$ and $\boldsymbol{\sigma}^{F_2}$—the first case; of only one of these partitions—the other two cases. We consider the first case. The other two can be reduced to this one if it is assumed that one of the pairs of functions considered below does not have a jump on $\mathscr{L}$. Let f_1 and f_2 be the functions in the tuple $F_1 \circ \rho$ that are defined in the domains of the partition $\sigma_* \Xi^{F_1}$ adjacent to $\mathscr{L}$; g_1 and g_2 are the analogous functions in the tuple F_2. Then the function in δF defined in the $(\boldsymbol{\sigma}, \varepsilon/2)$-neighborhood of $\mathscr{L}$ has the form

$$\begin{aligned} h &= (f_1 \circ (\rho + g_1) - f_1 \circ \rho) - (f_2 \circ (\rho + g_2) - f_2 \circ \rho) \\ &= [(f_1 - f_2) \circ (\rho + g_1) - (f_1 - f_2) \circ \rho] + [f_2 \circ (\rho + g_1) - f_2 \circ (\rho + g_2)]. \end{aligned}$$

We estimate separately each of the square brackets; denote the first by h_1, and the second by h_2. The difference $f_1 - f_2$ is a function in the tuple δF_1. By definition, there exists a $C > 0$ such that in the $(\sigma^{F_1}, \varepsilon)$-neighborhood of the line $\rho\mathscr{L}$

$$|f_1 - f_2| < \sum \exp(-C \exp \operatorname{Re} \rho_0),$$

where the summation is over those j such that $\mathscr{L}$ is a standard boundary line of the partition $\sigma_{j*}\Xi_0$; the diffeomorphisms σ_j are in the tuple σ^{F_1}. In the $(\boldsymbol{\sigma}, \varepsilon/2)$-neighborhood of the line $\mathscr{L}$

$$|(f_1 - f_2)'| < C' \sum \exp(-C'' \exp \operatorname{Re} \rho_j)$$

for some $C' > 0$ and $C'' \in (0, C)$. This follows from Proposition 1 in §2.1 and Cauchy's inequality. Consequently, if δ is sufficiently small, then in the $(\boldsymbol{\sigma}, \varepsilon/3)$-neighborhood of $\mathscr{L}$

$$\begin{aligned} |h_1| &< C' \left|\sum \exp(-C'' \exp \operatorname{Re} \rho_j \circ \rho)\right| |g_1| \\ &< C' \sum \exp(-C'' \exp \operatorname{Re} \rho_j \circ \rho) \end{aligned}$$

since $|g_1| < \delta < 1$. This is the required estimate on the function of the tuple forming the coboundary, but the estimate is only on the first term. Let us estimate the second term, omitting the details; the arguments are analogous to the preceding arguments. Using the exponential estimate for regular cochains and the Cauchy inequality, we obtain $|f'_2| < \exp \nu\xi$ in Ω_2^δ for some $\nu > 0$. The same holds for $f'_2 \circ \rho$, because $\operatorname{Re} \rho < \alpha\xi$ in Ω_2^δ. Consequently, we have that

$$|h_2| < C \exp(\nu\xi)|g_1 - g_2| < C \sum \exp \circ C_2(\nu\xi - \exp \operatorname{Re} \rho_j),$$

where the summation is over those j such that $\mathscr{L}$ is a boundary line of the partition $\sigma_{j*}\Xi_0$; this time the functions σ_j are from the tuple $\boldsymbol{\sigma}^{F_2}$. By the definition of an admissible diffeomorphism (part 4 of Definition 4 in 1.4),

$$\nu\xi - \varepsilon \exp \operatorname{Re} \rho_j \to -\infty$$

for arbitrary $\varepsilon > 0$. Consequently, there exist C' and C'' such that

$$|h_2| < \sum C' \exp(-C'' \exp \operatorname{Re} \rho_j).$$

The final estimate $|h| \leq |h_1| + |h_2|$ is what is required in Definition 10 in §1.6.

LEMMA 5. *Suppose that* $\boldsymbol{\Omega}$ *is a proper class of standard domains, and* $\mathscr{D}$ *is the set of admissible germs of class* $\boldsymbol{\Omega}$. *Let* F_1 *and* F_2 *be the germs of regular functional cochains, the second weakly decreasing*:

$$F_1 \in \mathscr{F}\mathscr{C}_{\mathrm{reg}}(\mathscr{D}), \qquad F_2 \in \mathscr{F}\mathscr{C}_{\mathrm{wr}}^{+}(\mathscr{D}).$$

Then

$$F = F_1 \circ (\mathrm{id} + F_2) \in \mathscr{F}\mathscr{C}_{\mathrm{reg}}(\mathscr{D}).$$

If in addition F_1 *is a weakly or rapidly decreasing cochain, then* F *is a weakly or rapidly decreasing cochain, respectively.*

PROOF. For arbitrary $\delta > 0$ take the domains Ω_2^δ and Ω_1 (independent of δ) in the same way as in the proof of Lemma 4. We prove that for sufficiently small δ the germ F has a representative on Ω_2^δ satisfying all the requirements of Definition 10 in §1.6. We remark that all the arguments of parts 1 and 2 in the proof of Lemma 4 go through for $\sigma = \rho = \mathrm{id}$ and $F_2 \in \mathscr{F}\mathscr{C}_{\mathrm{wr}}(\mathscr{D})$; recall that the condition $\operatorname{Re} \rho \prec \alpha\xi$, which is not satisfied for $\rho = \mathrm{id}$, is used only in part 3 of the proof. Assume that $\rho = \mathrm{id}$ in parts 1 and 2 of the proof of Lemma 4, and choose δ as in these parts. We get that F is a cochain defined in Ω_2^δ and corresponding to a regular partition of type $\boldsymbol{\sigma}^{F_1} \cup \boldsymbol{\sigma}^{F_2}$.

We estimate the modulus of the germ F. By a condition of the lemma, $|F_1| < C_1 \exp \nu\xi$ in the domain Ω_1^δ, and $|F_2| < \delta$ in Ω_2^δ. Consequently, in Ω_2^δ,

$$|F| < C' \exp \nu\xi, \qquad C' = C_1 \exp \nu\delta.$$

If F_1 is a weakly or rapidly decreasing cochain, then in Ω_1^δ we have the respective inequalities

$$|F_1| < C_1|\zeta|^{-5}$$

or $|F_1| < C_1 \exp(-\nu\xi)$, $\nu > 0$. Then, respectively,

$$|F| < C'|\zeta|^{-5}, \qquad C' = C \sup_{\Omega_2^\delta}(|\zeta| - \delta)^5|\zeta|^{-5},$$

or

$$|F| < C' \exp(-\nu\xi), \qquad C' = C_1 \exp \nu\delta.$$

This concludes the estimate of the modulus of F.

The coboundary is estimated just as in Lemma 4; the argument used in part 4 of the proof of Lemma 4 goes through with minimal changes for the difference

$$\begin{aligned} &f_1 \circ (\mathrm{id} + g_1) - f_2 \circ (\mathrm{id} + g_2) \\ &\quad = (f_1 \circ (\mathrm{id} + g_1) - f_2 \circ (\mathrm{id} + g_1)) + (f_2 \circ (\mathrm{id} + g_1) - f_2 \circ (\mathrm{id} + g_2)). \end{aligned}$$ ►

§2.6. Properties of special admissible germs of class n

The properties of the germs of class $\mathscr{D}_0^n$ and $\mathscr{D}_1^{n-1}$ are formulated in this section. They are then proved by induction on n. The induction base (the case $n = 1$) is analyzed in this section. The induction step—the passage from n to $n+1$—is realized in Chapter V. Therefore, all the lemmas below have a double number, with the digit 5 as the first. The index indicates the value of the parameter with respect to which the induction is being carried out. The value n of the index corresponds to the classes $\mathscr{D}_0^n$ and $\mathscr{D}_1^{n-1}$. The sets $\mathscr{D}_0^n$ and $\mathscr{D}_1^{n-1}$ are defined with the help of the groups G^{n-1} and G^{n-2}, for the elements of which we can assume as proved all the assertions that will then be proved for the elements of G^n.

LEMMA 5.3_n. a. *For each $g \in G^{n-1}$ the germ $A^{1-n}g$ maps the germ of any half-strip containing $(\mathbf{R}^+, \infty)$ biholomorphically into the germ of any sector $(S^\forall, \infty)$,*

$$S^\forall = \{|\arg \zeta| < \alpha,\ \alpha \in (\pi/2, \pi)\}.$$

b. *For each $g \in G^{n-1}$ in the domain $(\Pi^\forall, \infty)$, $\Pi^\forall = \{|\operatorname{Im} \zeta| < a,\ \xi \geq 0\}$, $a > 0$ arbitrary, the limit*

$$\mu_n(g) = \lim_{(\Pi^\forall, \infty)} (A^{1-n}g)'$$

exists, positive, zero, or infinite.

LEMMA 5.4_n. a. *The germs of the class $\mathscr{D}_0^n$ are admissible of class Ω_n.*

b. *The germs of the form $\sigma = A^{-n}g \circ \exp$ with $g \in G_{\text{slow}}^{n-1^+} \cup G_{\text{rap}}^{n-1}$ are nonessential (of class $\mathbf{\Omega}_n$).*

c. *The product of two germs of class* $\mathscr{D}_0^n$ *is a nonessential germ.*

LEMMA 5.5_n. a. *Germs of class* $\mathscr{D}_{\text{slow}}^{n-1} \circ \mathscr{L}^{n-1}$ *are admissible of class* n. *If* $g \in G_{\text{slow}}^{n-1^-}$, *then* $(A^{1-n} g\Pi, \infty) \subset (\Pi, \infty)$. *For an arbitrary germ* $g \in G^{n-1}$ *the following assertions are satisfied*:

b. *For arbitrary* $c > 0$ *and* $\varepsilon > 0$,

$$(\ln|\zeta|)^c \prec |A^{-n} g| \prec \exp|\zeta|^\varepsilon \quad \textit{in } (S^\forall, \infty),$$
$$(\ln|\zeta|)^c \prec |A^{1-n} g| \prec \exp \xi^\varepsilon \quad \textit{in } (\Pi, \infty).$$

c. *Let* $\sigma = A^{-n} g$. *Then* $\sigma'/\sigma \to 0$ *in* $(\mathbf{C}^+, \infty)$.

d. *Suppose that in addition* $g \in G_{\text{slow}}^{n-1^+}$, *and let* $\rho = \sigma^{-1}$. *Then there exists a standard domain of class* n *such that in this domain* $\operatorname{Re}\rho < \varepsilon\xi$ *for every* $\varepsilon > 0$.

In the remainder of this chapter all nonessential germs are of class $\mathbf{\Omega}_n$.

Lemmas 5.3_n–5.5_n below are assumed to be valid—induction hypotheses. In Chapter V Lemmas 5.3_{n+1}–5.5_{n+1} are proved—the induction step. Here these lemmas are proved for $n = 1$—the induction base.

It is proved in §1.6 that the germs of class $\mathscr{D}_0^1$ are equivalent in the sense of Definition 18 in §1.6E to germs of the form $\zeta \mapsto (\exp\mu\zeta)$, $\mu \in (0, 1)$, $C > 0$, while germs of class $\mathscr{D}_1^0$ are equivalent to affine germs $\zeta \mapsto \alpha\zeta + \beta$. The set $\mathscr{L}^0$ is empty.

In Lemma 5.3_1 the germ $A^{1-n} g$ belongs to G^0 and is almost regular (see the definition in §§0.3 and 1.7). The lemma is trivial for it.

In Lemma 5.4_1 the germ σ is equivalent to a germ $\zeta \to C\exp\circ\mu\zeta$. For such germs the lemma is also trivial (assertion (a) corresponds to the case $0 < \mu < 1$, and assertion (b) to the case $\mu \geq 1$).

In Lemma 5.5_1b–d we have that $\sigma \in A^{-1} g$, $g \in \mathscr{R}$. Consequently, $\sigma = C\zeta^{\mu+o(1)}$. For such germs all the assertions of the lemma are trivial. We mention only that the case $g \in G_{\text{slow}}^{0^-}$ corresponds to $\mu < 1$. By Proposition 1 in D of §1.6, the function σ given by the preceding formula in some standard domain of class 1 satisfies the inequality

$$\operatorname{Re}\sigma \prec |\sigma| \prec |\zeta|^{\mu+\delta} \prec \varepsilon\xi$$

for $\mu + \delta < 1$. In Lemma 5.5_1.a, the sets $\mathscr{D}_{\text{slow}}^0$ and $\mathscr{L}^0$ are empty, and the lemma is true as a property of the empty set.

This proves Lemmas 5.3_1–5.5_1 and finishes the induction base.

§2.7. The second shift lemma and the conjugation lemma, function-theoretic variant: $\mathrm{SL}\,2_{n,\text{reg}}$ and $\mathrm{CL}_{n,\text{reg}}$

A. Formulations. We recall the conventions in §1.10:

$$\mathscr{D}^m = \begin{cases} \mathscr{D}_1^m & \text{for } m \leq n-1, \\ \mathscr{D}_0^m & \text{for } m = n. \end{cases}$$

Let also

$$\mathscr{F}^{m^{(+)}}_{\mathrm{reg},\,g} = \mathscr{F}\mathscr{C}_{\mathrm{reg}}(\mathscr{D}^m)\circ \exp^{[m]}\circ g\,, \qquad g\in G^{m-1}$$

The plus in parentheses means that the equality is true both with the plus (without parentheses) and without it.

Lemma $\mathrm{SL}\,2_{n,\,\mathrm{reg}}$. *Suppose that* $m=n-1$ *or* $m=n$, $(k,f)\prec(m,g)$, *and* $\varphi\in\mathscr{F}^k_{\mathrm{reg},\,f}$. *Then*

$$\varphi\circ(\mathrm{id}+\mathscr{F}^{m^+}_{\mathrm{reg},\,g})-\varphi\subset\mathscr{F}^{m^+}_{\mathrm{reg},\,g}.$$

Lemma $\mathrm{CL}_{n,\,\mathrm{reg}}$. *Suppose that* $m=n-1$ *or* $m=n$, $1\le k\le m-1$, *and* $f\in G^k$. *Then*

$$\mathrm{Ad}(f)(\mathrm{id}+\mathscr{F}^{m^+}_{\mathrm{reg},\,g})\subset \mathrm{id}+\mathscr{F}^{m^+}_{\mathrm{reg},\,gf}.$$

The assertions of these lemmas for fixed k are denoted by $\mathrm{SL}\,2_{k,\,n,\,\mathrm{reg}}$ and $\mathrm{CL}_{k,\,n,\,\mathrm{reg}}$, respectively. Both lemmas can be proved simultaneously by induction on k. The induction base is $\mathrm{CL}_{-1,\,n,\,\mathrm{reg}}$, and the induction step is carried out according to the scheme

$$\mathrm{CL}_{k-1,\,n,\,\mathrm{reg}}\overset{\mathrm{A}}{\Rightarrow}\mathrm{SL}\,2_{k,\,n,\,\mathrm{reg}}\overset{\mathrm{B}}{\Rightarrow}\mathrm{CL}_{k,\,n,\,\mathrm{reg}}.$$

The implication A is proved for $k\le m$ in this and the next three subsections, and the implication B is proved for $k\le m-1$ in E.

Induction base: $\mathrm{CL}_{-1,\,n,\,\mathrm{reg}}$. Recall that $G^{-1}=-\mathscr{A}\!f\!f$. For $k=-1$ the $\mathrm{CL}_{k,\,n,\,\mathrm{reg}}$ follows immediately from the linearity of the germs $f\in G^k$.

B. Induction step: implication A. Is is required to prove that for each pair of germs

$$F_1\in\mathscr{F}\mathscr{C}_{\mathrm{reg}}(\mathscr{D}^k)\quad\text{and}\quad F_2\in\mathscr{F}\mathscr{C}_{\mathrm{reg}}(\mathscr{D}^m)$$

with $k\le m$ and $m=n-1$ or n and each pair of germs $f\in G^{k-1}$, $g\in G^{m-1}$ with $f\prec\!\prec g$ in G^{m-1} for $k=m$ there exist a germ $F\in\mathscr{F}\mathscr{C}^+_{\mathrm{reg}}(\mathscr{D}^m)$ and a standard domain Ω of class m such that

$$F_1\circ\exp^{[k]}\circ f\circ(\mathrm{id}+F_2\circ\exp^{[m]}\circ g)-F_1\circ\exp^{[k]}\circ f=F\circ\exp^{[m]}\circ g;$$

the equality holds in the domain $g^{-1}\circ\ln^{[m]}\Omega$. It is equivalent to the equality

$$F_1\circ[\mathrm{Ad}(f^{-1}\circ\ln^{[k]})\circ(\mathrm{id}+F_2\circ\exp^{[m]}\circ g)]-F_1=F\circ\sigma,$$

where $\sigma=\exp^{[m]}\circ h\circ\ln^{[k]}$, $h=g\circ f^{-1}$; the equality holds in the domain $\rho\Omega$, where $\rho=\sigma^{-1}$. We investigate the composition in square brackets.

By the $\mathrm{CL}_{k-1,\,n,\,\mathrm{reg}}$, which appears in the induction hypothesis, there exists a germ $F_3\in\mathscr{F}\mathscr{C}^+_{\mathrm{reg}}(\mathscr{D}^m)$ such that

$$\mathrm{Ad}(f^{-1})(\mathrm{id}+F_2\circ\exp^{[m]}\circ g)=\mathrm{id}+F_3\circ\exp^{[m]}\circ h.$$

Note further that $\exp^{[k]} \in \mathscr{F}_{\text{reg}}^{k-1}$. By the $\text{SL}\,2_{k-1, n, \text{reg}}$, which appears in the induction hypothesis, there exists a germ $F_4 \in \mathscr{F}\mathscr{C}_{\text{reg}}^{+}(\mathscr{D}_1^m)$ such that

$$\exp^{[k]} \circ (\text{id} + F_3 \circ \exp^{[m]} \circ h) = \exp^{[k]} + F_4 \circ \exp^{[m]} \circ h.$$

Consequently,

$$A^{-k} \circ (\text{id} + F_3 \circ \exp^{[m]} \circ h) = \text{id} + F_4 \circ \sigma.$$

Accordingly, it is required to prove that in some standard domain Ω of class m there exists a representative of the germ $F \in \mathscr{F}\mathscr{C}_{\text{reg}}^{+}(\mathscr{D}^m)$ such that in $\rho\Omega$

$$F_1 \circ (\text{id} + F_4 \circ \sigma) - F_1 = F \circ \sigma,$$

or, equivalently,

$$F_1 \circ (\rho + F_4) - F_1 \circ \rho = F. \tag{$*$}$$

REMARK. This computation does not use the inequalities $f \prec\prec g$ for $f, g \in G^{m-1}$; the result of it will be used in the proof of Lemma $\text{SL}\,3_{n, \text{reg}}$ for $f \succ\succ g$ or $f \circ g^{-1} \in G_{\text{rap}}^{m-1}$.

We prove $(*)$ by using Lemma 4 of §2.5. Below, condition 1 in it is verified separately for $m = n$ and $m = n - 1$ in Propositions 1 and 2, respectively, and conditions 2 and 3 are verified in Proposition 3.

C. Group properties of special admissible germs.

PROPOSITION 1. *Suppose that* $(k, f) \prec (n, g)$, $f \in G^{k-1}$, $g \in G^{n-1}$; $\sigma = \exp^{[n]} \circ g \circ f^{-1} \circ \ln^{[k]}$. *Then*

$$\sigma \circ \mathscr{D}^k \subset \mathscr{D}_0^n \cup \{\textit{the set of nonessential germs}\}.$$

PROOF. We consider three cases: $k \le n - 2$, $k = n - 1$, $k = n$.
In the first two cases it is required to prove that

$$\sigma \circ \mathscr{D}_1^k \subset \mathscr{D}_0^n \cup \{\text{the set of nonessential germs}\}. \tag{$**$}$$

CASE 1: $k \le n - 2$. For an arbitrary germ $\sigma_0 \in \mathscr{D}_1^k$ the composition $\sigma \circ \sigma_0$ is nonessential in this case. Indeed, the inverse germ has the form

$$\rho_0 \circ \rho = \rho_0 \circ \ln^{[n-k-1]} \circ A^{1-n} h \circ \ln, \qquad h = f \circ g^{-1}.$$

By assertion a of Lemma 5.3_n, which appears in the induction hypothesis, the germ $A^{1-n} h$ is holomorphic in the germ at infinity of an arbitrary horizontal half-strip containing $(\mathbf{R}^+, \infty)$ and carries it into the germ at infinity of an arbitrary sector. Denote by $\Pi^\forall$ an arbitrary horizontal half-strip, and by $S^\forall$ an arbitrary sector; both sets contain $(\mathbf{R}^+, \infty)$ by definition. The germ ρ acts as follows:

$$(\mathbf{C}^+, \infty) \xrightarrow{\ln} (\Pi, \infty) \xrightarrow{A^{1-n}h} (S^\forall, \infty) \xrightarrow{\ln^{[n-k-1]}} (\Pi^\forall, \infty).$$

Accordingly, the image $\rho(\mathbf{C}^+, \infty)$ belongs to any (arbitrarily narrow) half-strip containing $(\mathbf{R}^+, \infty)$. By the definition of an admissible diffeomorphism, the derivative ρ_0' is bounded, and ρ_0 carries $(\Pi^\forall, \infty)$ into $(C\Pi^\forall, \infty)$ for some $C > 0$. Consequently, the image $\rho_0 \circ \rho(\mathbf{C}^+, \infty)$ belongs to an arbitrary half-strip and, moreover, to the domain Π_* (see §1.5). This easily implies that the germ $\sigma \circ \sigma_0$ is nonessential in the sense of Definition 3 in §1.5.

CASE 2: $k = n - 1$. In this case

$$\sigma = \exp \circ A^{1-n} h^{-1}, \qquad h = f \circ g^{-1} \in G^{n-1}.$$

We prove (**) for $k = n - 1$. Recall that $\mathscr{D}_1^{n-1} = \mathscr{D}_0^{n-1} \cup \mathscr{D}_*^{n-1} \circ \mathscr{L}^{n-1}$, and $\mathscr{D}_*^{n-1} = A^{1-n}(G_{\text{rap}}^{n-2} \cup G_{\text{slow}}^{n-2^+})$. Below we prove

ASSERTION *.

$$\mathscr{L}^{n-1} \subset A^{1-n} G^{n-1}.$$

Let $\sigma_0 \in \mathscr{D}_1^{n-1}$. Suppose first that $\sigma_0 \in \mathscr{D}_0^{n-1}$. Then the germ $\sigma \circ \sigma_0$ is nonessential. Indeed,

$$\sigma_0 = A^{1-n} g_1 \circ \exp, \qquad g_1 \in G^{n-2},$$
$$\tilde{\rho} = (\sigma \circ \sigma_0)^{-1} = \ln \circ (A^{1-n}(g_1^{-1} \circ h)) \circ \ln, \qquad g_1^{-1} \circ h \in G^{n-1}.$$

The germ $\tilde{\rho}$ acts according to the scheme:

$$\tilde{\rho} : (\mathbf{C}^+, \infty) \xrightarrow{\ln} (\Pi, \infty) \xrightarrow{A^{1-n}(g_1^{-1} \circ h)} (S^\forall, \infty) \xrightarrow{\ln} (\Pi^\forall, \infty).$$

The proofs are the same as in Case 1.

Suppose now that $\sigma_0 \in \mathscr{D}_*^{n-1} \circ \mathscr{L}^{n-1}$, $\sigma_0 = A^{1-n} g_1 \circ \tilde{h}$, $g_1 \in G^{n-2}$, $\tilde{h} \in \mathscr{L}^{n-1}$, $\tilde{h} = A^{1-n} \tilde{g}$, for some $\tilde{g} \in G^{n-1}$, by assertion *. Then $\sigma \circ \sigma_0 \in \mathscr{D}_0^n \cup \{$the set of nonessential germs$\}$. Indeed,

$$\begin{aligned} \sigma \circ \sigma_0 &= \exp \circ A^{1-n}(h \circ g_1 \circ \tilde{g}) \\ &= A^{-n}(h \circ g_1 \circ \tilde{g}) \circ \exp, \qquad h \circ g_1 \circ \tilde{g} \in G^{n-1}. \end{aligned}$$

The statement in Case 2 now follows from Lemma 5.4_nb and the definition of the set $\mathscr{D}_0^n$.

CASE 3: $k = n$. In this case it is required to prove that $\sigma \circ \mathscr{D}_0^n \subset \mathscr{D}_0^n \cup \{$the set of nonessential germs$\}$, where $\sigma = A^{-n} h$, and $h = g \circ f^{-1} \succ\succ \text{id}$ in G^{n-1}. For an arbitrary germ $\sigma_0 \in \mathscr{D}_0^n$ we have that

$$\sigma_0 = (A^{-n} g_1) \circ \exp, \qquad g_1 \in G_{\text{slow}}^{n-1^-}.$$

Consequently,

$$\sigma \circ \sigma_0 = A^{-n}(h \circ g_1) \circ \exp, \qquad h \circ g_1 \in G^{n-1}.$$

If $h \circ g_1 \in G_{\text{slow}}^{n-1^-}$, then $\sigma \circ \sigma_0 \in \mathscr{D}_0^n$ by definition; if $h g_1 \in G_{\text{rap}}^{n-1} \cup G_{\text{slow}}^{n-1^+}$, then the germ $\sigma \circ \sigma_0$ is nonessential by Lemma 5.4_nb. ►

PROPOSITION 2. *Suppose that* $(k, f) \prec (n-1, g)$, $f \in G^{k-1}$, $g \in G^{n-2}$, *and* $\sigma = \exp^{[n-1]} \circ (g \circ f^{-1}) \circ \ln^{[k]}$.

Then

$$\sigma \circ \mathscr{D}_1^k \subset \mathscr{D}_1^{n-1} \cup \{\text{the set of nonessential germs}\}.$$

PROOF.

CASE 1: $k \leq n-3$. This is handled just like Case 1 of Proposition 1.

CASE 2: $k = n-2$. In this case $\sigma = \exp \circ A^{1-n} h$, $h = g \circ f^{-1}$. The proof that

$$\sigma \circ \mathscr{D}_1^{n-2} \subset \mathscr{D}_1^{n-1} \cup \{\text{the set of nonessential germs}\}$$

is the same as in Case 2 of Proposition 1.

CASE 3: $k = n-1$. In this case $\sigma = A^{1-n} h$, $h \in G_{\text{slow}}^{n-2^+}$. The proof that

$$\sigma \circ \mathscr{D}_1^{n-1} \subset \mathscr{D}_1^{n-1} \cup \{\text{the set of nonessential germs}\}$$

is the same as in Case 3 of Proposition 1, except that the definition of $\mathscr{D}_1^{n-1}$ instead of $\mathscr{D}_0^n$ is used.

PROOF OF ASSERTION $*$. Let $m = n-1$. A generating element of the group $\mathscr{L}^m$ has the form (Definition 15 in §1.6)

$$\tilde{h} = \operatorname{Ad}(A^{1-m} g \circ \ln) \tilde{f}, \qquad \tilde{f} = A(g_1) f,$$
$$f \in A^0, \quad g_1 \in \mathscr{H}^0, \quad \tilde{f} \in \mathscr{M}_{\mathbf{R}}, \quad g \in G^{m-1},$$
$$\mu_m(g) = 0.$$

In the proof of Assertion $*$ the last requirement is not used. A simple computation shows that

$$A^m \tilde{h} = \operatorname{Ad}(\exp^{[m-1]} \circ g) \tilde{f}.$$

Consequently,

$$A^m \tilde{h} = \operatorname{Ad}(g) A^{m-1} \tilde{f} = \operatorname{Ad}(g)[(A^m g_1) A^{m-1} f].$$

The germ in the square brackets belongs to the set

$$H^m \circ J^{m-1} \cap \mathscr{M}_{\mathbf{R}} \subset G^m.$$

Further, $g \in G^{m-1} \subset G^m$. Consequently,

$$\tilde{h} \in G^m = G^{n-1}. \quad ►$$

D. Estimate of the germ ρ. The following proposition verifies conditions 2 and 3 of Lemma 4.

PROPOSITION 3. *Suppose that* $m = n$ *or* $m = n-1$, *and* σ *is the same germ as in Proposition* 1 *or* 2 *for* $m = n$ *or* $m = n-1$, *respectively. Then for an arbitrary standard domain* Ω *of class* n *there exists a standard domain* $\tilde{\Omega}$ *such that* $\rho \tilde{\Omega} \subset \Omega$, *and in this domain*

$$\operatorname{Re} \rho < \varepsilon \xi \quad \textit{for arbitrary } \varepsilon > 0. \qquad (***)$$

PROOF. We use assertions obtained in the course of proving Propositions 1 and 2. We consider the same three cases as in these proofs.

CASE 1: $k \le m-2$. The germ ρ carries $(\mathbf{C}^+, \infty)$ into the germ at infinity of an arbitrary half-strip containing $(\mathbf{R}^+, \infty)$. By Definition 1 in §1.5, a standard domain contains not only the germ of an arbitrary horizontal half-strip, but also the germ of any sector of the form $|\arg\zeta| < \alpha < \pi/2$. Consequently, $\widetilde{\Omega}$ can be taken to be an arbitrary standard domain of class m "sufficiently far to the right." In this case the inequality $(***)$ follows easily from assertion d of Lemma 5.5_n; an analogous and somewhat more complicated argument is carried out in Case 2.

CASE 2: $k = m-1$. In this case the germ ρ acts according to the scheme

$$(\mathbf{C}^+, \infty) \overset{\ln}{\to} (\Pi, \infty) \overset{A^{1-m}h}{\to} (S^{\forall}, \infty).$$

The last inclusion follows from assertion a in Lemma 5.3_n, which appears in the induction hypothesis. The inequality $(***)$ follows in this case from the estimates

$$\operatorname{Re}\rho \le |\rho| \le |(A^{1-m}h)\circ \ln| \le \exp(-\varepsilon \ln|\zeta|) = |\zeta|^{\varepsilon}.$$

The last inequality follows from assertion b in Lemma 5.5_n. In a quadratic standard domain, in which

$$|\zeta| \prec \xi^3,$$

the preceding estimate on $\operatorname{Re}\rho$ implies $(***)$.

CASE 3: $k = m$. In this case $\sigma = \rho^{-1} = A^{-m}h$, $h \in G_{\text{slow}}^{m-1^+}$. Therefore, the germ σ is admissible of class m according to assertion a in Lemma 5.5_m, which appears in the induction hypothesis for $m \le n$. The inequality $(***)$ follows from assertion d in this lemma. ►

All the conditions in Lemma 4 have been verified. The use of it proves the equality $(*)$ and hence the implication A.

E. Implication B. By the ADT_k, which appears in the induction hypothesis for $k \le n-1$, there exists for each $f \in G^k$ an expansion

$$f^{-1} = a + \sum \varphi_j, \qquad \varphi_j \in \mathscr{F}_{\text{reg}, g_j}^{k_j^+}, \quad g_j \in G^{k_j - 1}.$$

Then

$$\begin{aligned}\operatorname{Ad}(f)(\mathrm{id} + \mathscr{F}_{\text{reg}, g}^{m^+}) &= [a \circ (\mathrm{id} + \mathscr{F}_{\text{reg}, g}^{m^+}) + \sum \varphi_j \circ (\mathrm{id} + \mathscr{F}_{\text{reg}, g}^{m^+})] \circ f \\ &\overset{①}{\subset} \left(a + \sum \varphi_j + \mathscr{F}_{\text{reg}, g}^{m^+}\right) \circ f = \mathrm{id} + \mathscr{F}_{\text{reg}, g\circ f}^{m^+}\end{aligned}$$

The inclusion ① follows from the affineness of a and Lemma $\mathrm{SL}\,2_{k,n,\text{reg}}$ proved above with implication A; here the latter appears in the induction hypothesis. To use this lemma we must set $\varphi = \varphi_j$ in its formulation. The implication B is proved, and with it $\mathrm{SL}\,2_{n,\mathrm{reg}}$. ►

§2.8. The third shift lemma, function-theoretic variant, $\mathrm{SL}\,3_{n,\mathrm{reg}}$

Recall that, by the convention in §1.10

$$\mathscr{D}^m = \begin{cases} \mathscr{D}_1^m & \text{for } m \le n-1, \\ \mathscr{D}_0^m & \text{for } m = n. \end{cases}$$

Let, for brevity, $\mathscr{F}\mathscr{C}_{\mathrm{reg}}^{m^{(+)}} = \mathscr{F}\mathscr{C}_{\mathrm{reg}}^{(+)}(\mathscr{D}^m)$.

LEMMA $\mathrm{SL}\,3_{n,\mathrm{reg}}$. (a) *Suppose that* $m \le n$, *and either* $f \succ\!\succ g$ *in* G^{m-1} *or* $f \circ g^{-1} \in G_{\mathrm{rap}}^{m-1}$. *Then*

$$\mathscr{F}_{\mathrm{reg},f}^{m^{(+)}} \circ (\mathrm{id} + \mathscr{F}_{\mathrm{reg},g}^{m^+}) \subset \mathscr{F}_{\mathrm{reg},f}^{m^{(+)}}.$$

(b)

$$(\mathrm{id} + \mathscr{F}_{\mathrm{reg},g}^{m^+})^{-1} = \mathrm{id} + \mathscr{F}_{\mathrm{reg},g}^{m^+}.$$

PROOF. It is required to prove that for arbitrary $F_1 \in \mathscr{F}\mathscr{C}_{\mathrm{reg}}^{(+)}(\mathscr{D}^m)$, $F_2 \in \mathscr{F}\mathscr{C}_{\mathrm{reg}}^{+}(\mathscr{D}^m)$, and f, g satisfying the condition of the lemma there exists a germ $F \in \mathscr{F}\mathscr{C}_{\mathrm{reg}}^{(+)}(\mathscr{D}^m)$ such that

$$F_1 \circ \exp^{[m]} \circ f \circ (\mathrm{id} + F_2 \circ \exp^{[m]} \circ g) = F \circ \exp^{[m]} \circ f.$$

We investigate the germ F given by this equality. It follows from the remark at the end of §2.7B that there exists a germ $\widetilde{F}_2 \in \mathscr{F}\mathscr{C}_{\mathrm{reg}}^{+}(\mathscr{D}^m)$ such that

$$F = F_1 \circ (\mathrm{id} + \widetilde{F}_2 \circ \rho),$$

where $\rho = A^{-m}h$, $h = g \circ f^{-1}$. By a condition of the lemma, $h \in G_{\mathrm{slow}}^{m-1^-} \cup G_{\mathrm{rap}}^{m-1}$.

We prove that $\widetilde{F}_2 \circ \rho$ is a weakly decreasing cochain of class $\mathscr{F}\mathscr{C}_{\mathrm{wr}}(\mathscr{D}^m)$. To do this it is necessary to prove that

$$\sigma \circ \mathscr{D}^m \subset \mathscr{D}^m \cup \{\text{the set of nonessential germs}\}, \quad \text{where } \sigma = \rho^{-1}, \tag{$*$}$$

along with the estimate

$$|\widetilde{F}_2 \circ \rho| < |\zeta|^{-5} \tag{$**$}$$

in some standard domain of class n.

Formula $(*)$ and Lemma 3 in §2.3 imply that $\widetilde{F}_2 \circ \rho \in \mathscr{F}\mathscr{C}_{\mathrm{reg}}(\mathscr{D}^m)$; this and the estimate $(**)$ imply that $\widetilde{F}_2 \circ \rho \in \mathscr{F}\mathscr{C}_{\mathrm{wr}}(\mathscr{D}^m)$. Assertion a of Lemma $\mathrm{SL}\,3_{n,\mathrm{reg}}$ is obtained from this and Lemma 5 in §2.5.

In the case $h \in G_{\mathrm{rap}}^{m-1}$, $(*)$ follows from Proposition 1 of §2.4B; in the case $h \in G_{\mathrm{slow}}^{m-1^-}$, it follows from Propositions 1 and 2 in §2.7 (Case 3).

We prove the estimate $(**)$. By the definition of a rapidly decreasing cochain, there exists a $\nu > 0$ such that $|\widetilde{F}_2| < \exp(-\nu\xi)$ in some standard domain Ω of class n. By Lemmas 5.5_na and 5.2_nb, there exists a domain $\widetilde{\Omega}$ of class n such that $\rho\widetilde{\Omega} \subset \Omega$. In view of the estimate in b of Lemma 5.5_n it can be assumed that the inequality $\operatorname{Re}\rho \succ c\ln|\zeta|$ holds in $\widetilde{\Omega}$ for all $c > 0$. Consequently, in this domain

$$\exp(-\nu \operatorname{Re}\rho) \prec |\zeta|^{-\nu c}.$$

Since c is arbitrary, it can be assumed that $\nu c > 5$. This proves $(**)$ and thus assertion a in Lemma $\mathrm{SL}\,3_{n,\mathrm{reg}}$.

Let us prove assertion b in that lemma: for $m = n$ or $m = n - 1$

$$(\mathrm{id} + \mathscr{F}^{m^+}_{\mathrm{reg},g})^{-1} = \mathrm{id} + \mathscr{F}^{m^+}_{\mathrm{reg},g}.$$

According to Lemma $\mathrm{CL}_{n-1,n,\mathrm{reg}}$, we have that for $g \in G^{n-1}$

$$\mathrm{id} + \mathscr{F}^{m^+}_{\mathrm{reg},g} = \mathrm{Ad}(g)(\mathrm{id} + \mathscr{F}^{m^+}_{\mathrm{reg},\mathrm{id}}).$$

Therefore, it suffices to prove assertion b for $g = \mathrm{id}$. Further, we have the following equality, proved below:

$$\mathrm{id} + \mathscr{F}^{m^+}_{\mathrm{reg},\mathrm{id}} = A^m(\mathrm{id} + \mathscr{F}\mathscr{C}^{m^+}_{\mathrm{reg}}). \qquad (***)$$

An equivalent assertion:

$$A^{-m}(\mathrm{id} + \mathscr{F}^{m^+}_{\mathrm{reg},\mathrm{id}}) = \mathrm{id} + \mathscr{F}\mathscr{C}^{m^+}_{\mathrm{reg}}.$$

We prove by induction on k the equality

$$A^{-k}(\mathrm{id} + \mathscr{F}^{m^+}_{\mathrm{reg},\mathrm{id}}) = \mathrm{id} + \mathscr{F}\mathscr{C}^{m^+}_{\mathrm{reg}} \circ \exp^{[m-k]}$$

$(0 \le k \le m)$. The needed assertion is obtained for $k = m$.

THE INDUCTION BASE: $k = 0$. In this case the equality to be proved becomes an obvious identity.

INDUCTION STEP. Suppose that the assertion has been proved for some $k \le m - 1$; we prove it for $k + 1$. By the induction hypothesis,

$$\begin{aligned} A^{-(k+1)}(\mathrm{id} + \mathscr{F}^{m^+}_{\mathrm{reg},\mathrm{id}}) &\subset A^{-1}(\mathrm{id} + \mathscr{F}\mathscr{C}^{m^+}_{\mathrm{reg}} \circ \exp^{[m-k]}) \\ &= \exp\circ(\ln + \mathscr{F}\mathscr{C}^{m^+}_{\mathrm{reg}} \circ \exp^{[m-k+1]}) = \zeta \cdot \exp\mathscr{F}\mathscr{C}^{m^+}_{\mathrm{reg}} \circ \exp^{[m-k+1]}. \end{aligned}$$

By Corollary 1 in §2.2,

$$\exp\mathscr{F}\mathscr{C}^{m^+}_{\mathrm{reg}} = 1 + \mathscr{F}\mathscr{C}^{m^+}_{\mathrm{reg}}.$$

Then

$$\exp\mathscr{F}\mathscr{C}^{m^+}_{\mathrm{reg}} \circ \exp^{[m-k-1]} = 1 + \mathscr{F}\mathscr{C}^{m^+}_{\mathrm{reg}} \circ \exp^{[m-k-1]}.$$

We next get by Lemma 1 in §2.1 that

$$\begin{aligned}\zeta\cdot\mathscr{F}\mathscr{C}_{\mathrm{reg}}^{m^+}\circ\exp^{[m-k-1]} &= [\ln^{[m-k-1]}\mathscr{F}\mathscr{C}_{l,\mathrm{reg}}^{m^+}]\circ\exp^{[m-k-1]}\\ &= \mathscr{F}\mathscr{C}_{\mathrm{reg}}^{m+1}\circ\exp^{[m-k-1]}.\end{aligned}$$

From this,

$$\zeta\cdot\exp\mathscr{F}\mathscr{C}_{\mathrm{reg}}^{m^+}\circ\exp^{[m-k-1]} = \mathrm{id}+\mathscr{F}\mathscr{C}_{\mathrm{reg}}^{m^+}\circ\exp^{[m-k-1]}.$$

The equality $(***)$ is proved.

Accordingly, assertion b of Lemma $\mathrm{SL}\,3_{m,\mathrm{reg}}$ is equivalent to the following:

$$[\mathrm{Ad}(\exp^{[m]}\circ g)(\mathrm{id}+\mathscr{F}\mathscr{C}_{l,\mathrm{reg}}^{m^+})]^{-1} = \mathrm{Ad}(\exp^{[m]}\circ g)(\mathrm{id}+\mathscr{F}\mathscr{C}_{l,\mathrm{reg}}^{m^+}),$$

or

$$(\mathrm{id}+\mathscr{F}\mathscr{C}_{\mathrm{reg}}^{m^+})^{-1} = \mathrm{id}+\mathscr{F}\mathscr{C}_{\mathrm{reg}}^{m^+}.$$

The last equality will be derived directly from the definitions. Let $\varphi\in\mathscr{F}\mathscr{C}_{\mathrm{reg}}^{m^+}$. By definition, there exists a standard domain Ω of class m and a number $\nu>0$ such that $|\varphi|<\exp(-\nu\xi)$ in the domain Ω and even in the ε-neighborhood Ω^ε of Ω for some $\varepsilon>0$. It can be assumed without loss of generality that $|\varphi|<\varepsilon/2$ in Ω^ε. The correction of the inverse map-cochain has the form $-\tilde\varphi$: $(\mathrm{id}+\varphi)^{-1}=\mathrm{id}-\tilde\varphi$ and satisfies the function equation

$$\tilde\varphi=\varphi\circ(\mathrm{id}-\tilde\varphi).$$

This implies the following a priori estimate: the correction $\tilde\varphi$ is bounded in the $(\varepsilon/2)$-neighborhood of the domain Ω, by the constant $\varepsilon/2$. We now obtain a sharper estimate on $\tilde\varphi$ from the estimate on φ:

$$|\tilde\varphi(\zeta)|<\exp(-\nu\xi+\nu\varepsilon/2)<\exp(-\nu\xi/2);$$

the last inequality is valid if the function ξ is sufficiently large in Ω.

Accordingly, the correction $\tilde\varphi$ decreases exponentially in the $(\varepsilon/2)$-neighborhood of the standard domain Ω. The cochain $\tilde\varphi$ corresponds to the same partition as φ: that is, to a regular partition of class $\mathscr{D}^m$. The cochain $\tilde\varphi$ is $(\sigma,\varepsilon/2)$-extendible, and extends to the $(\sigma,\varepsilon/2)$-neighborhood of the extended main domain of the corresponding partition, because the margin between the (σ,ε)-neighborhood and the $(\sigma,\varepsilon/2)$-neighborhood of domains of a regular partition and the extended main domain of the regular partition decreases no more rapidly than $|\zeta|^{-3}$, while the correction $\tilde\varphi$ decreases exponentially.

This verifies requirements 1 and 2 of Definition 10 in §1.6. Requirement 3 of this definition has already been verified above. Finally, we get an upper estimate of the coboundary of the cochain $\tilde\varphi$. In place of the difference coboundary of $\tilde\varphi$ we consider the correction of the composition coboundary of the cochain $\mathrm{id}+\tilde\varphi$. The correction of the composition coboundary of this map-cochain and its difference coboundary vanish simultaneously and differ by a factor $1+o(1)$. The composition coboundaries of mutually inverse map-cochains are mutually inverse. Consequently, the correction $-\psi$ of the composition coboundary of $\mathrm{id}-\tilde{\varphi}$ satisfies the equation

$$\psi = \delta_\circ \varphi \circ (\mathrm{id}-\psi),$$

where $\mathrm{id}+\delta_\circ\varphi$ is the composition coboundary of the cochain $\mathrm{id}+\varphi$. The subscript $\circ$ indicates composition. Therefore, ψ can be estimated from above by rigging functions differing from the rigging functions estimating φ_δ from above only by a shift of the argument:

$$\exp(-c\exp\operatorname{Re}\rho_j(\xi_j+\tilde{c})), \qquad |\tilde{c}|<\varepsilon.$$

These functions are majorized by rigging functions of the same class, but corresponding to another value of c, since the derivatives ρ_j' are bounded. This finishes the verification of all the requirements of Definition 10 in §1.6 and proves Lemma $\mathrm{SL}\,3_{n,\mathrm{reg}}$.

§2.9. Properties of the group $A^{-n}J^{n-1}$

The group J^{n-1} is generated by the germs in the set

$$\widetilde{J}^{n-1} \stackrel{\mathrm{def}}{=} \{\mathrm{Ad}(g)A^{n-1}f \mid g\in G^{n-1},\ f\in\mathscr{A}^0\}.$$

The set $\widetilde{J}^{n-1}$ can be divided into three subsets, depending on the value of $\mu_n(g)$;

$$\mathrm{Ad}(g)A^{n-1}f \in \begin{cases} J_\infty^{n-1} & \text{for } \mu_n(g)=\infty,\\ J_*^{n-1} & \text{for } \mu_n(g)\in(0,\infty),\\ J_0^{n-1} & \text{for } \mu_n(g)=0.\end{cases}$$

Analogous definitions with n replaced by k are used for any $k\le n-1$.

Recall that $J_0^{n-1}\subset L^n$ by definition. To prove Lemma $\mathrm{SL}\,4_{n,\mathrm{reg}}$ it is necessary to investigate "shifts of cochains by cochains from $A^{-n}J^{n-1}$." For this we need the lemmas that make up the main content of this section.

Lemma 6_n. $A^{-n}J_\infty^{n-1}\subset \mathrm{id}+\mathscr{F}\mathscr{C}_{\mathrm{wr}}(\mathscr{D}_0^n)$.

Lemma 7_n. $A^{-n}J_*^{n-1}\subset A^{-n}G_{\mathrm{rap}}^{n-1}\circ(\mathrm{id}+\mathscr{F}\mathscr{C}_{\mathrm{wr}}(\mathscr{D}_0^n))$.

Lemma 7_n is based on the next lemma.

Lemma 8_n. *Suppose that*

$$j=\mathrm{Ad}(g)A^{n-1}f\in J_*^{n-1},$$

$$\mu_n(g)=\mu, \qquad |f-\mathrm{id}|=o(1)\exp(-\gamma\xi),$$

and $\mu\nu\ge 5$. *Then*

$$h=A^{-n}+j\in\mathrm{id}=\mathscr{F}\mathscr{C}_{\mathrm{wr}}(\mathscr{D}_0^n).$$

Lemmas 6_n and 8_n can be proved in parallel; the proofs separate only at the very end of the argument.

PROOF. Suppose that $h \in A^{-n}J_\infty^{n-1}$ or h is the same germ as in Lemma 8_n. Then $h = \mathrm{Ad}(\rho)f$, where

$$f \in \mathscr{A}^0, \quad \rho = (A^{1-n}g) \circ \ln, \quad g \in G^{n-1}, \quad \mu_n(g) \in (0, \infty].$$

1. We prove that h is defined in some standard domain of class n. For this we take two standard domains Ω_1 and Ω_2 of class n in which representatives of the germ ρ are defined that satisfy the following additional requirements:

1°. $\Omega_2 \subset \Omega_1$, $\mathrm{dist}(\Omega_2, \partial\Omega_1) \geq 1$;

2°. the germ f extends to the domain $\rho\Omega_2$, and there

$$f = \mathrm{id} + \varphi, \qquad |\varphi| < \exp(-\nu\xi)$$

where the constant ν is positive in the proof of Lemma 6_n, and is the same as in the condition of the lemma in the proof of Lemma 8_n.

Such domains Ω_1 and Ω_2 can be chosen in view of the properness of the class Ω_n of standard domains (Lemma 5.2_nd).

Assume in addition that Ω_2 is located in a "far removed right half-plane" $\xi \geq a$; in what follows, the constant a will be taken large enough that the asymptotic inequalities of type $f \prec g$ in Ω_2 becomes pointwise inequalities: $f < g$.

We prove that the germ h extends to Ω_2 as a mapping $\Omega_2 \to \Omega_1$. Assume the opposite: there exists a point $\zeta \in \Omega_2$ to which the germ h does not extend. We bring this assumption to a contradiction, and at the same time estimate the correction of h, that is, we verify requirement 3 of Definition 10 in §1.6. Consider the composition

$$h_\theta = \mathrm{Ad}(\rho)(\mathrm{id} + \theta\varphi), \qquad \theta \in [0, 1].$$

Obviously, $h_0 = \mathrm{id}$ and $h_1 = h$. Therefore, the composition h_θ is defined at ζ for small θ. Let θ_* be the supremum of those θ such that $h_{\tilde\theta}(\zeta)$ is defined for all $\tilde\theta \in [0, \theta_*]$. If $\theta_* < 1$, then $h_{\theta_*}(\zeta) \in \partial\Omega_1$, otherwise θ_* would not be the supremum of the set described. If $\theta_* = 1$, then $h_{\theta_*}(\zeta) \in \partial\Omega_1$, otherwise the germ h would extend to the point ζ. In both cases $h_{\theta_*} \in \partial\Omega_1$, that is,

$$|h_{\theta_*}(\zeta) - \zeta| \geq 1.$$

Let us bring this inequality to a contradiction. Suppose, as above, that $\sigma = \rho^{-1} = \exp \circ A^{1-n}g^{-1}$. We prove that there exists a $C > 0$ such that for a sufficiently large value of $|\zeta|$

$$|h_{\theta_*}(\zeta) - \zeta| \leq C|\zeta|\,|\varphi \circ \rho(\zeta)|. \tag{$*$}$$

If we then get an upper estimate of the right-hand side of this inequality, we contradict the preceding. We have that

$$h_{\theta_*}(\zeta) = \sigma \circ (\mathrm{id} + \theta_*\varphi) \circ \rho = \sigma \circ (\rho + \theta_*\varphi \circ \rho).$$

Recall that the number θ_* is chosen so that the segment

$$\Sigma = [\rho(\zeta), \rho(\zeta) + \theta_* \varphi \circ \rho(\zeta)]$$

belongs to the domain of the function σ. Let us express the ratio h_{θ_*}/ζ in terms of the logarithmic derivative of σ:

$$h_{\theta_*}/\zeta = \sigma \circ (\rho + \theta_* \varphi \circ \rho)/\sigma \circ \rho = \exp \int_{\rho(\zeta)}^{\rho(\zeta)+\theta_*\varphi\circ\rho(\zeta)} \sigma'/\sigma.$$

By the definition of σ,

$$\sigma'/\sigma = (\ln \sigma)' = (A^{1-n} g^{-1})'.$$

The limit of this logarithmic derivative in the germ of the domain $\rho(\mathbf{C}^+, \infty)$ is equal to $1/\mu_n(g)$, that is, is equal to zero or μ^{-1} under the conditions of Lemma 6_n or 8_n, respectively. Therefore, for arbitrary $\delta > 0$ and for sufficiently large $|\zeta|$, $|\sigma'/\sigma| |_\Sigma \le \mu^{-1} + \delta$, $\mu^{-1} = 0$ for $\mu = \infty$. Then

$$|h_{\theta_*}(\zeta)/\zeta| \le |\exp(\mu^{-1} + \delta)\theta_* |\varphi \circ \rho(\zeta)|,$$

$$|h_{\theta_*}(\zeta) - \zeta| \le \zeta(\mu^{-1} + 2\delta)\theta_* |\varphi \circ \rho(\zeta)|$$

if $|\varphi \circ \rho(\zeta)|$ is sufficiently small.

We now get an upper estimate of $|\varphi \circ \rho|$. For some $\nu > 0$

$$|\varphi| < \exp(-\nu\xi).$$

in the domain $\rho\Omega_2$. Consequently,

$$|\varphi \circ \rho| \le \exp(-\nu \operatorname{Re} A^{1-n} g \circ \ln) = 1/|\exp \cdot A^{1-n} g \circ \ln|^\nu = 1/|A^{-n} g|^\nu,$$

$$|h_{\theta_*}(\zeta) - \zeta| \le C|\zeta|/|A^{-n} g|^\nu.$$

The rest of the arguments are carried out separately under the conditions of Lemma 6_n and Lemma 8_n.

The case $\mu_n(g) = \infty$, *Lemma* 6_n. In this case $A^{1-n} g$ increases more rapidly than any linear germ in an arbitrary right half-strip sufficiently far enough to the right, $A^{-n} g$ increases more rapidly than any power, and $(A^{-n} g)^{-\nu}$ decreases more rapidly than any power in an arbitrary right half-plane far enough to the right. This implies that if $|\zeta|$ is sufficiently large, then

$$|h_{\theta_*}(\zeta) - \zeta| < C_1 |\zeta|^{-N} < 1, \qquad (**)$$

which contradicts the assumption.

The case $\mu_n(g) \in (0, \infty)$, *Lemma* 8_n. In this case $|A^{-n} g|^{-\nu} = |\zeta^{\mu+o(1)}|^{-\nu} = |\zeta|^{-\mu\nu+o(1)}$. For a sufficiently large value of a

$$|h_{\theta_*}(\zeta) - \zeta| \le C|\zeta|^{-5}.$$

This again gives us the inequalities $(**)$ (when $N = 5$) and contradicts the assumption.

Accordingly, the germ h extends to a standard domain Ω_2, and there its correction decreases more rapidly than any power if $\mu_n(g) = \infty$, and no more slowly than $|\zeta|^{-5}$ if $\mu_n(g) \in (0, \infty)$. Thus, this correction decreases just as required in the definition of weakly decreasing regular cochains.

2. We verify the remaining requirements of Definition 10 in §1.6. The germ h has the form $\sigma \circ f \circ \rho$. Let us prove that the germ $f \circ \rho$ corresponds to a regular partition and is ε-extendible for some $\varepsilon > 0$. To do this we use Lemma 3. The functional cochain f is of type $(\nu_1, \dots, \nu_N)$, $\nu_j \subset \mathbf{R}$. It follows from the definition of the set $\mathscr{D}_0^n$ and the fact that $A^{n-1}\mathscr{A}\!f\!f \subset G^{n-1}$ that the germ $\sigma \circ \nu_j$ belongs to $\mathscr{D}_0^n$ or is nonessential by Lemma 5.4_nb. It follows from Lemma 3 that the regular functional cochain $f \circ \rho$ is of class $\mathscr{D}_0^n$. In particular, the first two requirements of the definition in §1.6 hold: the germ $f \circ \rho$ corresponds to a regular partition, is ε-extendible for some $\varepsilon > 0$, and admits all the necessary realizations. Moreover, part of requirement 4 holds: the mappings corresponding to the coboundary of $f \circ \rho$ extend to the generalized ε-neighborhood of the boundary lines of the same partition. Consequently, the germ $h = \sigma \circ f \circ \rho$ corresponds to the same partition and is also ε-extendible for some $\varepsilon > 0$. Similarly, the functions of the tuple δh are extendible. The correction of the coboundary can be estimated in the same way as the correction of the cochain itself. Consider the $\mathbf{R}$-realization of the cochain f; the k-realization is handled similarly. Obviously,

$$\delta_\circ h = \mathrm{Ad}(\rho)\delta_\circ f,$$

where $\delta_\circ$ is the composition coboundary; see the end of §2.8. Let $\mathscr{L}$ be a standard boundary line of the partition (see Definition 8 in §1.6) corresponding to the germ f, and let $\mathrm{id}+\psi$ be the function in $\delta_\circ f$ corresponding to $\mathscr{L}$. By definition, there exists a $C > 0$ such that $|\psi| < \sum \exp(-C \exp \nu_j^{-1}\xi)$, where the summation is over all j such that $\mathscr{L}$ is a line of the partition $\nu_{j*}\Xi_{\mathrm{st}}$. Arguing as in the beginning of the proof, we get an analogue of formula $(*)$:

$$|\,\mathrm{Ad}(\rho)(\mathrm{id}+\psi) - \mathrm{id}\,| < C_1|\zeta| \sum \exp(-C \exp \mathrm{Re}\, \nu_j \rho).$$

By decreasing C, that is, choosing $C' \in (0, C)$, and increasing a, that is, shifting the domain Ω_2 to the right, it is possible to ensure that in Ω_2

$$C_1|\zeta| \sum \exp(-C \exp \mathrm{Re}\, \nu_j \rho) < \sum \exp(-C' \exp \mathrm{Re}\, \nu_j \rho)$$

This is the required estimate on the coboundary. Lemmas 6_n and 8_n are proved. ▶

PROOF OF LEMMA 7_n. Let f, g, and h be the same as in Lemma 8_n, but do not require the inequality $\mu\nu \geq 5$, $\mu = \mu_n(g)$. It must be proved that there exist an $h_1 \in A^{-n}G^{n-1}$ and an $h_2 \in \mathrm{id} + \mathscr{F}\mathscr{C}_{\mathrm{wr}}(\mathscr{D}_0^n)$ such that $h = h_1 \circ h_2$. Take the constant $\nu > 0$ such that $\nu\mu_n(g) \geq 5$. By Definition 1 in §1.3, the germ $f \in \mathscr{A}^0$ can be expanded in a Dulac series with linear part the identity. Let f_1 be a partial sum of this series approximating the germ f to within $o(1)\exp(-\nu\xi)$ in some standard domain of class n. Then

$$f = f_1 \circ f_2, \qquad |f_2 - \mathrm{id}| = o(1)\exp(-\nu\xi).$$

Let: $h_j = \mathrm{Ad}(\rho) f_j$, $j = 1, 2$.

Let us prove that $h_1 \in A^{-n} G^{n-1}_{\mathrm{rap}}$. For this we verify the requirements of Definition 10 in §1.6. It will first be proved that $h_1 \in A^{-n} G^{n-1}$. We have that

$$A^n h_1 = \mathrm{Ad}(g) A^{n-1} f_1.$$

But $f_1 \in \mathscr{R}^0$. Consequently, $A^{n-1} f_1$ belongs to the group G^{n-1} and even to the group G_{n-1}; see Definition 1 in §1.3. Further, $g \in G^{n-1}$. Since G^{n-1} is a group, we get that $A^n h_1 \in G^{n-1}$.

We prove that the germ h_1 increases no more rapidly and no more slowly than a linear germ on $(\mathbf{R}^+, \infty)$. This is equivalent to the assertion that the correction of the germ Ah_1 is bounded on $(\mathbf{R}^+, \infty)$. Let us prove this. We have

$$Ah_1 = \mathrm{Ad}(A^{1-n} g \circ \ln \circ \exp) f_1 = \mathrm{Ad}(A^{1-n} g) f.$$

In this case the germ $A^{1-n} g$ has a bounded derivative together with the inverse on $(\mathbf{R}^+, \infty)$, since $\mu_n(g) \in (0, \infty)$. The germ f_1 has an exponentially small correction. Therefore, the germ Ah_1 has not only a bounded but even an exponentially small correction on $(\mathbf{R}^+, \infty)$. Lemma 8_n gives us immediately that $h_2 \in \mathrm{id} + \mathscr{F}\mathscr{C}^+_{\mathrm{wr}}(\mathscr{D}^n_0)$. This finishes the proof of Lemma 7_n. ►

§2.10. The fourth shift lemma, assertion a

Let $\mathscr{F}^{n-1}_{\mathrm{reg}} = \mathscr{F}\mathscr{C}^+_{\mathrm{reg}}(\mathscr{D}^{n-1}_1) \circ \exp^{[n-1]} \circ G^{n-2} = \bigcup_{g \in G^{n-2}} \mathscr{F}^{n-1^+}_{\mathrm{reg},\, g}$

LEMMA SL $4_{n,\mathrm{reg}}$a. $J^{n-1} \subset \mathrm{Gr}(\mathrm{id} + \mathscr{F}^{n-1^+}_{\mathrm{reg}})$.

PROOF. It suffices to carry out the proof for generating elements of the group J^{n-1}. These generators have the form

$$h = \mathrm{Ad}(g) A^{n-1}(\mathrm{id} + \varphi), \qquad \varphi \in \mathscr{F}\mathscr{C}^0_+, \ g \in G^{n-1}.$$

REMARK. This is not the definition of h, but a consequence of the definition; for h to be in J^{n-1} it is necessary to impose additional requirements on φ that are not used in the argument given.

Accordingly, our goal is to prove that

$$h \in \mathrm{id} + \mathscr{F}^{n-1^+}_{\mathrm{reg}}.$$

We prove this first for $g = \mathrm{id}$. In the case of general g, a fifth shift lemma is needed; it is formulated below.

Thus, we prove first that

$$A^{n-1} f \in \mathrm{id} + \mathscr{F}^{n-1^+}_{\mathrm{reg},\,\mathrm{id}}$$

for $f = \mathrm{id} + \varphi$, $\varphi \in \mathscr{F}\mathscr{C}^0_+$. By the definitions of the operation A and the set $\mathscr{F}^{n-1^+}_{\mathrm{reg},\,\mathrm{id}}$, this is equivalent to the condition

$$\ln^{[n-1]} \circ (\mathrm{id} + \varphi) \circ \exp^{[n-1]} - \mathrm{id} \in \mathscr{F}\mathscr{C}^{n-1^+}_{\mathrm{reg}} \circ \exp^{[n-1]},$$

or

$$\tilde{\varphi} \overset{\mathrm{def}}{=} \ln^{[n-1]} \circ (\mathrm{id} + \varphi) - \ln^{[n-1]} \in \mathscr{F}\mathscr{C}^{n-1^+}_{\mathrm{reg}}.$$

In fact, a stronger assertion is valid: $\tilde{\varphi} \in \mathscr{F}\mathscr{C}^{0^+}_{\mathrm{reg}}$ for $\varphi \in \mathscr{F}\mathscr{C}^{0^+}_{\mathrm{reg}}$. We prove this. All the requirements of the definition of simple functional cochains (Example 2 after Definition 13 in §1.6) are satisfied, because the derivative $(\ln^{[n-1]})'$ is bounded. Indeed the functional cochain $\tilde{\varphi}$ is defined in the same standard domain as φ, and corresponds to the same partition. Let σ be the type of this partition, and suppose that the cochain φ is (σ, ε)-extendible. Then the cochain $\tilde{\varphi}$ has this property, too, since the functions in the tuples φ and $\tilde{\varphi}$ are defined in the same domains. The estimates in requirements 3 and 4 of Definition 10 in §1.6 follow for $\tilde{\varphi}$ and $\delta\tilde{\varphi}$ from the inequalities

$$|\tilde{\varphi}| < C|\varphi|, \qquad |\delta\tilde{\varphi}| < C|\delta\varphi|,$$

where C is the supremum of the modulus of the derivative of $\ln^{[n-1]}$ in the domain of the cochain $\tilde{\varphi}$.

This proves assertion a of the lemma for $g = \mathrm{id}$.

Let us proceed to the case of an arbitrary $g \in G^{n-1}$. We need the following "fifth shift lemma" (function-theoretic variant).

LEMMA SL $5_{n,\,\mathrm{reg}}$. $\mathscr{F}^{n-1^{(+)}}_{\mathrm{reg}} \circ J^{n-2} \subset \mathscr{F}^{n-1^{(+)}}_{\mathrm{reg}}$.

This lemma is proved in the next section.

We prove that for arbitrary $g \in G^{n-1}$ and $\varphi \in \mathscr{F}^{n-1^{(+)}}_{\mathrm{reg},\,\mathrm{id}}$

$$\mathrm{Ad}(g)(\mathrm{id} + \varphi) \in (\mathrm{id} + \mathscr{F}^{n-1^+}_{\mathrm{reg}}).$$

According to the multiplicative decomposition theorem MDT_{n-1}, which appears in the induction hypothesis, we get that

$$g = g_1 \circ j \circ u, \qquad g_1 \in G^{n-2}, \ j \in J^{n-2}, \ u \in H^{n-1}.$$

By the conjugation lemma $\mathrm{CL}_{n,\,\mathrm{reg}}$ for $m = n-1$, we get that

$$\mathrm{Ad}(g_1)(\mathrm{id} + \varphi) = \mathrm{id} + \psi \in \mathrm{id} + \mathscr{F}^{n-1^+}_{\mathrm{reg},\,g_1}.$$

Arguing as in the proof of the conjugation lemma in §2.7 and using Lemma $\mathrm{SL}\,4_{n-1,\,\mathrm{reg}}$a appearing in the induction hypothesis and $\mathrm{SL}\,2_{n,\,\mathrm{reg}}$ (proved before) we get that

$$j^{-1} \circ (\mathrm{id} + \psi) = j^{-1} + \tilde{\psi}, \qquad \tilde{\psi} \in \mathscr{F}^{n-1^+}_{\mathrm{reg},\,g_1}.$$

From this,

$$j^{-1} \circ (\mathrm{id} + \psi) \circ j = \mathrm{id} + \tilde{\psi} \circ j, \qquad \tilde{\psi} \circ j \in \mathscr{F}^{n-1^+}_{\mathrm{reg},\, g_1}.$$

by Lemma $\mathrm{SL}\,5_{n,\,\mathrm{reg}}$. Accordingly,

$$\mathrm{Ad}(g_1 \circ j)(\mathrm{id} + \varphi) \in \mathrm{id} + \mathscr{F}^{n-1^+}_{\mathrm{reg},\, g_1}.$$

Finally, $u \in \mathrm{Gr}(\mathrm{id} + \mathscr{F}^{n-1}_{0,\,\mathrm{reg}})$. Together with Lemma $\mathrm{SL}\,2_n$ this implies assertion a of the lemma. ►

§2.11. The fourth and fifth shift lemmas, function-theoretic variant

By the convention recalled at the beginning of §2.7, Lemmas $\mathrm{SL}\,4_{n,\,\mathrm{reg}}\mathrm{b}$ and $\mathrm{SL}\,5_{n,\,\mathrm{reg}}$ can be formulated as follows.

LEMMA $*$. *Suppose that $m = n$ or $m = n - 1$. Then for arbitrary $g \in G^{m-1}$*

$$\mathscr{F}^{m^{(+)}}_{\mathrm{reg},\, g} \circ J^{m-1} = \mathscr{F}^{m^{(+)}}_{\mathrm{reg},\, g}.$$

A. Reduction to the properties of the group L^m.

PROOF. The proof is carried out simultaneously for both values of m. Let $F_1 \in \mathscr{F}\mathscr{C}^{(+)}_{\mathrm{reg}}(\mathscr{D}^m)$ be an arbitrary cochain, and let j be a generating element of the group J^{m-1} of the form

$$j = \mathrm{Ad}(g_1) A^{m-1} f, \qquad g_1 \in G^{m-1}, \ f \in \mathscr{A}^0.$$

It is required to prove that there exist a cochain $F \in \mathscr{F}\mathscr{C}^{(+)}_{\mathrm{reg}}(\mathscr{D}^m)$ and a standard domain Ω of class m such that

$$F_1 \circ \exp^{[m]} \circ g \circ j = F \circ \exp^{[m]} \circ g;$$

the equality holds in the domain $g^{-1} \circ \ln^{[m]} \Omega$. It is equivalent to the equality $F_1 \circ \kappa = F$ in the domain Ω, where

$$\kappa = A^{-m} \mathrm{Ad}(g^{-1}) j = A^{-m} \mathrm{Ad}(g_1 g^{-1}) A^{m-1} f \overset{\mathrm{def}}{=} A^{-m} h.$$

The element h is again a generating element in J^{m-1} of the form

$$h = \mathrm{Ad}(\tilde{g}) A^{m-1} f, \qquad \tilde{g} \in G^{m-1}, \ f \in A^0, \ \tilde{g} = g_1 \circ g^{-1}.$$

We consider three cases: $\mu_m(\tilde{g}) = \infty$; $\mu_m(\tilde{g}) \in (0, \infty)$; $\mu_m(\tilde{g}) = 0$.

CASE 1: $\mu_m(\tilde{g}) = \infty$. In this case $A^{-m} h \in +\mathrm{id}\mathscr{F}\mathscr{C}_{\mathrm{wr}}(\mathscr{D}^m)$ by Lemma 6_m, and $F_1 \circ \kappa = F_1 \circ A^{-m} h = F \in \mathscr{F}\mathscr{C}^{(+)}_{\mathrm{reg}}(\mathscr{D}^m)$ by Lemma 5 in §2.5.

CASE 2: $\mu_m(\tilde{g}) \in (0, \infty)$. In this case, by Lemma 7_m,

$$A^{-m} h = h_1 \circ h_2, \qquad h_1 \in A^{-m} G^{m-1}_{\mathrm{rap}}, \ h_2 \in \mathrm{id} + \mathscr{F}\mathscr{C}_{\mathrm{wr}}(\mathscr{D}^m).$$

According to Lemmas $\mathrm{SL}\,1_{m,\,\mathrm{reg}}$ and 5 in §2.5,

$$F_1 \circ A^{-m} h = F_1 \circ h_1 \circ h_2 \in \mathscr{F}\mathscr{C}^{(+)}_{\mathrm{reg}}(\mathscr{D}^m) \circ h_2 \subset \mathscr{F}\mathscr{C}^{(+)}_{\mathrm{reg}}(\mathscr{D}^m).$$

CASE 3: $\mu_m(\tilde{g}) = 0$. In this case $A^{-m}h \in L^m$, and to finish the proof it is necessary to use Lemma 3 bis. To verify its conditions for $m = n$ and $m = n - 1$ it suffices to prove the following propositions, respectively.

PROPOSITION 1. *For any germ $h \in L^n$ and any $\sigma_0 \in \mathscr{D}_0^n$ the germs σ_0 and $h\sigma_0$ are equivalent. Moreover, there exists a standard domain of class n in which*

$$\varepsilon\xi \prec \operatorname{Re} h \prec \varepsilon^{-1}\xi, \qquad |h| \prec \varepsilon^{-1}|\zeta|$$

for arbitrary $\varepsilon \in (0, 1)$.

The last two inequalities are not required under the condition of Lemma 3 bis, but it is natural to prove them all together.

PROPOSITION 2. *For any germ $h \in L^{n-1}$ and any $\sigma_0 \in \mathscr{D}_1^{n-1}$ there exists a germ $\sigma_1 \in \mathscr{D}_1^{n-1}$ such that the germs $h\sigma_0$ and σ_1 are equivalent. Moreover, the inequalities of Proposition 1 hold in some standard domain of class n.*

REMARK. The proof of the last assertion in Proposition 1 goes through for any $m \leq n$ instead of n, in particular, for $m = n - 1$; therefore, the last assertion of Proposition 2 does not need a separate proof.

B. Group properties of the sets L^n and $\mathscr{D}_0^n$.

PROOF OF PROPOSITION 1. By Definition 18 in §1.6, it must be proved that the germ $\tilde{h}$ determined by the equality $h \circ \sigma_0 = \sigma_0 \circ \tilde{h}$ in the germ of the domain $\sigma_0^{-1}(\mathbf{C}^+, \infty)$ is negligible. Suppose first that h is a generating element of the group L^n of the form

$$h = A^{-n}(\operatorname{Ad}(\tilde{g})A^{n-1}f), \qquad f \in \mathscr{A}^0, \ \tilde{g} \in G^{n-1}, \ \mu_n(\tilde{g}) = 0.$$

By assumption,

$$\sigma_0 = A^{-n}g \circ \exp, \qquad g \in G_{\text{slow}}^{n-1^-}.$$

Then

$$\begin{aligned} \tilde{h} &= \operatorname{Ad}(\sigma_0)h = Ah_1, \\ h_1 &= A^{-n}(\operatorname{Ad}(g_1)A^{n-1}f), \\ g_1 &= \tilde{g} \circ g. \end{aligned}$$

Since $\mu_n(\tilde{g}) = 0$ and $g \preceq \mathrm{id}$, we get that $\mu_n(g_1) = 0$; consequently, $\tilde{h} \in AL^n$. We prove

PROPOSITION 3 (7). *The correction of any germ $\tilde{h} \in AL^n$ decreases more rapidly than any power of ζ in the domain $(\sigma_0^{-1}S^\forall, \infty)$ for an arbitrary $\sigma_0 \in \mathscr{D}_0^n$.*

PROOF. Note that $(\sigma_0^{-1}S^\forall, \infty) = (\sigma\Pi^\forall, \infty)$, where $\sigma = A^{1-n}g^{-1}$. By Lemma 5.3_n, the derivative σ' has a limit μ in $(\Pi^\forall, \infty)$, and $\mu \geq 1$,

(7) In this section $S^\forall$ is any sector of the form $\{|\arg\zeta| < \alpha\}$, $\alpha \in (\pi/2, \pi)$, and $\Pi^\forall$ is any half-strip of the form $\{\xi \geq 0, |\eta| < \alpha\}$, $\alpha \in (\pi/2, \pi)$.

because $g^{-1} \succeq \mathrm{id}$. Consequently, the inverse mapping $\rho = \sigma^{-1}$ has a bounded derivative less or close to unity in $(\Pi^{\forall}, \infty)$. Therefore, $(\sigma\Pi^{\forall\varepsilon}, \infty)$ contains the δ-neighborhood of the germ of $(\sigma\Pi^{\forall}, \infty)$ for some $\delta > 0$. This remark implies that it suffices to prove the estimate $|\tilde{h} - \mathrm{id}| \prec |\zeta|^{-N}$ in $(\sigma\Pi^{\forall}, \infty)$ for arbitrary $N > 0$ in the case when h is a generating element of L^n: compositions of germs with decreasing corrections do not lead the germ of the domain $(\sigma\Pi, \infty)$ beyond the limits of the germ of $(\sigma\Pi^{\varepsilon}, \infty)$, where the corrections of all the factors are estimated. Below we prove

PROPOSITION 4.

$$A^{1-n}(\mathrm{id} + \mathscr{F}_{\mathrm{reg}}^{n-1^+}) \subset \mathrm{id} + \mathscr{F}_{\mathrm{reg}}^{n-1^+} \circ \ln^{[n-1]}.$$

It now suffices to prove that for an arbitrary germ $\varphi \in \mathscr{F}_{\mathrm{reg}}^{n-1^+}$ and for any $N > 0$

$$|\varphi \circ \ln^{[n-1]}| \prec |\zeta|^{-N} \quad \text{in } (\sigma\Pi^{\forall}, \infty).$$

This is a simple consequence of the definition of rapidly decreasing cochains. Namely, by the definition of the set $\mathscr{F}_{\mathrm{reg}}^{n-1^+}$, there exist germs $F \in \mathscr{F}\mathscr{C}_{\mathrm{reg}}^{+}(\mathscr{D}^{n-1})$ and $f \in G^{n-2}$ such that $\varphi = F \circ \exp^{[n-1]} \circ f$. About the germ F we need only know that there exist an $\varepsilon > 0$ and a standard domain Ω such that $|F| < \exp(-\varepsilon\xi)$ in Ω. Then

$$\varphi \circ \ln^{[n-1]} = F \circ A^{1-n} f,$$
$$(A^{1-n} f)\sigma = A^{1-n}(f \circ g^{-1}).$$

It follows from Lemma 5.3_na that

$$(A^{1-n}(f \circ g^{-1})\Pi^{\forall}, \infty) \subset (\Omega, \infty).$$

Therefore, in $(\sigma\Pi^{\forall}, \infty)$

$$|F \circ A^{1-n} f| \prec \exp(-\varepsilon \operatorname{Re} A^{1-n} f).$$

But, by Lemma 5.5_{n-1}b, for arbitrary $c > 0$,

$$|A^{1-n} f| \succ (\ln|\zeta|)^c \quad \text{in } (\mathbf{C}^+, \infty).$$

By Lemma 5.3_{n-1}a,

$$\arg(A^{1-n} f) \to 0 \quad \text{in } (\sigma\Pi^{\forall}, \infty)$$

(this is equivalent to the condition that $\arg A^{1-n}(f \circ g^{-1}) \to 0$ in $(\Pi^{\forall}, \infty)$, which follows from Lemma 5.3_na; recall that $\sigma = A^{1-n} g^{-1}$). Consequently,

$$\operatorname{Re} A^{1-n} f \succ (\ln|\zeta|)^c \succ N \ln|\zeta|$$

in $(\sigma\Pi^{\forall}, \infty)$ for all $c > 1$ and $N > 0$. Therefore, in $(\sigma\Pi^{\forall}, \infty)$

$$|\varphi \circ \ln^{[n-1]}| \prec \exp(-\varepsilon \operatorname{Re} A^{1-n} f) \prec \exp(-c \ln|\zeta|) = |\xi|^{-N}.$$

Proposition 3 is proved. ▶

This implies the first assertion in Proposition 1, modulo Proposition 4. Let us prove the second one.

From Proposition 3,

$$Ah = \mathrm{id} + \varphi\,, \qquad |\varphi| \prec |\zeta|^{-N} \quad \text{in } (\Pi^{\forall}, \infty).$$

This gives us that

$$h = A^{-1}(\mathrm{id} + \varphi) = \zeta(1 + o(\ln|\zeta|)^{-N}) \quad \text{in } (\mathbf{C}^{+}, \infty).$$

In this case $n \geq 2$, since $h \in L^n$ but $L^n = \varnothing$ for $n \leq 1$. It follows from Proposition 2 of Example 3 in §1.6 that for arbitrary $n \geq 2$ there exists a standard domain Ω of class n in which $|\zeta| \prec \xi(\ln|\zeta|)^3$. In this domain $|h - \mathrm{id}| \prec \alpha\xi$ for any $\alpha > 0$. This implies the second assertion of Proposition 1.

PROOF OF PROPOSITION 4. It suffices to prove that

$$\exp^{[n-1]} \circ (\mathrm{id} + \mathscr{F}_{\mathrm{reg}}^{n-1^+}) \subset \exp^{[n-1]} + \mathscr{F}_{\mathrm{reg}}^{n-1^+}.$$

But this follows immediately from the condition

$$\exp^{[n-1]} = \exp \circ \exp^{[n-2]} \in \mathscr{F}_{\mathrm{reg}}^{n-2}$$

and Lemma $\mathrm{SL}\,2_{n,\mathrm{reg}}$. This proves Proposition 4, and with it Proposition 1. ►►

In §4.9 we will use the following

COROLLARY. *For any germ* $j \in L^n$, $\lim_{(\mathbb{R}^+, \infty)} j/\xi = 1$.

PROOF. Let $\tilde{h} = Aj$. By Proposition 3 on $(\mathbb{R}^+, \infty)$ $\tilde{h} = \mathrm{id} + o(1)$. Then on $(\mathbb{R}^+, \infty)$

$$j = A^{-1}\tilde{h} = \exp \circ (\ln + o(1)) = \xi \cdot \exp o(1) = \xi \cdot (1 + o(1)). \quad ►$$

C. Properties of the sets L^{n-1} and $\mathscr{L}^{n-1}$.

For the proof of Proposition 2 we use the following lemma, which is proved by induction in §5.4.

LEMMA 5.6_n. a. $AL^{n-1} \subset \mathrm{id} + \mathscr{L}(\mathscr{F}_{+}^{n-2}) \circ \ln^{[n-2]}$.

b. *Each germ* $h \in L^{n-1}$ *extends biholomorphically to a germ on the infinite sector* S_α *for arbitrary* $\alpha \in (\pi/2, \pi)$; *moreover,* $h' \to 1$ *in* (S_α, ∞).

c. *For each germ* $\tilde{h} \in \mathscr{L}^{n-1}$ *there exists an equivalent germ* $h \in L^{n-1}$ *such that* $\tilde{h} - h = o(|\zeta|^{-N})$ *in* (S_α, ∞) *for any* $N > 0$.

d. *Each germ* $\tilde{h} \in \mathscr{L}^{n-1}$ *is biholomorphic in* (S_α, ∞), *and* $\tilde{h}' \to 1$ *in* (S_α, ∞).

Assertion a is used in the proof of assertions b–d, and they are used in the proof of Proposition 2.

INDUCTION BASE: $n = 1$. The set L^1 is empty, as mentioned above; therefore, Lemma 5.6_n is valid for $n = 1$, being a property of the empty set.

Assume that Lemma 5.6_m has been proved for $m \leq n$. The induction step is carried out in Chapter V, where Lemma 5.6_{n+1} is proved.

D. Group properties of the sets L^{n-1} and $\mathscr{D}_1^{n-1}$.

We prove Proposition 2. Let $h \in L^{n-1}$ be a generating element of the form

$$h = A^{1-n}(\mathrm{Ad}(\tilde{g})A^{n-2}f), \qquad \tilde{g} \in G^{n-2}, \ \mu_{n-1}(\tilde{g}) = 0, \ f \in \mathscr{A}^0,$$

and let $\sigma_0 \in \mathscr{D}_1^{n-1}$. In the case when $\sigma_0 \in \mathscr{D}_0^{n-1}$ Proposition 2 can be proved exactly like Proposition 1; just replace n by $n-1$.

Suppose now that

$$\sigma_0 = \sigma_1 \circ h_1, \qquad \sigma_1 = A^{1-n}g, \ g \in G_{\text{slow}}^{n-2^+} \cup G_{\text{rap}}^{n-2}, \ h_1 \in \mathscr{L}^{n-1}.$$

REMARK. Let us use the following general assertion. Suppose that σ_1 and σ_2 are two admissible germs of class $\mathbf{\Omega}$, and their composition quotient $\tilde{\sigma} = (\sigma_2)^{-1} \circ \sigma_1$ is extendible from $(\mathbf{R}^+, \infty)$ to some standard domain $\Omega \in \mathbf{\Omega}$ as a map-cochain, and there the correction of this quotient decreases faster than $|\zeta|^{-5}$. Then the germs σ_1 and σ_2 are $\mathbf{\Omega}$-equivalent.

In fact, there is a domain $\widetilde{\Omega}$ of class $\mathbf{\Omega}$ with the following properties: the germs $\rho_j = (\sigma_j)^{-1}$ are extendible to $\widetilde{\Omega}$ and $\rho_j \widetilde{\Omega} \subset \Omega$. The existence of $\widetilde{\Omega}$ follows from requirement 1 in the definition of admissibility (§1.5). In the domain $\rho_1 \widetilde{\Omega}$ the quotient is a diffeomorphism; it is negligible by the decreasing assumption about its correction. Thus, only the correction of $\tilde{\sigma}$ is estimated below.

We have that

$$(h\sigma)^{-1} \circ \sigma_1 = (h\sigma)^{-1} \circ \sigma \circ h_2 = (\mathrm{Ad}(\sigma)h^{-1}) \circ h_2.$$

Let $\tilde{h} = \mathrm{Ad}(\sigma)h^{-1}$. Then

$$\tilde{h} = A^{1-n}[\mathrm{Ad}(\tilde{g}g)A^{n-2}f^{-1}] = A^{1-n}\tilde{j}, \qquad \tilde{j} \in \tilde{J}^{n-2};$$

see the notation at the beginning of the subsection.

As before, let us consider three cases.

CASE 1: $\tilde{j} \in J_\infty^{n-2}$, $\mu_{n-1}(\tilde{g}g) = \infty$. In this case take $h_2 = \mathrm{id}$. By Lemma 6_{n-1}, $\tilde{j}$ is a weakly decreasing cochain, and its correction in some standard domain of class n is equal to $o(|\zeta|^{-5})$. The equivalence of the germs σ and $h\sigma$ now follows from the remark at the beginning of the proof.

CASE 2: $\tilde{j} \in J_*^{n-2}$, $\mu_{n-1}(\tilde{g}g) \in (0, \infty)$. In this case, by Lemma 7_{n-1} in §2.10, the following decomposition exists:

$$\tilde{h} = h_2 \circ h_3, \quad \text{where } h_2 \in \mathscr{D}_{\text{rap}}^{n-1} = A^{1-n}G_{\text{rap}}^{n-2} \text{ and } h_3 \in \mathscr{F}\mathscr{C}_{\text{wr}}(\mathscr{D}_1^{n-1}).$$

Take $\sigma_1 = \sigma \circ h_2$. Then $(h\sigma)^{-1} \circ \sigma_1 = h_3^{-1}$, and the germs $h\sigma$ and σh_2 are equivalent; the conclusion of the proof is the same as in Case 1.

CASE 3: $\tilde{j} \in J_0^{n-2}$, $\mu_{n-1}(\tilde{g}g) = 0$. In this case $\tilde{h} \in L^{n-1}$. The assertion d in Lemma 5.6_{n-1} implies the existence of a germ $h_2 \in \mathscr{L}^{n-1}$ such that $\tilde{h} = h_2 \circ h_3$, where h_3 is negligible in the sector S_α, $\alpha \in (\pi/2, \pi)$. The conclusion of the proof is the same as in Case 1.

This proves the proposition in the case when $h_1 = \mathrm{id}$. Suppose now that h_1 is an arbitrary germ in $\mathscr{L}^{n-1}$, and h and $\tilde{h}$ are the same germs as above.

In Case 1,

$$h \circ \sigma_1 \circ h_1 = \sigma_1 \circ \tilde{h} \circ h_1,$$

the germ $\tilde{h}$ is negligible, and $\sigma_1 \circ h_1 \in \mathscr{D}_1^{n-1}$.

In Case 2,

$$h \circ \sigma_1 \circ h_1 = \sigma_1 \circ h_2 \circ h_3 \circ h_1, \qquad h_2 \in \mathscr{D}_{\mathrm{rap}}^{n-1},$$

the germ h_3 is negligible, and $\sigma_1 \circ h_2 \circ h_1 \in \mathscr{D}_1^{n-1}$.

In Case 3,

$$h \circ \sigma_1 \circ h_1 = \sigma_1 \circ h_2 \circ h_3 \circ h_1, \qquad h_2 \in \mathscr{L}^{n-1},$$

the germ h_3 is negligible, and $\sigma_1 \circ h_2 \circ h_1 \in \mathscr{D}_1^{n-1}$.

In all three cases it remains to prove that if the germ h_0 is negligible and $h_1 \in \mathscr{L}^{n-1}$, then $\mathrm{Ad}(h_1)h_0$ is a negligible germ. In Case 1 this is applied to $h_0 = \tilde{h}$, and in Cases 2 and 3 to $h_0 = h_3$. It follows from Lemma 5.6_{n-1}d that the germ $\mathrm{Ad}(h_1)h_0$ is negligible. This finishes the proof of Proposition 2. ►

Propositions 1 and 2 and Lemma 3 bis imply Lemma $*$ in Case 3, the only one remaining after §A. ►

§2.12. The Regularity Lemma

We end this chapter with the following natural lemma.

REGULARITY LEMMA RL_n. *A partial sum of a STAR-m series with $m = n$ or $m = n - 1$ belongs to the set $\mathscr{F}_{\mathrm{reg,\,id}}^m$.*

PROOF. Suppose that $F \in \mathscr{F}\mathscr{C}^m$, Σ_∞ is the corresponding STAR-m for the composition $F \circ \exp^{[m]}$, and Σ is a partial sum of it. It is required to prove that

$$\Sigma \circ \ln^{[m]} \in \mathscr{F}\mathscr{C}^m.$$

The proof is by induction on n. For $n = 0$ a STAR-0 is a Dulac series; a partial sum of it is holomorphic in the whole plane and increases in $(\mathbf{C}^+, \infty)$ no more rapidly than an exponential. For $n = -1$ a STAR-n is a first-degree polynomial. Thus, the lemma is trivial for $n = 0$ (induction base).

INDUCTION STEP. Suppose that Lemma RL_{n-1} has been proved. We prove Lemma RL_n. Consider the case $m = n$; the proof is analogous in the case

$m = n - 1$. It suffices to consider one term of the series: $a \exp \mathbf{e}$, $a \in \mathscr{K}^n$, $\mathbf{e} \in E^n$. We have that $a \circ \ln^{[n]} \in \mathscr{F}\mathscr{C}_{\text{reg}}(\mathscr{D}^n)$. This follows from Propositions 1 and 2 in §2.7 and Lemma 3 in §2.3.

By the induction hypothesis, $\mathbf{e} \circ \ln^{[n-1]} \in \mathscr{F}\mathscr{C}_{\text{reg}}(\mathscr{D}^{n-1})$, since $\mathbf{e}$ is a partial sum of a STAR-$(n-1)$; see Definition 2^n in §1.7. By requirement 3° of this definition,

$$|\operatorname{Re} \circ \mathbf{e} \circ \ln^{[n]}| < \mu\xi$$

in some standard domain Ω of class n.

Consequently,

$$|\exp \circ \mathbf{e} \circ \ln^{[n]}| < \exp \mu\xi \quad \text{in } \Omega.$$

Finally, we estimate the coboundary of the cochain $\exp \circ \mathbf{e} \circ \ln^{[n]}$. By the induction hypothesis,

$$|\delta \mathbf{e} \circ \ln^{[n]}| < m,$$

where m is some rigging cochain of the partition corresponding to the cochain $\mathbf{e} \circ \ln^{[n]}$. Then

$$\begin{aligned} |\delta(\exp \circ \mathbf{e} \circ \ln^{[n]}| &< |\exp \circ \mathbf{e} \circ \ln^{[n]}|\,|\delta \mathbf{e} \circ \ln^{[n]}| \cdot O(1) \\ &< Cm \exp \mu\xi. \end{aligned}$$

But it was already proved above that the product of a rigging cochain of a regular partition by an exponential is majorized by another rigging cochain of the same partition.

This verifies the last requirement of Definition 10 in §1.6 and finishes the proof of Lemma RL_n. ►

CHAPTER III

The Phragmén-Lindelöf Theorem for Regular Functional Cochains

THEOREM. *Let F be a regular functional cochain of class $\mathscr{D}_0^n$ or $\mathscr{D}_1^{n-1}$ that decreases on $(\mathbf{R}^+, \infty)$ more rapidly than any exponential* $\exp(-\nu\xi)$, $\nu > 0$. *Then* $F^u \equiv 0$.

Here F^u is the main function of the tuple F (corresponding to the domain adjacent to $(\mathbf{R}^+, \infty)$ from above in the **R**-regular partition for F). Using the convention in §1.10, we write this theorem in the following form.

THEOREM. *Suppose that $m = n - 1$ or $m = n$, $F \in \mathscr{F}\mathscr{C}_{\text{reg}}(\mathscr{D}_1^m)$, and* $|F|_{(\mathbf{R}^+, \infty)}| < \exp(-\nu\xi)$ *for all* $\nu > 0$. *Then* $F^u \equiv 0$.

This theorem is proved by using as a model the classical "Phragmén-Lindelöf theorem for two quadrants," which is not new, but is known to me only from oral communication ([8]). §3.1 is devoted to its proof and to related questions.

§3.1. Classical Phragmén-Lindelöf theorems and modifications of them

A. THE PHRAGMÉN-LINDELÖF THEOREM FOR AN UNBOUNDED DOMAIN. *Suppose that $\mathscr{D}$ is a simply connected domain on the Riemann sphere that contains the point ∞ on its boundary. Assume that the function f is holomorphic in $\mathscr{D}$ and bounded and continuous on the closure of $\mathscr{D}$, which is taken in the topology of* **C** *and does not contain* ∞. *Then*

$$\sup_{\mathscr{D}} |f| = \sup_{\partial\mathscr{D}} |f|.$$

This theorem is a variant of the maximum principle for holomorphic functions.

B. THE PHRAGMÉN-LINDELÖF THEOREM FOR TWO QUADRANTS. *Suppose that the holomorphic function f: $\mathbf{C}^+ \to \mathbf{C}^+$ is bounded on the union of the*

([8]) After the manuscript of this book was written and a similar fragment of the text was published in [24], Y. Sibuja kindly communicated to me that this theorem was proved by J. N. Watson, "A theory of asymptotic series," Philos. Trans. Roy. Soc. London Ser. A **211** (1911), 279–313.

imaginary axis and the positive semi-axis and increases no more rapidly than an exponential in modulus: there exists a $\nu > 0$ *such that* $|f(\zeta)| < \exp \nu|\zeta|$. *Then* f *is bounded, and* $\sup_{\mathbf{C}^+} |f| = \sup_{\partial\mathbf{C}^+} |f|$.

PROOF. We prove first that f satisfies the maximum principle in each of the sectors S_1 and S_2; S_1 is the first coordinate quadrant, and S_2 the fourth. We consider the sector S_1 and a family of barrier functions g_ε that are holomorphic in S_1 and have modulus there greater than $\exp c|\zeta|^{1+\delta}$ for some $c > 0$ and $\delta > 0$. For example,

$$g_\varepsilon(\zeta) = \exp \varepsilon(\beta\zeta)^{1+\delta}, \qquad \beta = e^{i\pi/4}.$$

For an arbitrary positive value of the parameter ε the quotient f/g_ε is bounded in S_1, and by Theorem A

$$\sup_{S_1} |f/g_\varepsilon| = \sup_{\partial S_1} |f/g_\varepsilon|.$$

Passing to the limit as $\varepsilon \to 0$, we get that f is bounded in S_1. Similarly, f is bounded in S_2, and hence also in $\mathbf{C}^+$. Using Theorem A again, we get Theorem B. ►

C. COROLLARY 1. *If a holomorphic function* f: $\mathbf{C}^+ \to \mathbf{C}$ *is bounded and decreases on* $\mathbf{R}^+$ *faster than any exponential* $\exp(-\nu\xi)$, $\nu > 0$, *then* $f \equiv 0$.

PROOF. The function $f \exp \lambda\zeta$ satisfies the conditions of Theorem B for arbitrary $\lambda > 0$. Since $|\exp \lambda\zeta| = 1$ on $\partial\mathbf{C}^+$, this implies that

$$\sup_{\mathbf{C}^+} |f(\zeta)|\,|\exp \lambda\zeta| \leq \sup_{\partial\mathbf{C}^+} |f|.$$

Assume that there exists a point ζ: $\operatorname{Re}\zeta > 0$ and $f(\zeta) \neq 0$. Then $f(\zeta)\exp\lambda\zeta \to \infty$ as $\lambda \to \infty$, a contradiction. ►

REMARK. The Phragmén-Lindelöf theorem can be proved for cochains according to the same scheme as in Corollary 1. The proofs of this corollary and Theorem B were communicated to the author by E. A. Gorin.

COROLLARY 2. *Theorem* B *remains valid if in its formulation the real semi-axis is replaced by a curve* Γ *that joins some point on* $\partial\mathbf{C}^+$ *with the point* ∞ *on the Riemann sphere and has germ at infinity contained in the sector* S_α: $|\arg\zeta| \leq \alpha < \pi/2$. *The same applies to Corollary* 1.

The proof repeats almost word-for-word the proofs of Theorem B and Corollary 1.

COROLLARY 3. *Corollary* 2 *remains valid if in it* $\mathbf{C}^+$ *and* $\partial\mathbf{C}^+$ *are replaced by a standard domain* Ω *and* $\partial\Omega$.

PROOF. By the definition of a standard domain, there exists a conformal mapping ψ: $\Omega \to \mathbf{C}^+$ of the form $\zeta(1 + o(1))$. This implies that the germ at infinity of the curve $\Gamma = \psi\mathbf{R}^+$ belongs to an arbitrary sector S_α with $\alpha > 0$. Consequently, the function $f \circ \psi^{-1}$ satisfies all the conditions of Corollary 2. Corollary 2 thus implies Corollary 3. ►

COROLLARY 4. *Suppose that* Ω *is a subdomain of* $\mathbf{C}$ *containing the ray* $\mathbf{R}_a^+$: $\xi > a$ *for some* $a > 0$, *and assume that there exists a conformal mapping* ψ: $\Omega \to \mathbf{C}^+$ *carrying the germ of the ray* $(\mathbf{R}^+, \infty)$ *into a curve* (Γ, ∞) *whose germ at infinity belongs to some sector* S_α, $\alpha \in (0, \pi/2)$. *Suppose that the holomorphic function* f: $\Omega \to \mathbf{C}$ *increases no more rapidly than* $\exp \nu|\psi|$ *for some* $\nu > 0$, *and that* f *is bounded on* $\partial\Omega$ *and on* $\mathbf{R}_a^+$. *Then* f *is bounded in* Ω, *and*

$$\sup_{\Omega} |f| = \sup_{\partial\Omega} |f|.$$

If, moreover, f *decreases on* $\mathbf{R}_a^+$ *more rapidly than* $\exp(-\lambda \operatorname{Re} \psi)$ *for all* $\lambda > 0$, *then* $f \equiv 0$.

PROOF. Apply Corollary 2 to the function $f \circ \psi^{-1}$. ▶

§3.2. A preliminary estimate and the scheme for proving the Phragmén-Lindelöf theorem

A. Heuristic arguments. We describe the plan of proof of the Phragmén-Lindelöf theorem, replacing the exact statements by ones that are approximate but intuitive; the exact presentation is contained in B and C.

Let F be the cochain in the condition of the Phragmén-Lindelöf theorem. All realizations of F coincide in the upper half-plane, and hence they all decrease on $(\mathbf{R}^+, \infty)$ more rapidly than any exponential. Let F be a cochain of type $(\sigma_1, \dots, \sigma_N)$, $\sigma_1 \succ \cdots \succ \sigma_N$. Then the k-realization $F_{(k)}$ of F in the domain

$$\widetilde{\Omega}_k = \sigma_k(\Pi_{\text{main}}) \cap \Omega$$

(called the main domain [9] corresponding to the germ σ_k) is a cochain corresponding to an $\mathbf{R}$-regular partition of type $(\sigma_{k+1}, \dots, \sigma_N)$ (approximate statement). By the definition of a regular cochain, the coboundary of the cochain $F_{(k)}|_{\widetilde{\Omega}_k}$ admits the following upper estimate:

$$|\delta F_{(k)}|_{\widetilde{\Omega}_k}| < \exp(-c \exp \rho_{k+1})$$

for some $c > 0$.

The proof of the Phragmén-Lindelöf theorem is based on a preliminary estimate that, roughly speaking, asserts that:

I. If a cochain F on a standard domain of class n decreases on $(\mathbf{R}^+, \infty)$ more rapidly than any exponential, then for some $c > 0$

$$|F| < \exp(-c \exp \rho_1)$$

on $(\mathbf{R}^+, \infty)$.

II. If the k-realization $F_{(k)}$ of F in the domain $\widetilde{\Omega}_k$ satisfies the estimate

$$|F| < \exp(-c \exp \rho_k) \quad \text{on } (\mathbf{R}^+, \infty),$$

([9]) Recall that Π_{main} was defined in Part A of §0.4; see Figure 16.

then we have the stronger estimate

$$|F| < \exp(-c \exp \rho_{k+1}) \quad \text{on } (\mathbf{R}^+, \infty).$$

The main function F^u of the cochain F extends holomorphically to the domain $\widetilde{\Omega}_N$. It can be concluded from the classical Phragmén-Lindelöf theorem (Corollary 4 in §3.1) and the estimate

$$|F^u| < \exp(-c \exp \rho_N).$$

(proved by induction on k) that $F^u \equiv 0$.

The inequalities in A are not proved for all σ_j; only analogues of them are proved for some σ_j. Let us proceed to the exact statements.

B. A preliminary estimate.

As above, n is fixed, and the convention in §1.10 acts.

DEFINITION. A germ σ of class $\mathscr{D}_0^m$ is said to be m-rapid if

$$\sigma = \exp \circ A^{1-m} g, \qquad g \in G_{\text{slow}}^{m-1},$$
$$(A^{1-m} g)' \to 1 \quad \text{on } (\mathbf{R}^+, \infty),$$

and m-sectorial if

$$(A^{1-m} g)' \to \mu \in (0, 1) \quad \text{on } (\mathbf{R}^+, \infty).$$

All the remaining germs of the class $\mathscr{D}^m$ are said to be m-slow.

The name "sectorial" is due to the fact that the image of the strip $|\eta| < \pi/2$ under the action of a sectorial germ contains a sector and is contained in a sector of opening less than π. All germs of class $\mathscr{D}_0^1$ are sectorial; there are no m-rapid germs for $m = 1$. We will speak on the rapid, sectorial and slow germs of class $\mathscr{D}_0^n$ (respectively, $\mathscr{D}_1^{n-1}$), meaning n-rapid, etc. ($(n-1)$-rapid, etc.) germs respectively.

THE PRELIMINARY ESTIMATE. I. *Let Ω be a standard domain of class n and let F be a regular functional cochain of type $\boldsymbol{\sigma} = (\sigma_1, \dots, \sigma_N)$, $\sigma_1 \succ \dots \succ \sigma_N$, and of class $\mathscr{D}_0^n$ or $\mathscr{D}_1^{n-1}$. Suppose that F decreases on $(\mathbf{R}^+, \infty)$ more rapidly than any exponential, and let σ_1 be a rapid germ. Then there exists a $c > 0$ such that on $(\mathbf{R}^+, \infty)$*

$$|F| < \exp(-c \exp \rho_1), \qquad \rho_1 = \sigma_1^{-1}.$$

II. *Let the germs σ_j be the same as in the estimate* I, *and let $l = l(k) > k$ be the minimal number such that the composition quotient $\sigma_k^{-1} \circ \sigma_l$ $(l > k)$ has an unbounded correction on $(\mathbf{R}^+, \infty)$. Suppose that a bounded $\mathbf{R}$-regular functional cochain F of type $(\sigma_l, \dots, \sigma_N)$ is given in the domain $\widetilde{\Omega}_k = \sigma_k(\Pi_{\text{main}}) \cap \Omega$. Let the cochain F satisfy on $(\mathbf{R}^+, \infty)$ one of the following estimates:*

$$|F| < \exp(-c \exp \rho) \tag{$*$}$$

for $\rho = \sigma_k^{-1}$ *if* σ_k *is a rapid germ;*

$$|F| \prec \exp(-\exp(1-\delta)\rho) \qquad (**)$$

for $\rho = \sigma_k^{-1}$ *and any* $\delta > 0$ *if* σ_k *is not a rapid germ.*

Then $(*)$ *or* $(**)$ *holds on* $(\mathbf{R}^+, \infty)$ *for* $\rho = \sigma^{-1}$ *when* σ_l *is or is not a rapid germ, respectively.*

The preliminary estimate corresponding to the germ $\sigma = \rho^{-1}$ is defined to be estimate $(*)$ if σ is a rapid germ and $(**)$ if not.

C. The derivation of the Phragmén-Lindelöf theorem from the preliminary estimates goes according to the plan described in A. Unfortunately, the domains $\widetilde{\Omega}_k$ in the preliminary estimate are not in general ordered by inclusion, although there is a connection between the rates of growth of the germs σ_k and σ_l and the question of whether one of $\widetilde{\Omega}_k$ and $\widetilde{\Omega}_l$ belongs to the other. We proceed to describe this connection.

DEFINITION. Two germs σ_1 and σ_2 of class $\mathscr{D}_0^n$ or $\mathscr{D}_1^{n-1}$ are said to be weakly equivalent if their composition quotient $\sigma_1^{-1} \circ \sigma_2$ has a bounded correction on $(\mathbf{R}^+, \infty)$. Notation: $\sigma_1 \overset{w}{\sim} \sigma_2$.

REMARK 1. Germs equivalent in the sense of Definition 18 in §1.6 are weakly equivalent.

2. If $\sigma_1 \succ \sigma_2 \succ \sigma_3$ and $\sigma_1 \overset{w}{\sim} \sigma_3$, then $\sigma_1 \overset{w}{\sim} \sigma_2$ and $\sigma_2 \overset{w}{\sim} \sigma_3$.

3. If σ_1 and σ_2 are weakly equivalent, then the corresponding preliminary estimates for them are equivalent. In other words, the ratio

$$\exp(1-\delta)\rho_1 / \exp(1-\delta)\rho_2$$

is bounded for arbitrary $\delta > 0$; here $\rho_j = \sigma_j^{-1}$.

Indeed, $\rho_2^{-1} \circ \rho_1 = \mathrm{id} + O(1)$. Since the derivative of ρ_2 is bounded and $\rho_1 = \rho_2 \circ (\mathrm{id} + O(1))$, we get that $\rho_1 = \rho_2 + O(1)$ which implies the remark.

The connection mentioned at the beginning of the subsection is given by

LEMMA 5.7_na. *Suppose that* Ω *is an arbitrary standard domain of class* n, *and* $\sigma_1, \sigma_2 \in \mathscr{D}_0^n$ *or* $\sigma_1, \sigma_2 \in \mathscr{D}_1^{n-1}$, $\sigma_1 \succ \sigma_2$. *Then*

$$(\sigma_1 \Pi_{\mathrm{main}} \cap \Omega, \infty) \supset (\sigma_2 \Pi_{\mathrm{main}} \cap \Omega, \infty)$$

if the germs σ_1 *and* σ_2 *are not weakly equivalent;* σ_1 *and* σ_2 *can be renumbered so that for arbitrary* ε *and* δ *with* $0 < \delta < \varepsilon < 1$

$$(\sigma_1 \Pi_{\mathrm{main}}^{(\delta)} \cap \Omega, \infty) \subset (\sigma_2 \Pi_{\mathrm{main}}^{(\varepsilon)} \cap \Omega, \infty)$$

if σ_1 *and* σ_2 *are weakly equivalent.*

Recall that the domain Π_{main} and its generalized ε-neighborhood $\Pi_{\mathrm{main}}^{(\varepsilon)}$ were defined in §1.6 A; see Figures 12 and 16.

The lemma is proved in Chapter V, §7 by induction on n. The induction step is contained in §5.7, and the induction base here.

For $n = 1$ a partition of class $\mathscr{D}_0^n$ is equivalent to the sectorial partition $(\exp\circ\mu)_*\Omega_{\text{st}}$, and a partition of class $\mathscr{D}_1^{n-1}$ is equivalent to the stretched out standard one $\mu_*\Xi_{\text{st}}$. The lemma is obvious for such partitions.

REMARK 4. For weakly equivalent germs the order of growth is not connected with the question of one of the main domains belonging to the other. For example, for $\sigma_1 = \text{id}$ and $\sigma_2^{\pm} = \text{id} + 1 \pm \zeta^{-2}$ we have, on the one hand, that $\sigma_1 \prec \sigma_2^{\pm}$; on the other hand,

$$(\sigma_2^+\Pi_{\text{main}}, \infty) \subset (\sigma_1\Pi_{\text{main}}, \infty) \subset (\sigma_2^-\Pi_{\text{main}}, \infty).$$

We now use the preliminary estimates to describe an induction process that in the final analysis proves the inequality $(*)$ or $(**)$ for $\rho = \rho_N = \sigma_N^{-1}$.

PROPOSITION 1. *Let F be a regular functional cochain of type $\boldsymbol{\sigma} = (\sigma_1, \dots, \sigma_N)$, $\sigma_1 \succ \cdots \succ \sigma_N$, $\sigma_j \in \mathscr{D}_0^n$ or $\sigma_j \in \mathscr{D}_1^{n-1}$.*

Then for each germ $\sigma_k \in \boldsymbol{\sigma}$ there exists a weakly equivalent germ $\sigma_l \in \boldsymbol{\sigma}$ such that the l-realization of F in the intersection of the main domains corresponding to the germs $\sigma_k \in \boldsymbol{\sigma}$ that are weakly equivalent to σ_l extends to the germ of the domain $(\sigma_l\Pi_{\text{main}}, \infty)$. The extended cochain F corresponds to an $\mathbf{R}$-regular partition of type $(\sigma_m, \dots, \sigma_N)$, where σ_m is the first germ in the sequence $\boldsymbol{\sigma}$ that is not weakly equivalent to σ_l.

This is an immediate consequence of Lemma 5.7_n.a.

PROOF. We split the sequence $\boldsymbol{\sigma}$ into classes of weakly equivalent germs. By Remark 2, each such class contains germs that are in succession. By Lemma 5.7_n.a, each such class contains a germ σ_l having the following properties. For each germ σ_k of the same class and each $\varepsilon > 0$

$$(\sigma_k\Pi_{\text{main}}^{(\varepsilon)}, \infty) \supset (\sigma_l\Pi_{\text{main}}, \infty).$$

The proposition now follows directly from the ε-extendibility of the cochain F. ►

It is possible to add "fictitious" germs of corresponding class $\mathscr{D}_0^n$ or $\mathscr{D}_1^{n-1}$ to the sequence $\boldsymbol{\sigma}$ while preserving order. The cochain F remains regular, but its coboundary will be equal to zero on the new lines of the partition. Therefore, it will be assumed that a rapid germ is first in the sequence $\boldsymbol{\sigma}$, and there is necessarily a slow germ σ_* of the form

$$\sigma_* = \exp\circ A^{1-m}g, \qquad (A^{1-m}g)' \to 0 \quad \text{on } (\mathbf{R}^+, \infty),$$
$$m = n \quad \text{or} \quad m = n-1$$

(the last requirement is used in the last section of this chapter).

We now employ the preliminary estimate. By estimate I, the inequality $(*)$ holds for $\rho = \sigma_1^{-1}$. Further, by Proposition 1, in the sequence $\boldsymbol{\sigma}$ there exists a germ σ_l equivalent to σ_1 such that the l-realization of the cochain F on the corresponding main domain has the type $(\sigma_m, \dots, \sigma_N)$, $\sigma_l \overset{w}{\nsim} \sigma_m$,

$m \geq l+1$. By the preliminary estimate II, on $(\mathbf{R}^+, \infty)$ we have the inequality $(*)$ or $(**)$ corresponding to the germ σ_m.

We use induction on the number of classes of weak equivalence to prove $(*)$ or $(**)$ for the germ $\rho = \sigma_N^{-1}$. This finishes the induction process.

The following construction is needed to prove the preliminary estimate and to apply it at the last step to the function F^u (to the main function of F, which is holomorphic in the main domain of a partition of type σ_j, where σ_j is one of the germs in $\boldsymbol{\sigma}$ that is weakly equivalent to σ_N).

For each germ $\sigma \in \boldsymbol{\sigma}$ we must construct a domain Ω_σ that belongs to the corresponding main domain $\widetilde{\Omega}_\sigma = \sigma(\Pi_{\text{main}}) \cap \Omega$ and admits a conformal mapping $\psi\colon \Omega_\sigma \to \mathbf{C}^+$ having the following properties. First, there should not be a nonzero holomorphic function F on Ω_σ satisfying the preliminary estimate corresponding to the germ σ. The classical Phragmén-Lindelöf theorem (Corollary 4 in §3.1) should be applicable to the mapping ψ and the function F. For this, ψ should carry the ray $(\mathbf{R}^+, \infty)$ into the sector S_α (see Corollary 4), and the preliminary estimate corresponding to σ must be "effective"—it must imply that the modulus of F decreases on $(\mathbf{R}^+, \infty)$ more rapidly than $\exp(-\lambda \operatorname{Re} \psi)$ for arbitrary $\lambda > 0$. The construction of such a domain Ω makes it possible to pass from the preliminary estimate corresponding to σ_N to the equality $F^u \equiv 0$.

To prove the preliminary estimate II we must endow the domain Ω_σ with additional properties of "properness" to be described in §3.3.

In §§3.4–3.6 we prove a "general preliminary estimate" used to derive the preliminary estimates I and II: the estimate I immediately in §3.7, and the estimate II after the construction and investigation of the domains Ω_σ, §§3.8–3.10.

§3.3. Proper conformal mappings

We describe some geometric characteristics of conformal mappings of domains containing the germ of the positive semi-axis onto a half-plane.

A. Definitions.

DEFINITION 1. Let Ω be a connected simply connected domain belonging to a quadratic standard domain and containing $(\mathbf{R}^+, \infty)$. The degree of convexity for a conformal mapping $\psi\colon \Omega \to \mathbf{C}^+$ is defined to be the function

$$\mathbf{R}^+ \cap \Omega \to \mathbf{R}^+\colon\ \widetilde{\psi}(\xi) = \max_{\operatorname{Re}\zeta = \xi} \operatorname{Re} \psi(\zeta)$$

(see Figure 17, next page).

DEFINITION 2. A conformal mapping ψ of the domain Ω described in Definition 1 onto a half-plane is said to be proper if it extends to the ε-neighborhood of Ω for some $\varepsilon > 0$, and:

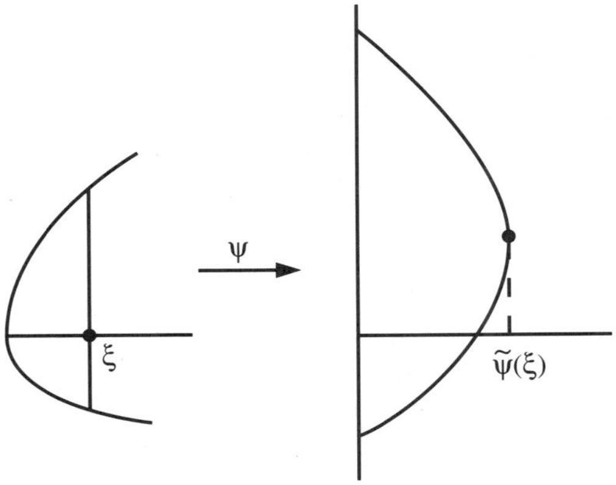

FIGURE 17

1°. ψ carries the germ of the ray $(\mathbf{R}^+, \infty)$ into the germ of the sector (S_α, ∞),

$$S_\alpha = \{\zeta \mid |\arg \zeta| < \alpha\}$$

for some $\alpha < \pi/2$.

2°. $\operatorname{Re}\psi' / \operatorname{Re}\psi \geq (1 + o(1))/\xi$ on $(\mathbf{R}^+, \infty)$.

3°. The degree of convexity of ψ increases no more rapidly than its real part on $(\mathbf{R}^+, \infty)$, that is,

$$\widetilde{\psi}(\xi) < c \operatorname{Re}\psi(\xi), \qquad \xi \in \mathbf{R}^+ \cap \Omega,$$

for some $c \geq 0$.

REMARK. As shown below, requirement 2° means that $\operatorname{Re}\psi$ grows no more slowly than $\xi^{1+o(1)}$, and perhaps faster than any power.

B. Examples. The simplest proper mapping is a conformal mapping of the sector $\arg \zeta \in (\alpha, \beta)$, $-\pi/2 < \alpha < 0 < \beta < \pi/2$, onto $\mathbf{C}^+$. The first two requirements in Definition 2 can be verified in an obvious way. The last one follows from the homogeneity of the power function.

The next example is given by

PROPOSITION 1. *For an arbitrary standard domain Ω belonging to a quadratic standard domain there exists a proper (in the sense of Definition 2) conformal mapping of this domain onto a half-plane.*

PROOF. By the definition of a standard domain, Definition 1 in §1.5, Ω is symmetric with respect to $\mathbf{R}^+$, and there exists a conformal mapping $\psi: \Omega \to \mathbf{C}^+$ whose correction on $(\mathbf{R}^+, \infty)$ is real and grows more slowly than any linear function on Ω:

$$\psi(\zeta) = \zeta(1 + \varphi(\zeta)), \qquad \varphi = o(1) \quad \text{in } \Omega. \tag{$*$}$$

The mapping ψ extends holomorphically to the ε-neighborhood of Ω in view of the same definition.

The first requirement of Definition 2 on ψ follows from the formula $(*)$. Further, by the same formula, a standard domain contains a sector with the bisector $(\mathbf{R}^+, \infty)$. Consequently, by Cauchy's inequality,

$$\psi'/\psi = (1+o(1))/\xi \quad \text{on } (\mathbf{R}^+, \infty).$$

This verifies requirement 2° of the definition of properness, because $\psi = \operatorname{Re}\psi$ on $\mathbf{R}^+$.

Finally, since Ω belongs to a quadratic standard domain, we get: 1. The function $\widetilde{\psi}$ is well defined; 2. The inequality $\operatorname{Re}\psi < \xi$ holds on $\partial\Omega$, since $\operatorname{Re}\psi = 0$ on $\partial\Omega$. This implies that the function $u_\varepsilon = \operatorname{Re}\psi - (1+\varepsilon)\xi$ increases no more rapidly than $|\zeta|$, is negative on $\partial\mathbf{C}^+$, and is bounded from above on $(\mathbf{R}^+, \infty)$ for all $\varepsilon > 0$. The function $u_\varepsilon \circ \psi^{-1}$ is harmonic on $\mathbf{C}^+$, is negative on $\partial\mathbf{C}^+$, is bounded from above on $(\mathbf{R}^+, \infty)$, and increases no more rapidly than a linear function: $|u_\varepsilon \circ \psi^{-1}| \prec C|\zeta|$ on $(\mathbf{C}^+, \infty)$. Let f be a holomorphic function such that $\operatorname{Re} f = u_\varepsilon \circ \psi^{-1}$. Then $|\exp f| < 1$ on $\partial\mathbf{C}^+$, and $|\exp f|$ is bounded on $(\mathbf{R}^+, \infty)$. By the Phragmén-Lindelöf theorem for two quadrants, $|\exp f| < 1$ everywhere in $\mathbf{C}^+$. Consequently, $\operatorname{Re} f < 0$ in $\mathbf{C}^+$, and $u_\varepsilon < 0$ in Ω. Passing to the limit as $\varepsilon \to 0$, we get that $\operatorname{Re}\psi - \xi < 0$ everywhere in Ω. Consequently, for every $\delta > 0$

$$\widetilde{\psi}(\xi) \overset{\text{def}}{=} \max_{\operatorname{Re}\zeta = \xi} \operatorname{Re}\psi(\zeta) < \xi \prec (1+\delta)\psi(\xi)$$

on $(\mathbf{R}^+, \infty)$. This proves requirement 3° in the definition of properness.

C. Properties of proper mappings. We list the properties of proper mappings needed to prove the preliminary estimate. Everywhere below in this section ψ is a proper mapping in the sense of Definition 2 unless otherwise stated.

PROPERTY I. *On* $(\mathbf{R}^+, \infty)$ *the function* $\operatorname{Re}\psi$ *is nonnegative and monotonically increasing*: $\operatorname{Re}\psi' \succ 0$ *on* $(\mathbf{R}^+, \infty)$.

PROOF. The first assertion follows from requirement 1° in the definition of properness, and the second from the first and requirement 2° in the same definition. ▶

Before formulating Property II, we prove

PROPOSITION 2. *Suppose that* f: $(\mathbf{R}^+, \infty) \to (\mathbf{R}^+, \infty)$ *is a monotonically increasing function, and* $f'/f \geq A/x$. *Then for arbitrary* $C > 0$ *the relation* $f \circ C \geq C^A \cdot f$ *holds for* $C \geq 1$, *and* $f \circ C \leq C^A \cdot f$ *for* $C \leq 1$.

PROOF. Obviously, $f(Cx)/f(x) = e^I$, where $I = \int_x^{Cx} (f'(\xi)/f(\xi))\, d\xi$. Let $C \leq 1$. Then $I \leq A \ln C$. Consequently, $f(Cx) \leq C^A f(x)$. The case $C \geq 1$ is treated similarly. ▶

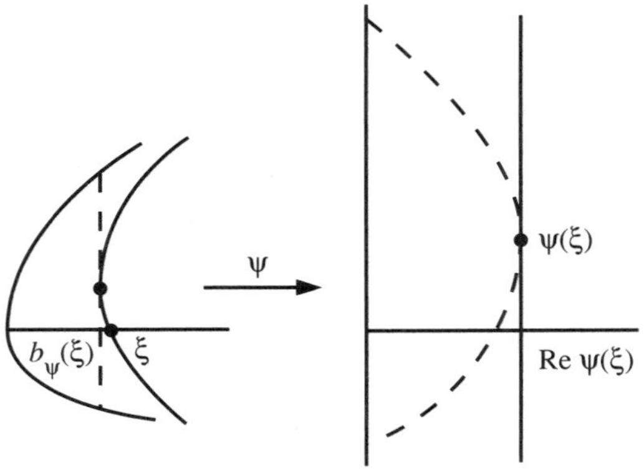

FIGURE 18

PROPERTY II. *If* $C < 1$, *then* $\operatorname{Re}\psi \circ C \prec C^{1-\delta} \cdot \operatorname{Re}\psi$ *for arbitrary* $\delta > 0$, *and for any sufficiently small* $\delta > 0$

$$\operatorname{Re}\psi \circ (1-\delta) \prec (1-\delta/2)\operatorname{Re}\psi \quad \text{on } (\mathbf{R}^+, \infty).$$

PROOF. This is an immediate consequence of requirement 2° in Definition 2, Property I, and Proposition 2. ►

Before stating Property III, we give

DEFINITION 3. The degree of nonrealness of the mapping ψ in Definition 1 is defined to be the function b_ψ: $\mathbf{R}^+ \cap \Omega \to \mathbf{R}^+$,

$$b_\psi(\xi) = \min_{\operatorname{Re}\psi(\zeta) = \operatorname{Re}\psi(\xi)} \operatorname{Re}\zeta.$$

REMARK 1. It follows from Definitions 1 and 3 that

$$\widetilde{\psi}(b_\psi(\xi)) = \operatorname{Re}\psi(\xi);$$

see Figure 18.

PROPERTY III. *The degree of nonrealness of a proper mapping increases no more slowly than a linear function: for some* $C > 0$

$$b_\psi(\xi) > C\xi.$$

PROOF. Let $\xi/b_\psi(\xi) = C(\xi)$. We assume ξ is large enough that the asymptotic inequalities of Definition 2 become pointwise inequalities, in particular,

$$\operatorname{Re}\psi'(\xi)/\operatorname{Re}\psi(\xi) \geq (1-\delta)/\xi.$$

Then, by Remark 1, we get from requirement 3° of Definition 2 that

$$\operatorname{Re}\psi(\xi) = \widetilde{\psi}(b_\psi(\xi)) \leq C\operatorname{Re}\psi(b_\psi(\xi)).$$

Let us apply Proposition 2 to the function $f = \operatorname{Re}\psi$, considering that $C(\xi) \geq 1$, since $b_\psi(\xi) \leq \xi$:

$$\operatorname{Re}\psi(\xi) = \operatorname{Re}\psi(C(\xi)b_\psi(\xi)) \geq C(\xi)^{1-\delta},$$

$$\operatorname{Re}\psi(b_\psi(\xi)) \geq C^{-1}C(\xi)^{1-\delta}\operatorname{Re}\psi(\xi).$$

Consequently,

$$1 \le C(\xi)^{1-\delta} \le C, \qquad C(\xi) \le C^{1/(1-\delta)}. \quad ▶$$

PROPERTY IV. *If requirement* 2° *of Definition 2 is strengthened to*

$$\xi \operatorname{Re} \psi' / \operatorname{Re} \psi \to \infty \quad on\ (\mathbf{R}^+, \infty),$$

then Property III *is also strengthened, to*

$$b_\psi(\xi) = (1 + o(1))\xi.$$

PROOF. Arguing as in the proof of Property III, we get that

$$1 \le C(\xi) \prec C^{1/A}$$

for arbitrary $A > 0$. ▶

§3.4. Trivialization of a cocycle

The goal of this section is to find for a given (not necessarily bounded) cochain a bounded cochain with the same coboundary, and to get an upper estimate of its modulus. This permits us then to estimate the original cochain by applying the classical Phragmén-Lindelöf theorems in §3.1 to the difference between the original cochain and the one found, which difference is a holomorphic function.

A. Properties of regular partitions.

DEFINITION 1. Suppose that Ω is a domain in the plane $\mathbf{C}$ that intersects each vertical in a segment, perhaps empty. Let Ξ be a partition of Ω whose boundary lines are the graphs of functions $\eta = \eta(\xi)$. For each $\xi \in \Omega \cap \mathbf{R}$ we define the total slope $C_{\Xi,\Omega}(\xi)$ of the lines of the partition as the sum of the secants of the slope angles of the partition lines, taken at the points of intersection of all these lines with the segment $\{\operatorname{Re} \zeta = \xi\} \cap \Omega$.

The next lemma contains a summary of the properties of the diffeomorphisms of class $\mathscr{D}_0^n$ and $\mathscr{D}_1^{n-1}$ needed for proving the preliminary estimate.

LEMMA 5.8_n. *Suppose that* $m = n - 1$ *or* $m = n$, *and* σ *is an arbitrary germ of class* $\mathscr{D}^m$ (*see the convention in* §1.10). *Then there exists a standard domain* Ω *of class* n *such that*:

(a) *the boundary lines of the partition* $\sigma_* \Xi_{\mathrm{st}}$ *of the domain* Ω *are the graphs of the functions* $\eta = \eta(\xi)$;

(b) *the total slope of the lines of this partition does not exceed* $C\xi^4$;

(c) *the real part of the germ* $\rho = \sigma^{-1}$ *in* (Ω, ∞) *is nondecreasing upon moving away from the real axis along a vertical in the case when* $\sigma \in \mathscr{D}_0^n \cup \mathscr{D}_0^{n-1} \cup \mathscr{D}_{\mathrm{slow}}^{n-1} \circ \mathscr{L}^{n-1}$, *and satisfies for any* $\delta > 0$ *the inequality*

$$\operatorname{Re} \rho \succ (1 - \delta)\rho \circ \operatorname{Re},$$

if $\sigma \in \mathscr{D}_{\mathrm{rap}}^{n-1} \circ \mathscr{L}^{n-1}$.

B. Lemma on trivialization.

LEMMA 1 (on trivialization of a cocycle). *Suppose that* Ω *is a domain belonging along with its* ε*-neighborhood* Ω^ε *to a standard domain* Ω_{st}. *Let* F *be an* ε*-extendible functional cochain defined in* Ω *and corresponding in this domain to the partition* $\Xi = \prod_1^N \sigma_{j*}\Xi_{\mathrm{st}}$, *where* $\sigma_j \in \mathscr{D}_0^n$ *or* $\mathscr{D}_1^{n-1}$. *Suppose that* Ω_{st} *is such that in it assertions* (a) *and* (b) *of Lemma* 5.8_{n+1} *are satisfied for the diffeomorphisms* σ_j. *Suppose that the coboundary of* F *in the* ε*-neighborhood* Ω^ε *of* Ω *can be estimated from above by a function* m, *and*

$$\max_{\partial\Xi\cap\Omega^\varepsilon} m \leq m_0, \qquad \int_{\partial\Xi\cap\Omega^\varepsilon} m\, ds = I < \infty$$

(*the integration is with respect to arclength*). *Then there exists a functional cochain* Φ *on* Ω, *corresponding to the same partition* Ξ, *such that*

$$\delta\Phi = \delta F \text{ on } \partial\Xi \cap \Omega \quad \text{and} \quad \max|\Phi| \leq C\varepsilon^{-1}(m_0 + I),$$

where C *is a constant depending on the partition* $\Xi^F = \Xi$, *but not on* F.

PROOF. The proof is based on an explicit formula for Φ. When computing a function in the tuple δF on a boundary line of the partition Ξ, one first considers in the formula the domain lying to the left of the line, and then the domain lying to the right; the orientation on the partition lines is induced from the natural orientation of the ray $\mathbf{R}^+$ under the projection $\zeta \mapsto \xi = \operatorname{Re}\zeta$. Let

$$\Phi = \frac{1}{2\pi i}\int_{\partial\Xi\cap\Omega^\varepsilon} \frac{\delta f(\tau)}{\tau - \zeta}\, d\tau.$$

By a theorem of Plemelj [26], $\delta\Phi = \delta f$. Let us estimate $|\Phi(\zeta)|$ for $\zeta \in \Omega$. We consider two cases.

1°. $\operatorname{dist}(\zeta, \partial\Xi) \geq \varepsilon/C$, where $C = \sup_{j,\Omega^\varepsilon}|\rho_j'(\zeta)|$, $\rho_j = \sigma_j^{-1}$. Then $|\Phi(\zeta)| \leq C\varepsilon^{-1}I$.

2°. $\operatorname{dist}(\zeta, \partial\Xi) < \varepsilon/C$. Suppose that ζ is in the domain $\mathscr{D}$ of the partition Ξ. Then the disk K about ζ with radius ε/C is entirely contained in the generalized ε-neighborhood of $\mathscr{D}$ (see Proposition 1 in §2.1). In the formula for Φ we replace the integrals along "curvilinear chords" (the connected components of the intersection $\partial\Xi \cap K$) by integrals along the arcs of the circle ∂K with the same endpoints that are separated from ζ in K by the corresponding curvilinear chords. The moduli of the integrals along arcs belonging to ∂K can be estimated from above by the constant $2C\pi m_0/\varepsilon$, and the integral along the remaining part of the contour by the constant CI/ε. This proves the lemma. ►

The cochain Φ in Lemma 1 is called a trivialization of the cocycle δF.

§3.5. The maximum principle for functional cochains

LEMMA 2. *Suppose that under the conditions of Lemma* 1 *the domain* Ω *is standard, and the functional cochain* F *increases no more rapidly than an exponential* $\exp \nu\xi$ *in* Ω *and is bounded on* $\partial\Omega$ *and on* $\mathbf{R}^+$. *Then* F *is bounded in* Ω, *and*

$$\sup_{\Omega} |F| \leq \sup_{\partial\Omega} |F| + 2C\varepsilon^{-1}(m_0 + I),$$

where C, ε, m_0, *and* I *are the same as in Lemma* 1.

PROOF. Suppose that Φ is the trivialization of δF given by Lemma 1. Then $F - \Phi$ is a holomorphic function in a standard domain Ω and satisfies the condition of Corollary 2 in §3.1. In view of this corollary,

$$\sup_{\Omega} |f - \Phi| = \sup_{\partial\Omega} |f - \Phi|.$$

Further,

$$\sup_{\Omega} |f| \leq \sup_{\Omega} |f - \Phi| + \sup_{\Omega} |\Phi| \leq \sup_{\partial\Omega} |f| + 2 \sup_{\Omega} |\Phi|.$$

The estimate in Lemma 2 follows from the estimate in Lemma 1 on $|\Phi|$. ►

LEMMA 3. *Suppose that the conditions of Lemma* 1 *on trivialization of a cocycle hold, and that there exists a conformal mapping* $\psi \colon \Omega \to \mathbf{C}^+$ *carrying the germ of the ray* $(\mathbf{R}^+, \infty)$ *into the germ of the sector* (S_α, ∞) *for some* $\alpha \geq 0$, *where* $S_\alpha = \{|\arg \zeta| < \alpha < \pi/2\}$. *Assume that the cochain* F *increases no faster than* $\exp \nu|\psi|$ *for some* $\nu > 0$ *and is bounded on* $\partial\Omega$ *and on* $(\mathbf{R}^+, \infty)$. *Then the estimate of Lemma* 2 *is valid.*

PROOF. The proof repeats word-for-word that of Lemma 2, except that Corollary 2 is replaced by Corollary 4. ►

§3.6. A general preliminary estimate

The estimate in this section is the first step in the proof of the preliminary estimate in §3.2. In its conditions it is now assumed that the functional cochain is defined in a domain admitting a proper conformal mapping onto $\mathbf{C}^+$. Such domains with the properties described in §3.2 will be constructed below.

LEMMA 4 (preliminary estimate). *Suppose that* Ω *is a domain belonging together with its* ε*-neighborhood* Ω^ε *to a standard domain* Ω_{st} *and admitting a proper conformal mapping* ψ *onto* $\mathbf{C}^+$ *(see Definition* 2 *in* §3.3); *in particular,* ψ *is extendible to the part of* Ω^ε *outside some compact set. Suppose that* Ω_{st} *is contained in some quadratic standard domain and the derivative* $\widetilde{\psi}'$ *is bounded away from zero by some positive constant on* $(\mathbf{R}^+, \infty)$. *Let* $m = n$ *or* $m = n - 1$, *and let* F *be the restriction to* Ω *of a regular functional cochain defined and* ε*-extendible in the domain* Ω_{st}. *Suppose that the restriction* F *to* Ω *corresponds to an* $\mathbf{R}$*-regular partition* Ξ *of* Ω *of type* $\boldsymbol{\sigma}$:

$$\boldsymbol{\sigma} = (\sigma_1, \ldots, \sigma_N), \qquad \sigma_1 \succ \sigma_2 \succ \cdots \succ \sigma_N, \quad \sigma_j \in \mathscr{D}^m.$$

Assume that the partitions $\sigma_{j*}\Xi_{\text{st}}$ *and the germs* $\rho_j = \sigma_j^{-1}$ *on* Ω_{st} *satisfy the conclusions of Lemma* 5.8_n. *Moreover, assume that:*

1°. *For each* $\lambda > 0$,

$$|F| \prec \exp(-\lambda \operatorname{Re} \psi)$$

on $(\mathbf{R}^+, \infty)$.

2°. *For* $\rho_1 = \sigma_1^{-1}$,

$$f \stackrel{\text{def}}{=} (\exp \rho_1)' / \widetilde{\psi}' \to \infty \quad \textit{on } (\mathbf{R}^+, \infty),$$

when $\sigma_1 \in \mathscr{D}_0^n \cup \mathscr{D}_0^{n-1} \cup (\mathscr{D}_{\text{slow}}^{n-1} \circ \mathscr{L}^{n-1})$ *(case* I*), and*

$$f_\delta \stackrel{\text{def}}{=} (\exp(1-\delta)\rho_1)' / \widetilde{\psi}' \to \infty \quad \textit{on } (\mathbf{R}^+, \infty)$$

when $\sigma_1 \in \mathscr{D}_{\text{rap}}^{n-1} \circ \mathscr{L}^{n-1}$ *(case* II*).*

Then in case I *there exist a* $C > 0$ *and a* $c > 0$ *such that for an arbitrary monotone function* $\widetilde{f} < f$ *with* $\widetilde{f} \nearrow \infty$ *on* $(\mathbf{R}^+, \infty)$

$$|F(\xi)| \prec \exp(-C \operatorname{Re} \psi(\xi) \cdot \widetilde{f}(c\xi)) \quad \textit{on } (\mathbf{R}^+, \infty).$$

In case II *there exists a* $C > 0$ *such that for every sufficiently small* $\delta > 0$ *and every monotone function* $\widetilde{f}_\delta < f_\delta$ *with* $\widetilde{f}_\delta \nearrow \infty$ *on* $(\mathbf{R}^+, \infty)$

$$|F(\xi)| \prec \exp(-C \operatorname{Re} \psi(\xi) \cdot \widetilde{f}_\delta((1-\delta)\xi))) \quad \textit{on } (\mathbf{R}^+, \infty).$$

The two assertions are proved in parallel, the proofs diverging at one place: namely, at the use of assertion (c) of Lemma 5.8_n, which is different in Cases I and II. The existence of the derivative $\widetilde{\psi}'$ follows from the definition; see the remark in §3.7 below.

REMARK. To the cochain F on Ω_{st} there can correspond a modified partition Ξ^F of the type $(\widetilde{\boldsymbol{\sigma}}, k)$:

$$\widetilde{\boldsymbol{\sigma}} = (\sigma_{-k+1}, \dots, \sigma_0, \sigma_1, \dots, \sigma_N), \qquad \sigma_j \succ \sigma_{j+1},$$

$$\Xi^F = \bigcap_{-k+1}^{0} \sigma_{j*}\Xi_M \bigcap_{1}^{N} \sigma_{j*}\Xi_{\text{st}}.$$

The conditions of the lemma are satisfied if the lines of the partition $\sigma_{j*}\Xi_M$ do not intersect Ω for $j = -k+1, \dots, 0$.

PROOF. It can be assumed without loss of generality that F is bounded. Indeed, by the definition of regular cochains, there exist $C > 0$ and $\nu > 0$ such that $|F(\zeta)| < C \exp \nu\xi$ in Ω_{st}. Then $F \exp(-\nu\zeta)$ is a bounded regular functional cochain of the same type as F. Accordingly, let $|F| \le 1$. For each $a \in \mathbf{R}^+ \cap \Omega$ let

$$\Omega_a = \{\zeta \mid \operatorname{Re} \psi(\zeta) > \operatorname{Re} \psi(a)\}.$$

For sufficiently large a the mapping $\zeta \mapsto \psi(\zeta) - \psi(a)$ is defined in the ε-neighborhood Ω_a^ε of the domain Ω_a and carries Ω_a into $\mathbf{C}^+$, because ψ

extends to the part of the ε-neighborhood of Ω outside a compact set, by assumption. For arbitrary a and $\lambda > 0$ we consider the cochain

$$F_{\lambda,a} = F \exp \circ \lambda(\psi(\zeta) - \psi(a))$$

on the domain Ω_a^ε.

By condition 1° in Lemma 4, it is bounded on $(\mathbf{R}^+, \infty)$. Our goal is to apply the maximum principle to the cochain $F_{\lambda,a}$ and prove that if a is sufficiently large, then there is a "large but not too large" value $\lambda(a)$ such that $|F_{\lambda,a}| < 2$. This gives us the inequality

$$|F(\xi)| < 2\exp(-\lambda \operatorname{Re}(\psi(\xi) - \psi(a))).$$

Fixing ξ and choosing $a = a(\xi)$ (considering that $\lambda = \lambda(a)$ is "large"), we shall get from this the inequality in Lemma 4.

Note that $|F_{\lambda,a}|_{\partial\Omega_a}| \le 1$, because $|\exp \lambda(\psi(\zeta) - \psi(a))||_{\partial\Omega_a} = 1$.

We now estimate the constants m_0 and I in Lemma 1, replacing the cochain F and the domain Ω by $F_{\lambda,a}$ and Ω_a in the condition of the lemma; we express them in terms of a and λ, regarding a as arbitrary and choosing λ. By choosing a sufficiently large we can make all the asymptotic inequaltiies in §§3.3–3.5 become pointwise inequalities in Ω_a^ε; assume that this has already been done.

Since the partition Ξ of Ω is $\mathbf{R}$-regular, the coboundary of the cochain F on Ω can be estimated by the rigging cochain on all the boundary lines of the partition. For us the following estimate suffices:

$$|\delta F||_{\partial\Xi\cap\Omega} < m = \sum_1^N C_1 \exp(-C_2 \exp \rho_j), \qquad \rho_j = \sigma_j^{-1}.$$

By assertion (c) in Lemma 5.8_n,

$$\begin{aligned} m(\zeta) \le & \sum_1^{N_1} C_1 \exp(-C_2 \exp \operatorname{Re} \rho_j(\xi)) \\ & + \sum_{N_1+1}^{N} C_1 \exp(-C_2 \exp \operatorname{Re}(1-\delta)\rho_j(\xi)), \qquad \xi = \operatorname{Re}\zeta. \end{aligned}$$

In the first sum, j runs through the values for which $\sigma_j \in \mathscr{D}_0^n \cup \mathscr{D}_0^{n-1} \cup \mathscr{D}_{\text{slow}}^{n-1} \circ \mathscr{L}^{n-1}$, and in the second sum it runs through the values for which $\sigma_j \in \mathscr{D}_{\text{rap}}^{n-1} \circ \mathscr{L}^{n-1}$.

Let us consider Case I of Lemma 4 first. For $j \ge 2$ we have that $\rho_1 \prec \rho_j$ on $(\mathbf{R}^+, \infty)$, because $\sigma_j \succ \sigma_{j+1}$ and $\rho_j = \sigma_j^{-1}$. Consequently,

$$m(\zeta) \le C_1' \exp(-C_2 \exp \rho_1(\xi)), \qquad C_1' = NC_1.$$

Now let

$$m_{\lambda,a}(\zeta) = C_1' \exp(-C_2 \exp \rho_1(\xi) + \lambda \operatorname{Re}(\psi(\zeta) - \psi(a))).$$

Obviously, $|\delta F_{\lambda,a}| < m_{\lambda,a}$ in Ω_a^ε. According to requirement 3° in the definition of proper mappings (§3.3), for some $C > 0$

$$\operatorname{Re}\psi(\zeta) < C \operatorname{Re}\psi(\xi), \qquad \xi = \operatorname{Re}\zeta.$$

We proceed directly to an estimate of the constants

$$m_0 = \max_{\partial\Xi\cap\Omega_a^\varepsilon} m_{\lambda,a}, \qquad I = \int_{\partial\Xi\cap\Omega_a^\varepsilon} m_{\lambda,a}\,dx.$$

Let us begin with the integral and pass to integration over a ray of the positive semi-axis. By Definition 3 in §3.3, the domain Ω_a^ε belongs to the half-plane $\xi > b \overset{\text{def}}{=} b_\psi(a) - \varepsilon$ (see Figure 18). By assertion (b) in Lemma 5.8_n, the total slope of the lines of the partition Ξ in Ω_a^ε does not exceed $C_3\xi^4$, where C_3 does not depend on a. Therefore,

$$I \le \int_b^\infty C''\xi^4 \exp(-C_2 \exp\rho_1(\xi) + \lambda(\widetilde{\psi}(\xi) - \operatorname{Re}\psi(a))), \qquad b = b_\psi(a) - \varepsilon.$$

It is this inequality that motivates the definition of the degree of convexity of a mapping ψ in §3.2 A.

Denote the integrand by m_+. The integral I is easy to estimate if $m_+'/m_+ \le -1$ for $\xi \ge b$. Then

$$m_+(\xi) \le m_+(b)\exp(b - \xi), \qquad I(a) \le m_+(b).$$

We have that

$$m_+'/m_+ = -C_2(\exp\rho_1)' + \lambda\widetilde{\psi}' + o(1).$$

Suppose that the functions f and $\widetilde{f}$ are the same as in Case I of Lemma 4.

Now let

$$\lambda(a) = \frac{C_2}{2}\widetilde{f}(b), \qquad b = b_\psi(a) - \varepsilon.$$

Then

$$m_+'/m_+(\xi) \le C_2\widetilde{\psi}'(\xi)\left(\frac{1}{2}\widetilde{f}(b) - \widetilde{f}(\xi)\right) + o(1).$$

Note that $b \to \infty$ as $a \to \infty$, by Property III of proper maps. Taking into account that the derivative $\widetilde{\psi}'$ is bounded away from zero by a positive constant in view of the assumption of the lemma and that the function $\widetilde{f}$ is monotone and tends to infinity, we conclude that $m_+'/m_+ < -1$ for $\xi \ge b$ if a is large enough. Then

$$I \le m_+(b) \le C''b^4 \exp(-C_2 \exp\rho_1(b))$$

because in the expression for $m_+(b)$

$$\begin{aligned}\widetilde{\psi}(b) - \operatorname{Re}\psi(a) &= \widetilde{\psi}(b_\psi(a) - \varepsilon) - \operatorname{Re}\psi(a)\\ &\le \widetilde{\psi}(b_\psi(a)) - \operatorname{Re}\psi(a) = 0;\end{aligned}$$

the last equality follows from Remark 1 in §3.3C. Since $\exp\rho_1$ tends to infinity together with ρ_1 on $(\mathbf{R}^+, \infty)$, we get that $I \le m_+(b) \to 0$ as $a \to 0$. Similarly, $m_0 \le m_+(b_\psi(a) - \varepsilon) \to 0$ as $a \to \infty$.

Applying the maximum principle to the cochain $F_{\lambda,a}$, we get that

$$\sup_{\Omega_a} |F_{\lambda,a}| \leq \sup_{\partial\Omega_a} |F| + o(1) \quad \text{as } a \to \infty.$$

For sufficiently large a,

$$\sup_{\Omega_a} |F_{\lambda,a}| \leq 2.$$

Consequently, for arbitrary $\xi > a$,

$$|F(\xi)| < 2\exp(-\lambda \operatorname{Re}(\psi(\xi) - \psi(a))).$$

By definition, $\lambda = C\widetilde{f}(b)$ for some $C > 0$. We now fix a real ξ and let δ be small enough that

$$\operatorname{Re}(\psi(\xi) - \psi((1-\delta)\xi)) > \frac{\delta}{2} \operatorname{Re}\psi(\xi)$$

(it holds for all sufficiently small δ and all sufficiently large ξ; see Property II of proper mappings). Setting $a = (1-\delta)\xi$ and using Property II of proper mappings, we get that

$$|F(\xi)| \leq 2\exp(-C\operatorname{Re}\psi(\xi)\widetilde{f}(c\xi)) \tag{$*$}$$

for some $C > 0$ and $c > 0$. This proves Lemma 4 in Case I.

The proof is analogous in Case II, except that in the estimate of $m(\zeta)$ the exponent ρ_1 is replaced by $(1-\delta)\rho_1$ for arbitrary $\delta > 0$:

$$m(\zeta) \leq C_1' \exp(-C_2 \exp(1-\delta)\rho_1).$$

The function $\exp(1-\delta)\rho_1$ increases exponentially, since ρ_1 increases linearly by the definition of $\mathscr{D}^{n-1}_{\text{rap}}$ and by property $(*)$ in §2.11C of the group $\mathscr{L}^{n-1}$. Consequently, all the computations in Case I go through with ρ_1 replaced by $(1-\delta)\rho_1$. This changes the estimate on $b = b_\psi(a) - \varepsilon$: $b = (1+o(1))(1-\delta/2)\xi$ for $a = (1-\delta/2)\xi$, by Property IV in §3.3, $b > (1-\delta)\xi$. The function f is replaced by f_δ, and $\widetilde{f}$ by $\widetilde{f}_\delta$. We get instead of $(*)$ the inequality

$$|F(\xi)| \leq 2\exp(-C\psi(\xi))\widetilde{f}_\delta((1-\delta)\xi). \tag{$**$}$$

This concludes the proof of Lemma 4 in Case II. ►

§3.7. A preliminary estimate of cochains given in a standard domain

The first step in the proof of the preliminary estimate is the use of Lemma 4 in the case when Ω is a standard domain belonging to a quadratic standard domain, and $\rho_1^{-1} = \sigma_1$ is a rapid germ. We consider the case $\sigma_1 \in \mathscr{D}_0^n$; the case $\sigma_1 \in \mathscr{D}_1^{n-1}$ may be repeated word-for-word replacing n by $n-1$.

Remark. For $n = 1$ there are no rapid germs of class $\mathscr{D}_0^n$. The arguments in §§3.7–3.9 apply to the case $n > 1$. However, §3.7 is not hard to modify for the case when the germ σ_1 is sectorial. Therefore, the arguments in Chapter

III work also for $n = 1$. An essentially simpler proof of the Phragmén-Lindelöf theorem for $n = 1$ is given in the first part [24].

We verify the conditions of Lemma 4.

1°. By a condition of the Phragmén-Lindelöf theorem, $|F| \prec \exp(-\lambda\xi)$ on $(\mathbf{R}^+, \infty)$ for arbitrary $\lambda > 0$. According to the definition of a standard domain, there exists a conformal mapping ψ: $\Omega \to \mathbf{C}^+$ equal to $\xi(1 + o(1))$ on $(\mathbf{R}^+, \infty)$. This proves the first condition of the lemma:

$$|F| \prec \exp(-\lambda \operatorname{Re} \psi) \quad \text{on } (\mathbf{R}^+, \infty) \text{ for any } \lambda > 0.$$

By Proposition 1 in §3.3, ψ can be chosen to be proper. The inequality $\widetilde{\psi}' \succ a \succ 0$ is proved after the next remark. Consequently, Lemma 4 is applicable to the cochain F and the mapping ψ. The germ σ_1 is rapid, and hence the lemma must be used in Case I. By Lemma 4, there exist $C > 0$ and $c > 0$ such that on $(\mathbf{R}^+, \infty)$

$$|F(\xi)| \prec \exp(-C \operatorname{Re} \psi(\xi))\widetilde{f}(c\xi),$$

where $\widetilde{f}$ is an arbitrary monotone function tending to ∞ and less than f:

$$f = (\exp \rho_1)'/\widetilde{\psi}'.$$

We prove that $(1/2\xi)\exp \rho_1$ can be taken as the function $\widetilde{f}$. By the definition of a rapid germ,

$$\exp \rho_1 = A^{-n} g, \qquad g \in G_{\text{slow}}^{n-1^+}, \ \mu_n(g) = 1.$$

Let

$$A^{1-n} g = \operatorname{id} + \varphi_{n-1}.$$

Since $\mu_n(g) = \lim_{(\mathbf{R}^+, \infty)} (A^{1-n} g)'$, we get that $\varphi'_{n-1} \to 0$ on $(\mathbf{R}^+, \infty)$. Then on $(\mathbf{R}^+, \infty)$

$$(\exp \rho_1)' = (\xi \exp \circ \varphi_{n-1} \circ \ln)' = (\exp \circ \varphi_{n-1} \circ \ln)(1 + o(1)).$$

REMARK. For any ξ define the set $M_\xi = \{\zeta_{\max}(\xi)\}$ by the equality $\widetilde{\psi}(\xi) = \operatorname{Re} \psi(\zeta_{\max}(\xi))$. It follows from the definition of $\widetilde{\psi}$ that M_ξ is not empty for large ξ and is finite because the function ψ is analytic. Moreover, $\widetilde{\psi}'(\xi) = \max_{\zeta \in M_\xi} \operatorname{Re} \psi'(\zeta)$.

Further, $\operatorname{Re} \psi' = 1 + o(1)$ in (Ω, ∞), by the definition of a standard domain. Consequently, $\widetilde{\psi}' \succ a \succ 0$, as required in Lemma 4. Moreover,

$$f = (\exp \rho_1)'/\widetilde{\psi}' = (\exp \circ \varphi_{n-1} \circ \ln)(1 + o(1)) \succ \widetilde{f}.$$

We prove that the function $\widetilde{f} = \frac{1}{2} \exp \circ \varphi_{n-1} \circ \ln$ tends monotonically to $+\infty$ on $(\mathbf{R}^+, \infty)$. This follows immediately from the following proposition:

PROPOSITION 5.1_n. *Suppose that* $g \in G_{\text{slow}}^{n-1^+}$, $\mu_n(g) = 1$, *and* $A^{1-n} g = \text{id} + \varphi_{n-1}$. *Then the germ* φ_{n-1} *tends monotonically to* $+\infty$ *on* $(\mathbf{R}^+, \infty)$.

For $n = 1$ this proposition is trivial (it relates to an empty set of objects, since for $g \in G^0$ the equality $\mu_1(g) = 1$ implies that $g \in G^0_{\text{rap}}$). Assume that the proposition (the property of the group G^{n-1}) has been proved (induction hypothesis). In § 5.5_nc we prove Proposition 5.1_{n+1}—the induction step. This proves that the chosen function $\widetilde{f}$ has the necessary properties.

It follows from the monotonicity of ψ on $\mathbf{R}^+ \cap \Omega$ that $\psi(c\xi) < \psi(\xi)$ for $c < 1$. Moreover, $\psi(\xi) = \xi(1 + o(1))$ on $(\mathbf{R}^+, \infty)$. Therefore, Lemma 4 gives us that

$$|F(\xi)| \prec \exp(-C\psi \cdot \widetilde{f}) \circ (c\xi) \prec \exp(-\tfrac{C}{3} \exp \rho_1(c\xi)).$$

But

$$\begin{aligned} \exp \rho_1(c\xi) &= c\xi \cdot \exp \circ \varphi_{n-1} \circ \ln c\xi \\ &= c\xi \cdot \exp \circ \varphi_{n-1} \circ (\ln \xi + \ln c) \\ &= c\xi \cdot (\exp \circ \varphi_{n-1} \circ \ln \xi)(1 + o(1)) \\ &= (c + o(1)) \exp \rho_1, \end{aligned}$$

because $\varphi'_{n-1} \to 0$ on $(\mathbf{R}^+, \infty)$. This proves the preliminary estimate I in the case under consideration.

* * *

Accordingly, in this section we have taken the first step in proving the Phragmén-Lindelöf theorem. Namely, let F be a regular functional cochain of class $\mathscr{D}_0^n$ or $\mathscr{D}_1^{n-1}$ and of type $\boldsymbol{\sigma} = (\sigma_1, \dots, \sigma_N)$ given in some standard domain and decreasing faster than any exponential on $(\mathbf{R}^+, \infty)$, and let σ_1 be a rapid germ. Then there exists a $c > 0$ such that the inequality $(*)$ in §3.2 holds on $(\mathbf{R}^+, \infty)$ with $\rho = \sigma_1^{-1}$.

§3.8. The Warschawski formula and corollaries of it

We formulate a theorem of Warschawski [37]. Consider a curvilinear strip S:

$$\widetilde{\gamma}_-(\xi) < \eta < \widetilde{\gamma}_+(\xi), \qquad -\infty < \xi < \infty.$$

Let $\theta = \widetilde{\gamma}_+ - \widetilde{\gamma}_-$, $\psi = (\widetilde{\gamma}_+ + \widetilde{\gamma}_-)/2$. The slope of the boundary of the strip is equal to zero at infinity if

$$\frac{\widetilde{\gamma}_\pm(u_1) - \widetilde{\gamma}_\pm(u_2)}{u_1 - u_2} \to 0 \quad \text{as } u_1, u_2 \to \infty, u_1 < u_2.$$

THEOREM. *Suppose that the strip* S *described above has zero slope of the boundary at infinity, and the derivatives* $\widetilde{\gamma}'_+$ *and* $\widetilde{\gamma}'_-$ *are continuous and have bounded variation in a neighborhood of* $+\infty$. *Assume that the integral* $\int \theta'^2/\theta$

converges; in view here and below are convergence of the integral at $+\infty$ *and convergence to zero at* $+\infty$. *Then a function* Φ *mapping* S *onto the right strip* $|\eta| < \pi/2$ *has the form*

$$\Phi(\zeta) = \lambda + \pi \int_{\xi_0}^{\xi} \frac{1+\psi'^2}{\theta} + i\pi \frac{\eta - \psi(\xi)}{\theta} + o(1).$$

DEFINITION 1. A half-strip of type W is a half-strip of the form

$$\Pi_W = \left\{ \zeta \mid \xi \geq a > 0,\ \eta \in \left[-\frac{\pi}{2} + \gamma_-(\xi),\ \frac{\pi}{2} - \gamma_+(\xi)_- \right] \right\},$$

where a is some positive constant, and γ_- and γ_+ are differentiable functions tending to zero together with their first derivatives as $\xi \to \infty$, with the convergence monotone from some point on. The functions γ_+ and γ_- are called the boundary functions of the half-strip of type W.

We apply the Warschawski theorem to the investigation of conformal mappings of half-strips of type W. A conformal mapping of a half-strip of type W onto the part of the right half-strip $|\eta| < \pi/2$ outside some compact set is said to be rectifying. The boundary functions γ_+ and γ_- can be extended to functions that are C^1-smooth on the whole axis and identically equal to zero on the negative semi-axis. Denote the extended functions by the same symbols. For the resulting strip $|\widetilde{\gamma}_\pm| = \pi/2 - \gamma_\pm$. We get from the conditions $\gamma'_\pm \to 0$ as $\xi \to 0$ that a half-strip of type W has zero slope at infinity. The monotonicity of the derivatives $\gamma'_\pm$ and the conditions $\gamma'_\pm \to 0$ imply that they are of bounded variation. Further, $\theta = \pi - (\gamma_+ + \gamma_-)$ and $\theta' = -(\gamma'_+ + \gamma'_-)$. We prove that the integral $\int \theta'^2/\theta$ always converges for half-strips of type W. Indeed, θ is bounded from above and is bounded away from zero; therefore, the convergence of the preceding integral is equivalent to the convergence of the integral $\int \theta'^2$. Furthermore,

$$\int \theta'^2 \leq \max|\theta'| \int |\theta'| \leq \max|\theta'| \int |\gamma'_-| + |\gamma'_+|.$$

The last integral converges because the functions $\gamma_\pm$ are monotone and the derivative θ' is bounded. Accordingly, we have

COROLLARY 1. *For an arbitrary half-strip of type W there exists a rectifying mapping, which is described by the Warschawski formula.*

We simplify the Warschawski formula for half-strips of type W. Note that $\int \psi'^2/\theta$ converges; this can be proved just like the convergence of $\int \theta'^2/\theta$. Therefore, $\int \psi'^2/\theta = C + o(1)$, $C > 0$. Further, $\psi(\xi) = -(\gamma_- + \gamma_+)/2 \to 0$ as $\xi \to \infty$. Therefore, $\psi/\theta = o(1)$. Finally,

$$\frac{\pi}{\theta} = \frac{\pi}{\pi - (\gamma_- + \gamma_+)} = 1 + \frac{\gamma_+ + \gamma_-}{\pi - (\gamma_+ + \gamma_-)}.$$

We get Corollary 2 from this.

COROLLARY 2. *For any half-strip of type W there exists a rectifying mapping of the form*

$$\Phi(\zeta) = \zeta + \int_{\xi_0}^{\xi} \frac{\gamma_+ + \gamma_-}{\pi - (\gamma_+ + \gamma_-)} + \lambda + o(1), \qquad \lambda \in \mathbf{R}.$$

REMARK. The integral on the right-hand side converges or diverges simultaneously with the integral $\int(\gamma_+ + \gamma_-)$ if $\gamma_+ > 0$ and $\gamma_- > 0$.

The subsequent material is added for completeness and is not used in the proof of the Phragmén-Lindelöf theorem.

COROLLARY 3. *A half-strip of type W admits a rectifying mapping with bounded correction if and only if the integral $\int(\gamma_+ + \gamma_-)$ converges.*

PROOF. Suppose that $\int(\gamma_+ + \gamma_-)$ converges. Then the boundedness of the correction of a rectifying mapping follows from Corollary 2.

We prove that if one rectifying mapping of a half-strip of type W has bounded correction, then all rectifying mappings of this half-strip have the same property. Indeed, suppose that Φ and Φ_1 are two rectifying mappings, and let Φ have a bounded correction. The composition $H = \Phi_1 \circ \Phi^{-1}$ carries the part of a right half-strip outside some compact set into the part of the same half-strip outside some other compact set. By the symmetry principle, the correction of this composition is $2\pi i$-periodic, and hence differs from a constant by an exponentially decreasing function. Therefore, the mapping $\Phi_1 = H \circ \Phi$ has bounded correction, being the composition of two mappings with the same property.

If now the integral $\int(\gamma_+ + \gamma_-)$ diverges, then the original half-strip admits a rectifying mapping with unbounded correction, and by what has been proved, it does not admit a rectifying mapping with bounded correction. ▶

§3.9. A preliminary estimate for cochains given in domains corresponding to rapid germs

In this section the preliminary estimate II in §3.2 is proved in the case when $\sigma = \sigma_k$ is a rapid germ (the definition is recalled below), and $\sigma_{\text{next}} = \sigma'_{k+1}$ is also a rapid germ or a sectorial germ (the definition is recalled at the beginning of D). We recall that in the formulation of the preliminary estimate II the germ σ_l is the first germ after σ_k in $\boldsymbol{\sigma}$ that is not weakly equivalent to σ_k.

Accordingly, σ is a rapid germ everywhere in this section:

$$\sigma = \exp \circ A^{1-n} g^{-1}, \qquad g \in G_{\text{slow}}^{n-1^+}, \quad \mu_n(g) = 1.$$

In this section $\sigma, \sigma_{\text{next}} \in \mathscr{D}_0^n$. The arguments carry over word-for-word to the case when $\sigma, \sigma_{\text{next}} \in \mathscr{D}_0^{n-1}$.

A. Construction of the domain Ω_σ.

LEMMA 5. *For an arbitrary rapid germ σ and an arbitrary standard domain Ω of class n there exists a domain Ω_σ with the following properties:*

1°. $(\Omega_\sigma, \infty) \subset (\sigma\Pi_{\text{main}} \cap \Omega, \infty)$;

2°. *there exists a proper (see Definition 2 in* §3.3) *conformal mapping* $\psi\colon \Omega_\sigma \to \mathbf{C}^+$.

Here the domain Ω_σ and the mapping ψ are constructed; the first assertion of the lemma is proved in subsection A simultaneously with the construction, and the second is proved in subsections F and G.

All the constructions are based on the properties of the half-strip $\widetilde{\Pi} = A^{1-n}g^{-1}\Pi$; as above, Π is a right half-strip (see §1.5). Lemma 5.7_n below asserts that this half-strip is of type W. The Warschawski formula relates the geometry of a half-strip $\widetilde{\Pi}$ with the asymptotic properties of the germ $A^{1-n}g$ rectifying $\widetilde{\Pi}$. This connection is used in two ways. First, by modifying $\widetilde{\Pi}$ we construct a new half-strip Π_1 whose image under the action of the mapping exp gives the desired domain Ω_σ. The Warschawski formula enables us to investigate the properties of the conformal mapping $\psi\colon \Omega_\sigma \to \mathbf{C}^+$. Second, $\exp\rho = A^{-n}g$. Therefore, the Warschawski formula gives information also about the second asymptotic expression of interest to us—that of the mapping $\exp\rho$. In the conclusion of this introductory part we mention that if $\Phi\colon \Pi_1 = \ln\Omega_\sigma \to \Pi$ is a rectifying mapping for the half-strip Π_1, then $A^{-1}\Phi\colon \Omega_\sigma \to \mathbf{C}^+\backslash K$ is a conformal mapping onto the part of $\mathbf{C}^+$ outside a compact set K. By the symmetry principle and the theorem on removable singularities, a conformal mapping $\mathbf{C}^+\backslash K \to \mathbf{C}^+$ extends holomorphically to a neighborhood of the point at infinity. Therefore, there exists a conformal mapping

$$\psi = A^{-1}\Phi + O(1/A^{-1}\Phi)\colon \Omega_\sigma \to \mathbf{C}^+.$$

We proceed to a description of the half-strip $\widetilde{\Pi}$.

LEMMA 5.7_n. *Suppose that* $\widetilde{g} \in G_{\text{slow}}^{n-1^-}$, $\mu_n(\widetilde{g}) = 1$, $\widetilde{\rho} = A^{1-n}\widetilde{g} = \mathrm{id} + \widetilde{\varphi}$, *and* $\rho_\lambda = \mathrm{id} + \lambda\widetilde{\varphi}$. *Then*: (10)

b. $\rho_\lambda\Pi$ *is a half-strip of type* W *for arbitrary* $\lambda > 0$;

c. *the germ of this half-strip at infinity belongs to the germ at infinity of the logarithm of an arbitrary standard domain of class* $\mathbf{\Omega}_n$;

d. *if* γ_λ *is the boundary function of the half-strip* $\rho_\lambda\Pi$, $\lambda > 0$, $\mu > 0$, *then*

$$\gamma_\lambda \searrow 0,\ \gamma_\lambda/\gamma_\mu \to \lambda/\mu,\ |\xi\gamma_\lambda'/\gamma_\lambda| \prec 2 \text{ and } |\xi\gamma_\lambda''/\gamma_\lambda'| \prec 3 \quad \text{on } (\mathbf{R}^+, \infty).$$

REMARK. The germ $\widetilde{g}$ is real; thus, the half-strip $\rho_\lambda\Pi$ is symmetric with respect to the real axis, and its boundary functions are the same.

Lemma 5.7_n is proved by induction on n. Induction base: $n = 1$. In this case $G_{\text{slow}}^0 = \{g \mid \mu_1(g) < 1\}$, and there are no rapid germs of class $\mathscr{D}_0^n$. Lemma 5.7_nb–5.7_nd are true, being properties of the empty set. Assertion 5.7_na is trivial, because germs of classes $\mathscr{D}_0^1$ and $\mathscr{D}_1^0$ are equivalent to exponentials and dilations, respectively; see §1.6. Due to the absence of

(10) Recall that Lemma 5.7_na was formulated in §3.2.

rapid germs, the proof of the Phragmén-Lindelöf theorem simplifies sharply for $n=1$; see part 1 [24]. The induction step in the proof of Lemma 5.7_n (passage from n to $n+1$) is made in §5.7. Lemma 5.7_n appears in the induction hypothesis.

We return to the description of the half-strip $\widetilde{\Pi} = A^{1-n}g^{-1}\Pi$, $g^{-1} \in G_{\text{slow}}^{n-1^-}$. By assumption, $\mu_n(g^{-1}) = 1$. Setting $\tilde{g} = g^{-1}$ in Lemma 5.7_n, we get that $\widetilde{\Pi}$ is a half-strip of type W.

Let us proceed to the construction of the domain Ω_σ. Let $\tilde{\rho}$, ρ_λ, and γ_λ be the same as in Lemma 5.7_n for $\tilde{g} = g^{-1}$; for brevity we write $\gamma_1 = \gamma$. Consider the half-strip Π_1 with boundary functions $\gamma_+ = \gamma_{1/2}$ and $\gamma_- = \gamma$, and let $\Omega_\sigma = \exp \Pi_1$. This domain will be the desired one. The inclusion $\Omega_\sigma \subset \Omega_{\text{st}}$ for any $\Omega_{\text{st}} \in \mathbf{\Omega}_n$ follows from Lemma 5.7_n c.

B. The conformal mapping $\psi\colon \Omega_\sigma \to \mathbf{C}^+$, and a proof that the preliminary estimate is effective in the domain Ω_σ. According to the Warschawski formula, there exists a rectifying mapping $\Phi\colon \Pi_1 \to \Pi$,

$$\Phi = \mathrm{id} + \varphi, \qquad \varphi = I + o(1),$$

$$I = \int^{\xi} \frac{\gamma + \gamma_{1/2}}{\pi - (\gamma + \gamma_{1/2})}.$$

By Lemma 5.7_n d, $\gamma_{1/2} = (\frac{1}{2} + o_{\mathbf{R}}(1))\gamma$. Here and below, $o_{\mathbf{R}}(1)$ is a real function tending to zero on $(\mathbf{R}^+, \infty)$. The lower limit of integration in the formula for I is chosen arbitrarily in the domain of the integrand and fixed; it is not essential for what follows; therefore, we do not write it.

The function $\tilde{\varphi} = A^{1-n}g^{-1} - \mathrm{id} = \tilde{\rho} - \mathrm{id}$ is defined in the ε-neighborhood L_+^ε of the arc $L_+\colon \xi \geq a$, $\eta = \pi/2$ for sufficiently large a according to Lemma 5.3_n. By the symmetry principle, points symmetric with respect to the ray L_+ pass into points that are conformally symmetric with respect to the curve $\tilde{\rho}L_+$ and lie in the image $\tilde{\rho}L_+^\varepsilon$. Since $\rho' \to 1$ in $(\Pi^\forall, \infty)$ (the limit exists by Lemma 5.3_n and is equal to 1 by assumption: $\mu_n(g) = 1$), we get that the set $\tilde{\rho}L_+^\varepsilon$ contains the $(\varepsilon/2)$-neighborhood of the curve $\tilde{\rho}L_+$ for sufficiently large a, and a "conformal symmetry" with respect to this curve is defined in it. By the symmetry principle, we get that an arbitrary conformal mapping defined in the domain $\tilde{\rho}L_+^\varepsilon$ on one side of the curve $\tilde{\rho}L_+$ and carrying it into the ray L_+ extends biholomorphically to the $(\varepsilon/2)$-neighborhood of $\tilde{\rho}L_+$. There is an analogous assertion for sufficiently large a and small ε for the curve $(\mathrm{id} + \tilde{\phi}/2)L_-$, where $L_- = \{\xi \geq a, \eta = -\pi/2\}$. This implies that the mapping Φ, which rectifies the half-strip Π_1, extends to the $\varepsilon/2$-neighborhood of the part of the half-strip outside some compact set.

We now construct a proper conformal mapping $\psi\colon \Omega = \Omega_\sigma \to \mathbf{C}^+$.

This will verify one of the conditions imposed on the domain Ω in Lemma 4 of §3.6. The mapping $\psi_0 = A^{-1}\Phi$ carries Ω_σ into the part of $\mathbf{C}^+$ outside some compact set. By the symmetry principle, there exists a mapping $\zeta \mapsto \zeta + O(1/\zeta)$ holomorphic in a neighborhood of infinity on the Riemann sphere that carries the domain $\psi_0 \Omega_\sigma$ into $\mathbf{C}^+$. Finally, the conformal mapping $\psi\colon \Omega_\sigma \to \mathbf{C}^+$ can be chosen in the form

$$\begin{aligned} \psi &= \psi_0 + O(1/\psi_0), & \psi_0 &= A^{-1}\Phi = \zeta \cdot \exp \circ \varphi \circ \ln, \\ \varphi &= I + o(1) \quad \text{in } (\ln \Omega_\sigma, \infty), & I &= \int^\xi \frac{\gamma + \gamma_{1/2}}{\pi - (\gamma + \gamma_{1/2})}. \end{aligned} \tag{$*$}$$

Since $\ln \psi = \ln \psi_0 + o(1)$ in (Ω_σ, ∞), we get that

$$\psi = \zeta \cdot \exp \circ (I \circ \ln + R), \qquad R = o(1).$$

The domain Ω_σ contains the germ of an arbitrary sector (S_α, ∞) for $0 < \alpha < \pi/2$. By the Cauchy inequality, the convergence of R to zero in (Ω_σ, ∞) implies that $(I \circ \ln + R)' = o(1)\zeta^{-1}$ in (S_α, ∞) for an arbitrary $\alpha \in (0, \pi/2)$. Therefore,

$$\psi' = (\exp \circ I \circ \ln)(1 + o(1)) \quad \text{in } (S_\alpha, \infty) \tag{$**$}$$

for arbitrary $\alpha \in (0, \pi/2)$. Everywhere in Ω_σ

$$\begin{aligned} \psi' &= (1 + O(\psi_0^{-2})) \cdot \psi_0', \\ \psi_0' &= (\exp \circ \varphi \circ \ln) \cdot (1 + \varphi' \circ \ln). \end{aligned} \tag{$***$}$$

It was proved above that the mapping Φ extends to the ε-neighborhood of the part of Π_1 outside some compact set. Consequently, the mapping ψ has the same property with respect to the domain $\Omega = \Omega_\sigma$. The rest of the requirements in the definition of properness are verified for ψ in subsections F and G; a positive lower estimate for $\widetilde{\psi}'$ is given in section D. To use Lemma 4 it remains to verify that the preliminary estimate is effective; this is done in the next subsection.

C. Effectiveness of the preliminary estimate. Let the mapping σ be the same as in the beginning of the section, and let $\rho = \sigma^{-1}$. It is required to prove that

$$\exp \rho / \operatorname{Re} \psi \to \infty \quad \text{on } (\mathbf{R}^+, \infty).$$

By definition, $\exp \rho = A^{1-n} g$. This mapping rectifies the half-strip $\widetilde{\Pi}$, is real, and hence is given by the Warschawski formula:

$$A^{1-n} g = \mathrm{id} + \varphi_{n-1}, \qquad \varphi_{n-1} = 2 \int^\xi \frac{\gamma}{\pi - 2\gamma} + \lambda + o(1), \quad \lambda \in \mathbf{R}.$$

The integral on the right-hand side diverges, since $g \in G_{\text{slow}}^{n-1^+}$, and thus the correction φ_{n-1} of the germ $A^{1-n} g$ is unbounded (otherwise the germ $A^{-n} g$ would be comparable with a linear germ). Consequently, the integral $\int^\xi \gamma$

also diverges, and with it the integral I. This consideration plays a central role in the estimates in subsections C–E. We have that

$$\exp\rho = \xi \cdot \exp\circ(\widetilde{I} + \lambda + o(1)) \circ \ln, \qquad \widetilde{I} = \int \frac{2\gamma}{\pi - 2\gamma}. \qquad ({*}{*}{*})$$

Consequently,

$$\exp\rho / \operatorname{Re}\psi = \exp\circ(\varphi_{n-1} - \varphi) \circ \ln,$$

$$\varphi_{n-1} - \varphi \succ \left(\frac{1}{2} - \delta\right) \int^{\xi} \frac{\gamma}{\pi} \to \infty \quad \text{on } (\mathbf{R}^+, \infty)$$

for arbitrary $\delta > 0$. This implies that the preliminary estimate is effective.

D. The preliminary estimate in the case when the germ σ_{next} is rapid. Suppose that the germs σ and σ_{next} are the same as in the preliminary estimate II, that is, $\sigma = \sigma_k$, and $\sigma_{\text{next}} = \sigma_l$, and both germs are rapid. Then

1°. $\sigma_{\text{next}} = \exp\circ A^{1-n} g_{\text{next}}^{-1}$, $g_{\text{next}} \in G_{\text{slow}}^{n-1^+}$, $\mu_n(g_{\text{next}}) = 1$, by the definition of a rapid germ.

2°. The germ σ_{next} is not weakly equivalent to the germ σ and $\sigma \succ \sigma_{\text{next}}$ by the condition of the preliminary estimate II.

To use Lemma 4 in §3.6, we compute the function $\widetilde{f}$ in this lemma.

STEP I. Let γ_{next} be a boundary function of the half-strip $\Pi_{\text{next}} = A^{1-n} g_{\text{next}}^{-1}\Pi$. This half-strip is of type W, by Lemma 5.7_nb; in particular, $\gamma_{\text{next}} \to 0$ on $(\mathbf{R}^+, \infty)$ by Definition 1 in §3.8. Then, by formula $({*}{*}{*})$, in which ρ, $\widetilde{I}$, and γ are replaced by ρ_{next}, $\widetilde{I}_{\text{next}}$, and γ_{next},

$$\exp\circ\rho_{\text{next}} = \xi \cdot \exp\circ(\widetilde{I}_{\text{next}} + \tilde{\lambda} + o(1)) \circ \ln,$$

$$\widetilde{I}_{\text{next}} = \int \frac{2\gamma_{\text{next}}}{\pi - 2\gamma_{\text{next}}}, \qquad \tilde{\lambda} \in \mathbf{R}.$$

Consequently, on $(\mathbf{R}^+, \infty)$,

$$(\exp\circ\rho_{\text{next}})' = \exp\circ(\widetilde{I}_{\text{next}} + o(1)) \circ \ln.$$

STEP II. We get an upper and a lower estimate for $\widetilde{\psi}'$. As in §3.7, suppose that the point $\zeta = \zeta_{\max}(\xi)$ is such that $\widetilde{\psi}(\xi) = \operatorname{Re}\psi(\zeta)$, $\operatorname{Re}\zeta = \xi$. Then $\arg\psi'(\zeta) = 0$ and $\widetilde{\psi}'(\xi) = \psi'(\zeta)$. In G we prove

PROPOSITION 1. *For the mapping ψ constructed in part* B *the equation $\arg\psi' = 0$ gives a set lying in the sector $S_{\pi/2-\varepsilon}$: $|\arg\zeta| < \pi/2 - \varepsilon$ for some $\varepsilon > 0$.*

The domain Ω_σ contains the germ at infinity of an arbitrary sector $S_{\pi/2-\delta}$, $\delta > 0$. The Koebe distortion theorem [26] implies that for some $C > 0$

$$C^{-1}|\psi'(\xi)| \prec \psi'(\zeta_{\max}(\xi)) \prec C|\psi'(\xi)| \quad \text{on } (\mathbf{R}^+, \infty).$$

Consequently,

$$C^{-1}|\psi'(\xi)| \prec \widetilde{\psi}'(\xi) \prec C|\psi'(\xi)| \quad \text{on } (\mathbf{R}^+, \infty).$$

The formula $(**)$ now implies that $\widetilde{\psi}' \succ \alpha \succ 0$, as is required in Lemma 4.

STEP III. We prove that $\gamma_{\text{next}} \succ \gamma$. This follows from condition 2° on the germ σ_{next}, the definition of weak equivalence of germs (subsection C in §3.2), and Lemma 5.5_n a. In greater detail, the correction of the germ $h = \sigma_{\text{next}}^{-1} \circ \sigma$ is unbounded by definition. But

$$h = A^{1-n} g_{\text{next}} \circ g^{-1}.$$

By Lemma 5.5_n a, it follows from the unboundedness of the correction of h and the inequality $h \succ \text{id}$ that $h^{-1} \in G_{\text{slow}}^{n-1}$, $(h^{-1}\Pi, \infty) \subset (\Pi, \infty)$, and

$$A^{1-n} g^{-1}(\Pi, \infty) \supset A^{1-n} g_{\text{next}}^{-1}(\Pi, \infty).$$

This means that $\gamma_{\text{next}} \succ \gamma$.

We now proceed directly to the computation of $\widetilde{f}$. We have that

$$\begin{aligned} f &= (\exp \rho_{\text{next}})'/\widetilde{\psi}' = \exp \circ (\widetilde{I}_{\text{next}} + \lambda + o(1)) \circ \ln / \widetilde{\psi}' \\ &\succ C_1 \exp \circ (\widetilde{I}_{\text{next}} - I) \circ \ln \overset{\text{def}}{=} \widetilde{f}. \end{aligned}$$

The last inequality uses the formula $(**)$ and the upper estimate for $\widetilde{\psi}'$ obtained in Step III. The function $\widetilde{f}$ is monotone, since the integrand of the first integral exceeds the integrand of the second, because $\gamma_{\text{next}} \succ \gamma$. For the same reason,

$$\widetilde{I}_{\text{next}} = I = \int^{\xi} \left(\frac{2\gamma_{\text{next}}}{\pi - 2\gamma_{\text{next}}} - \frac{\gamma + \gamma_{1/2}}{\pi - (\gamma + \gamma_{1/2})} \right) \succ \left(\tfrac{1}{2} - \delta\right) \int^{\xi} \gamma$$

for any $\delta > 0$. This proves that $\widetilde{f} \to +\infty$ on $(\mathbf{R}^+, \infty)$.

We now proceed immediately to the proof of the preliminary estimate II. Note that $\operatorname{Re}\psi(\xi) \succ \operatorname{Re}\psi(c\xi)$ for $0 < c < 1$, since $\operatorname{Re}\psi$ is monotone on $(\mathbf{R}^+, \infty)$. Therefore, $|F| \prec \exp \circ (-c \operatorname{Re}\psi \cdot \widetilde{f}) \circ (c\xi)$. But it follows from formula $(*)$ for ψ and the definition of $\widetilde{f}$ that

$$\operatorname{Re}\psi \circ \widetilde{f} = (C_1 + o(1))\xi \cdot \exp \widetilde{I}_{\text{next}} \circ \ln \succ C' \exp \circ \rho_{\text{next}}.$$

Further, $\widetilde{I}_{\text{next}} \circ \ln(c\xi) = \widetilde{I}_{\text{next}} \circ (\ln \xi + \ln c) = \widetilde{I}_{\text{next}} \circ \ln \xi + o(1)$, since the integrand in the integral $\widetilde{I}_{\text{next}}$ tends to zero. Consequently,

$$(\operatorname{Re}\psi \cdot \widetilde{f}) \circ (c\xi) \succ C''' \exp \rho_{\text{next}}.$$

This proves the preliminary estimate II in the case when the germs σ and σ_{next} are both rapid.

E. Proof of the preliminary estimate II in the case when the germ σ is rapid and σ_{next} is sectorial. In this case

$$\sigma_{\text{next}} = A^{-n} g_{\text{next}} \circ \exp,$$

$$g_{\text{next}} \in G_{\text{slow}}^{n-1^{-}}, \quad \lim_{(\Pi, \infty)} (A^{1-n} g_{\text{next}})' = \mu^{-1}, \quad 1 < \mu < \infty.$$

Then

$$\exp \rho_{\text{next}} = A^{-n} g_{\text{next}}^{-1} = \zeta^{\mu + R \circ \ln(\zeta)}, \qquad R = o(1) \quad \text{in } (\Pi, \infty).$$

The preliminary estimate II is deduced from Lemma 4 according to the same plan as in subsection D:

$$(\exp \rho_{\text{next}})' = \xi^{\mu - 1 + o(1)} \quad \text{on } (\mathbf{R}^{+}, \infty).$$

Using the formula $(**)$ and the equivalence of the germs $|\psi'|$ and $\widetilde{\psi}'$ on $(\mathbf{R}^{+}, \infty)$ (proved in D), we get that

$$f = (\exp \rho_{\text{next}})' / \widetilde{\psi} = \xi^{\mu - 1 + o(1)} \exp \circ (-I \circ \ln).$$

For arbitrary $\delta > 0$ the function

$$\widetilde{f} = \xi^{\mu - 1 - \delta} \exp \circ (-I \circ \ln)$$

tends monotonically to ∞ and is asymptotically less than f. The monotonicity of $\widetilde{f}$ follows from the positivity of the logarithmic derivative:

$$(\ln \widetilde{f})' = \xi^{-1} [\mu - 1 - \delta - I' \circ \ln].$$

The quantity in brackets is asymptotically positive, since the integrand of the integral I tends to zero and $\mu > 1$.

Consequently, by the preceding formula for $\widetilde{f}$ and the formula $(*)$ for ψ,

$$\operatorname{Re} \psi \cdot \widetilde{f} = \xi^{\mu - \delta + o(1)} \succ \exp \left(1 - \frac{2\delta}{\mu}\right) \rho_{\text{next}}.$$

Next, there exists a $c' > 0$ such that

$$(\operatorname{Re} \psi \cdot \widetilde{f}) \circ (c\xi) \succ c' \xi^{\mu - \delta + o(1)} \succ c' \exp \left(1 - \frac{2\delta}{\mu}\right) \rho_{\text{next}}.$$

Since δ is arbitrary, this proves the preliminary estimate II in the case under consideration.

F. Properness of the mapping ψ. Let us verify the requirements 1°–3° of Definition 2 in §3.3. We have that

$$\psi(\zeta) = \zeta \exp \circ (I \circ \ln |\zeta| + o(1)).$$

1°. The image of the ray $(\mathbf{R}^{+}, \infty)$ under the action of ψ is a curve along which the argument tends to zero:

$$\arg \psi(\xi) = \operatorname{Im} o(1) = o(1).$$

Consequently, the germ of $\psi(\mathbf{R}^+, \infty)$ lies in the germ at infinity of an arbitrary sector containing $(\mathbf{R}^+, \infty)$ strictly in its interior.

2°. It follows from the formulas $(*)$ and $(**)$ that

$$\xi \operatorname{Re} \psi' / \operatorname{Re} \psi = 1 + o(1) \quad \text{on } (\mathbf{R}^+, \infty).$$

This verifies requirement 2°.

3°. There exists a $C > 0$ such that

$$\widetilde{\psi}(\xi) \stackrel{\text{def}}{=} \max_{\operatorname{Re} \zeta = \xi} \operatorname{Re} \psi(\zeta) \succ C\psi(\xi).$$

This follows from Proposition 1 in D.

In more detail, it follows from the definition of $\widetilde{\psi}$ that $\arg \psi' = 0$ at the point $\zeta = \zeta_{\max}(\xi)$, where $\widetilde{\psi}(\xi) = \operatorname{Re} \psi(\zeta)$. Consequently, by Proposition 1, $\zeta \in S_\alpha$, $\alpha = \pi/2 - \varepsilon$ for some $\varepsilon > 0$. Moreover,

$$\widetilde{\psi}(\xi) = \operatorname{Re}[\zeta(\exp \circ I \circ \ln |\zeta| + o(1))].$$

There is a $C > 0$ such that in the sector S_α

$$|\zeta| < C \operatorname{Re} \zeta = C\xi, \qquad \ln |\zeta| < \xi + \ln C,$$
$$I \circ \ln |\zeta| = I \circ \ln \xi + o(1).$$

Consequently,

$$\widetilde{\psi}(\xi) \prec 2C\psi(\xi).$$

This proves requirement 3°.

It remains to prove Proposition 1. This is a consequence of Lemma 5.7_n and a simple fact in the theory of harmonic functions.

G. Proof of Proposition 1 in subsection D. We list the requirements on ψ that will be used to derive Proposition 1:

$$\psi = \psi_0 + O(\psi_0^{-1}), \qquad \psi_0 = A^{-1}(\mathrm{id} + \varphi);$$

the function $\mathrm{id} + \varphi$ rectifies the half-strip Π_1 with the boundary functions $\gamma_+ = \gamma_{1/2}$ and $\gamma_- = \gamma_1 = \gamma$. Let $\mathrm{id} - \widetilde{\varphi}\colon \Pi \to (\mathrm{id} + \varphi)\Pi = \Pi_1$ be the mapping inverse to $\mathrm{id} + \varphi$. This implies that

$$\gamma_{1/2}\left(\xi - \operatorname{Re} \widetilde{\varphi}\left(\xi + i\frac{\pi}{2}\right)\right) = \operatorname{Im} \widetilde{\varphi}\left(\xi + i\frac{\pi}{2}\right),$$
$$\gamma\left(\xi - \operatorname{Re} \widetilde{\varphi}\left(\xi - i\frac{\pi}{2}\right)\right) = -\operatorname{Im} \widetilde{\varphi}\left(\xi - i\frac{\pi}{2}\right).$$

The properties of the functions $\gamma_{1/2}$ and γ are listed in Lemma 5.7_nd, subsection A. Moreover, $\operatorname{Re} \varphi \to \infty$ and $\varphi' \to 0$ in (Π_1, ∞). These and only these properties are used in the proof of Proposition 1.

By the formula for ψ, in which the second term is holomorphic in a neighborhood of infinity,

$$\psi' = (1 + O(\psi_0^{-2})) \cdot \psi_0'.$$

PROPOSITION 1 BIS. *For sufficiently small* ε *and* C *the function* $|(\arg\psi_0')\circ\exp|$ *exceeds* $C\xi^{-2}$ *in the part* $\mathcal{U}$ *of the half-strip* Π_1 *outside the half-strip* $|\eta|<\pi/2-\varepsilon$.

We derive Proposition 1 from Proposition 1 bis. In Π_1 the function $\psi_0^{-2}\circ\exp$ can be estimated from above by a decreasing exponential, since it follows from $(*)$ that $|\psi_0|\succ|\zeta|$ in Ω_σ. Consequently, the assertion of Proposition 1 bis is valid for the functions $(\arg\psi_0')\circ\exp$ and $(\arg\psi')\circ\exp$ simultaneously. The desired derivation is complete.

PROOF OF PROPOSITION 1 BIS. 1°. By the formula $(***)$ for ψ_0',

$$(\arg\psi_0')\circ\exp=\operatorname{Im}\varphi+\arg(1+\varphi').$$

The correction of the mapping $\mathrm{id}-\widetilde{\varphi}=(\mathrm{id}+\varphi)^{-1}$ satisfies the equation $\varphi=\widetilde{\varphi}\circ(\mathrm{id}+\varphi)$. In the half-strip $\Pi_1=(\mathrm{id}+\varphi)\Pi$ the function $\arg\psi_0'\circ\exp$ takes the same values that the function

$$u\overset{\mathrm{def}}{=}\operatorname{Im}\widetilde{\varphi}-\arg(1-\widetilde{\varphi}')$$

takes in the half-strip Π. It suffices to prove that $|u|$ exceeds $C\xi^{-2}$ in the "boundary" half-strips $|\eta|\in(\frac{\pi}{2}-\varepsilon,\frac{\pi}{2})$ for some $\varepsilon>0$. This will imply Proposition 1 bis, since $\operatorname{Im}\widetilde{\varphi}\to0$ in (Π,∞).

2°. We prove that on the boundary of Π the function u takes the values $\gamma_{1/2}\cdot(1+o(1))$ and $-\gamma(1+o(1))$ on the upper and lower rays, respectively. Consider the upper ray. By the definition of $\gamma_{1/2}$,

$$\gamma_{1/2}'\circ(\xi-\operatorname{Re}\widetilde{\varphi}\circ(\xi+i\tfrac{\pi}{2}))=\tan\arg(1-\widetilde{\varphi}'(\xi+i\tfrac{\pi}{2})).$$

The derivative $\widetilde{\varphi}'$ exists on $\partial\Pi$, because the mapping $\mathrm{id}-\widetilde{\varphi}$ extends analytically to the ε-neighborhood of the boundary of Π. We note that if the logarithmic derivative of the function f, which is of constant sign and tends to zero, exceeds $-C\xi^{-1}$, then $|f|$ decreases no more rapidly than ξ^{-C} and, moveover, $f\circ(\xi\cdot(1+o(1))=(1+o(1))\cdot f(\xi)$. Indeed, for $\xi\ge\xi_0$

$$|f(\xi)|=|f(\xi_0)|\exp\int_{\xi_0}^{\xi}\frac{f'(\theta)}{f(\theta)}\,d\theta\succ|f(\xi_0)|\left(\frac{\xi_0}{\xi}\right)^C$$

By the Warschawski formula, $\operatorname{Re}\widetilde{\varphi}=o(1)\xi$. Consequently, by Lemma 5.7_nd,

$$\operatorname{Im}\widetilde{\varphi}\circ\left(\xi+i\frac{\pi}{2}\right)=\gamma_{1/2}(\xi)\cdot(1+o(1)),$$

$$\arg\left(1-\widetilde{\varphi}'\circ\left(\xi+i\frac{\pi}{2}\right)\right)=\gamma_{1/2}'(\xi)\cdot(1+o(1)).$$

But $\gamma_{1/2}'/\gamma_{1/2}\to0$. Consequently,

$$\arg\left(1-\widetilde{\varphi}'\circ\left(\xi+i\frac{\pi}{2}\right)\right)=o\left(\operatorname{Im}\widetilde{\varphi}\circ\left(\xi+i\frac{\pi}{2}\right)\right).$$

Thus,

$$u\left(\xi + i\frac{\pi}{2}\right) = \gamma_{1/2}(\xi)(1 + o(1)).$$

Similarly,

$$u\left(\xi - i\frac{\pi}{2}\right) = -\gamma(\xi)(1 + o(1)).$$

3°. Proposition 1 bis now follows from the next lemma.

LEMMA OF LANDIS. *Let u be a bounded harmonic function on the half-strip Π, and let the function γ on $(\mathbf{R}^+, \infty)$ have the properties that $\gamma > 0$, $\gamma \to 0$, $\gamma'/\gamma \to 0$, and $|\xi\gamma'/\gamma| \prec 2$. Let $u(\xi + i\frac{\pi}{2}) = \gamma(\xi)(\frac{1}{2} + o(1))$ and $u(\xi - i\frac{\pi}{2}) = -\gamma(\xi)(1 + o(1))$.*

Then there exists a constant $\beta < \pi/2$ such that the set of zeros of u near infinity lies in the half-strip $|\eta| < \beta$ and the function itself in the completion to this half-strip exceeds by modulus $C\xi^{-2}$ for some $C > 0$.

REMARK. It can be proved that this set approaches the ray $\eta = \pi/6$ asymptotically as $\xi \to \infty$.

PROOF. For each $\xi = \xi_0$ consider the rectangle Π_{ξ_0}: $\xi_0 \le \operatorname{Re}\zeta \le 2\xi_0$, $|\eta| \le \pi/2$. In it we construct a harmonic function v_{ξ_0} that will be less than u and will exceed $C\xi_0^{-2}$ in the rectangle $\Pi_{\xi_0+} : \eta \in [\pi/2 - \varepsilon, \pi/2]$, $4\xi_0/3 \le \xi \le 5\xi_0/3$ for any sufficiently large ξ_0 ($\xi_0 > \alpha$) and some $C > 0$, $\varepsilon > 0$ not depending on ξ_0. This will imply the desired esimate for u in the half-strip $\eta \in [\pi/2 - \varepsilon, \pi/2]$, $\xi \ge \alpha$. It is possible similarly to construct a function $w_{\xi_0} \ge u$ estimating $|u|$ from below for $\eta \in [-\pi/2, -\pi/2 + \varepsilon]$, $\xi \ge \alpha$. This proves the lemma. Of the two analogous constructions we describe only the first. Suppose that

$$a = \min_{\partial\Pi^+_{\xi_0}} u, \qquad b = \min_{\partial\Pi^-_{\xi_0}} u,$$

and $\partial\Pi^+_{\xi_0}$ and $\partial\Pi^-_{\xi_0}$ are the upper and lower sides of the rectangle Π_{ξ_0}. Let

$$v_{\xi_0} = v, \quad v|_{\partial\Pi^+_{\xi_0}} = a, \quad v|_{\partial\Pi^-_{\xi_0}} = b, \quad \Delta v = 0 \quad \text{in } \Pi_{\xi_0}.$$

On the vertical sides we set v equal to a continuous function f:

$$|f| \le C = \sup_{\Pi} |u|, \quad f(\tfrac{\pi}{2}) = a, \quad f(-\tfrac{\pi}{2}) = b,$$

and $f \le u|_{\{\operatorname{Re}\zeta = \zeta_0\} \cup \{\operatorname{Re}\zeta = 2\xi_0\}}$. Then in the domain $\widetilde{\Pi}_{\xi_0}$: $4\xi_0/3 \le \operatorname{Re}\zeta \le 5\xi_0/3$, $|\eta| \le \pi/2$ the function $v \le u$ will be exponentially close to the linear function $l = \alpha\eta + \beta$, $l(\pi/2) = a$, $l(-\pi/2) = b$. Indeed, the difference $w = v - l$ is a harmonic function:

$$|w| < 2C, \qquad w|_{\partial\Pi^+_{\xi_0}} = w|_{\partial\Pi^+_{\xi_0}} = 0.$$

Solving the Dirichlet problem for w by the method of separation of variables, we get that

$$w = \sum \frac{a_n \sin n\eta \cosh n\left(\xi - \frac{3}{2}\xi_0\right)}{\cosh(n\xi_0/2)}.$$

Here a_n are the Fourier coefficients of the function $f - l$ on $[-\pi/2, \pi/2]$. In the domain $\widetilde{\Pi}_{\xi_0}$ this function is less than $C' \exp(-\xi_0/6)$ for some $C' > 0$ independent of ξ_0, since $|f| \leq C$, where C is independent of ξ_0. As indicated above, it follows from the condition $|\xi\gamma'/\gamma| \prec 2$ that $|\gamma| \succ C\xi^{-2}$ for some $C > 0$. Therefore, $|a| > C'\xi_0^{-2}$ and $|b| > C''\xi_0^{-2}$. Consequently, in the rectangle Π_{ξ_0+} for some $\delta > 0$

$$l > \delta\xi_0^{-2}; \quad \delta\xi_0^{-2}/2 > C' \exp(-\xi_0/6)$$

for sufficiently large ξ_0. In this rectangle $v > \frac{\delta}{2}\xi_0^{-2}$ ►

Proposition 1 in subsection D and Lemma 5 in subsection A have been proved simultaneously.

* * *

In this section we have taken the next step in the proof of the Phragmén-Lindelöf theorem. Let F and $\boldsymbol{\sigma}$ be the same as in the condition of the theorem (see also the end of §3.7). Then for the first sectorial germ σ in the tuple $\boldsymbol{\sigma}$ and for any $\delta > 0$

$$|F| \prec \exp(-C \exp(1 - \delta)\rho), \qquad \rho = \sigma^{-1},$$

on $(\mathbf{R}^+, \infty)$.

§3.10. Preliminary estimates in domains corresponding to sectorial and slow germs

In this section the preliminary estimate II is proved in the case when the germs σ and σ_{next} are both not rapid.

A. Construction of domains corresponding to germs that are not rapid. The construction presented here takes care of all nonrapid germs in the sequence

$$\boldsymbol{\sigma} = (\sigma_1, \dots, \sigma_N), \qquad \sigma_1 \succ \cdots \succ \sigma_N, \ \sigma_j \in \mathscr{D}^m, \qquad (*)$$
$$m = n \quad \text{or} \quad m = n - 1.$$

Let $\Pi_{a,b}$ be the half-strip

$$\Pi_{a,b} = \{\xi \geq \xi_0, \ \eta \in [-a\pi, b\pi]\},$$
$$0 < a < \tfrac{1}{2}, \ \tfrac{1}{2} < b < 1, \ a + b > 1.$$

For all sectorial germs of the sequence $(*)$ of the form $\sigma = \exp \circ A^{1-m} g$, $\mu_m(g) \in (0, 1)$, we take the maximal value of $\mu_m(g)$, denote it by $\mu_{\max}$, and require that

$$\mu_{\max} b < \tfrac{1}{2}.$$

Obviously, for sufficiently large ξ_0 the half-strip $\Pi_{a,b}$ belongs to the main domain of an arbitrary modified standard partition:

$$\Pi_{a,b} \subset \Pi_{\text{main}}.$$

For any nonrapid germ σ of the sequence $(*)$ let

$$\Omega_\sigma = \sigma\Pi_{a,b}.$$

For all germs in $(*)$ that increase more slowly than sectorial germs

$$(\sigma\Pi_{\text{main}}, \infty) \subset (\Omega_{\text{st}}, \infty),$$

by Lemmas 5.3_n and 5.6_n.

Therefore, for them

$$(\Omega_\sigma, \infty) \subset (\sigma\Pi_{\text{main}} \cap \Omega_{\text{st}}, \infty).$$

For the sectorial germs of the sequence this inclusion is ensured by the inequality $\mu_{\max} b < 1/2$. Accordingly,

$$(\Omega_\sigma, \infty) \subset (\sigma\Pi_{\text{main}} \cap \Omega_{\text{st}}, \infty)$$

for all slow and sectorial germs σ in the sequence $(*)$.

B. The conformal mapping ψ: $\Omega_\sigma \to \mathbf{C}^+$. This mapping is very simple to construct. We first construct a mapping ψ_0: $\Omega \to \mathbf{C}^+ \backslash K$ where K is some disk. Let a_0: $\zeta \mapsto \alpha\zeta + \beta i$, a conformal mapping of the half-strip $\Pi_{a,b}$ onto the half-strip Π: $\xi \geq \xi_1$, $|\eta| \leq \pi/2$, where ξ_1 is chosen so that $\operatorname{Re} a_0(\xi_0) = \xi_1$;

$$\alpha = \frac{1}{a+b} < 1, \qquad \beta = \frac{\pi(a-b)}{2(a+b)} \in \left(0, -\frac{\pi}{2}\right).$$

The mapping $\psi_0 = \exp \circ a_0 \circ \rho$, $\rho = \sigma^{-1}$, acts as follows:

$$\psi_0\colon \Omega_\sigma \xrightarrow{\rho} \Pi_{a,b} \xrightarrow{a_0} \Pi \xrightarrow{\exp} \mathbf{C}^+ \backslash K.$$

More briefly,

$$\psi_0 = \omega \exp \alpha\rho, \qquad \omega = \exp i\beta, \ \alpha < 1, \ -\tfrac{\pi}{2} < \beta < 0. \tag{$**$}$$

The conformal mapping ψ: $\Omega_\sigma \to \mathbf{C}^+$ can be taken in the form

$$\psi = \psi_0 + O(1/\psi_0). \tag{$***$}$$

(Extendibility of ψ to the ε-neighborhood of the part of Ω_σ outside a compact set follows from Lemma 5.3_n, according to which the germ ρ is defined in the germ of the half-plane $(\mathbf{C}^+, \infty)$.)

C. Effectiveness of the preliminary estimate. We proceed to the proof of the preliminary estimate II. Let us begin with a proof that the preceding preliminary estimate is effective. In other words, we must prove that

$$\exp \rho / \operatorname{Re} \psi \to \infty \quad \text{on } (\mathbf{R}^+, \infty).$$

This is an immediate consequence of the definition of ψ: on $(\mathbf{R}^+, \infty)$

$$\exp \rho / \operatorname{Re} \psi = (\cos \beta + o(1))^{-1} \exp(1-\alpha)\rho \to \infty,$$

since $\alpha < 1$ and $\rho \to \infty$ on $(\mathbf{R}^+, \infty)$.

D. Proof that ψ is proper. We verify the requirements of Definition 2 in §3.3

1°. On $(\mathbf{R}^+, \infty)$,

$$\psi(\xi) = \omega \exp \alpha\rho(\xi) + o(1), \qquad \rho(\xi) \in \mathbf{R}^+.$$

Consequently, the curve $\psi(\mathbf{R}^+, \infty)$ tends asymptotically to the ray $\arg\zeta = \beta$, $|\beta| < \pi/2$.

2°. The logarithmic derivative of $\operatorname{Re}\psi$ on $(\mathbf{R}^+, \infty)$ is equal to $\alpha\rho' + o(1)$. The requirement 2° of Definition 2 in §3.3 will be proved if we prove that the limit $\lim_{(\mathbf{R}^+, \infty)} \xi\rho'$ is either more than 1 or equal to plus infinity.

PROPOSITION 1. *Suppose that* $\sigma \in \mathscr{D}_0^n \cup \mathscr{D}_0^{n-1} \cup \mathscr{D}_*^{n-1} \circ \mathscr{L}^{n-1}$ σ *is nonrapid germ,* $\rho = \sigma^{-1}$. *Then* $\rho'\xi \to \infty$ *on* $(\mathbf{R}^+, \infty)$ *if the germ* σ *is slow, and* $\rho'\xi \to \mu \in (1, \infty)$ *if* σ *is sectorial.*

PROOF. We consider the two basic cases: $\sigma \in \mathscr{D}_0^{n-1}$ and $\sigma \in \mathscr{D}_*^{n-1} \circ \mathscr{L}^{n-1}$. The case $\sigma \in \mathscr{D}_0^n$ is investigated like the first case with $n-1$ replaced by n.

Consider first the second case:

$$\rho = h \circ A^{1-n} g, \qquad h \in \mathscr{L}^{n-1}, \; g \in G_{\text{slow}}^{n-2^-} \cup G_{\text{rap}}^{n-2}.$$

Let $\widetilde{\sigma} = A^{1-n} g^{-1}$ and $\widetilde{\rho} = \widetilde{\sigma}^{-1}$. By Lemma 5.5_{n-1}c, $\widetilde{\sigma}'/\widetilde{\sigma} \to 0$ on $(\mathbf{R}^+, \infty)$. But

$$\widetilde{\sigma}' \circ \widetilde{\rho}/\widetilde{\sigma} \circ \widetilde{\rho} = 1/\widetilde{\rho}' \cdot \xi \to 0 \quad \text{on } (\mathbf{R}^+, \infty).$$

Consequently, $\widetilde{\rho}' \cdot \xi \to \infty$ on $(\mathbf{R}^+, \infty)$.

Further, by Lemma 5.6_n, $h' = (1 + o(1))$ for arbitrary $h \in \mathscr{L}^{n-1}$. Therefore,

$$\rho' \cdot \xi = h' \circ \widetilde{\rho} \cdot \widetilde{\rho}' \cdot \xi \to \infty \quad \text{on } (\mathbf{R}^+, \infty).$$

We now consider the first case: the germ σ is sectorial,

$$\sigma^{-1} = \rho = A^{2-n} g \circ \ln, \qquad g \in G_{\text{slow}}^{n-2^+}.$$

Then $\lim_{(\mathbf{R}^+, \infty)} \rho' \cdot \xi = \lim_{(\mathbf{R}^+, \infty)} (A^{2-n} g)'$. By Lemma 5.3_nb, the limit of the right-hand side always exists. It is not less than 1, since $g \succ \mathrm{id}$ and larger than 1 indeed, as σ is nonrapid. Its finiteness or nonfiniteness distinguishes rapid and sectorial germs from slow germs of class $\mathscr{D}_0^n$, by definition. ►

This verifies requirement 2° of Definition 2 in §3.3.

3°. We estimate the degree of convexity of the mapping ψ.

Suppose that $\max_{\operatorname{Re}\zeta = \xi} \operatorname{Re}\psi(\zeta) = \widetilde{\psi}(\xi)$ is attained at the point ζ. Then

$$\widetilde{\psi}(\xi) \leq \exp \circ \alpha \operatorname{Re}\rho(\zeta) + o(1).$$

Let Γ denote the segment $\Omega_\sigma \cap \{\operatorname{Re}\zeta = \xi\}$.

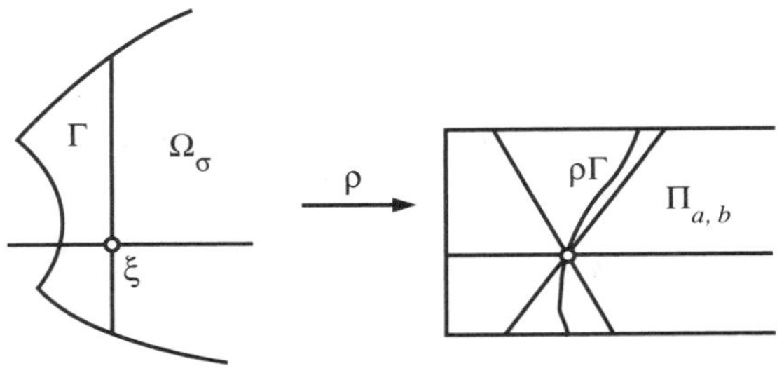

FIGURE 19

PROPOSITION 2_m. $\arg\sigma' \to 0$ *in* $(\Pi^\forall, \infty)$ *for an arbitrary germ* $\sigma \in \mathscr{D}_*^{m-1} \circ \mathscr{L}^{m-1}$.

For $m = 1$ this proposition deals with real dilations, hence is trivial. For $m \leq n$ it is based on the properties of the group G^k, $k \leq n-1$, and appears in the induction hypothesis. The induction step is made in Chapter V, where Proposition 2 of §5.8B (which is stronger than Proposition 2_{n+1}) is proved.

If σ is a sectorial germ, then $|\arg\sigma'|$ is separated from $\pi/2$ in the germ of the half-strip $(\Pi_{a,b}, \infty)$. This follows from the choice of a and b, see A.

This proves that the curve $\rho\Gamma$ lies in the cone obtained from the cone $|\arg\zeta - \pi/2| < \gamma < \pi/2$ by translating the vertex to the point $\rho(\xi)$. Consequently,

$$\operatorname{Re}\rho(\zeta) - \rho(\xi) < (a+b)\pi\tan\gamma, \qquad a+b < \tfrac{3}{2}$$

(see Figure 19). From this, $\widetilde{\psi}(\xi) < C\psi(\xi)$ for $C = \exp\frac{3}{2}\alpha\pi\tan\gamma$. It is proved that ψ is proper.

E. Investigation of the estimate given by Lemma 4: the function $\widetilde{f}_\delta$. It follows from the formulas (**) and (***) for ψ that on $(\mathbf{R}^+, \infty)$

$$\begin{aligned} f &= (\exp\rho_{\text{next}})'/\widetilde{\psi}' \\ &= (\exp(\rho_{\text{next}} - \alpha\rho)) \cdot (\rho'_{\text{next}}/\rho'(C + o(1)))(\operatorname{Re}\psi'(\xi)/\psi'(\zeta_{\max}(\xi))). \end{aligned}$$

But $\zeta = \zeta_{\max}(\xi)$ is the point where $\operatorname{Re}\zeta = \xi$, $\widetilde{\psi}(\xi) = \operatorname{Re}\psi(\zeta)$ and lies in the sector S_γ: $|\arg\zeta| < \gamma$ for some $\gamma < \pi/2$. By Koebe's distortion theorem, the last quotient in the formula for f is bounded below by a constant $C(\gamma) > 0$. Note that $\psi' = (\omega\alpha + o(1))(\exp\alpha\rho)\rho' \to \infty$ on $(\mathbf{R}^+, \infty)$ by Proposition 1 and the admissibility of the germ σ. This now proves that the germ $\widetilde{\psi}'$ is bounded away from zero by a positive constant, as required in Lemma 4. We prove that for sufficiently small C and δ the germ at infinity of the function $\widetilde{f}_\delta = C\exp((1-\delta)\rho_{\max} - \alpha\rho)$ tends monotonically to plus infinity and satisfies the inequality ([11]) $\widetilde{f}_\delta \prec \widetilde{f}_{\delta/2} \prec f$. For this it suffices to prove that $\rho' \preccurlyeq \rho'_{\text{next}}(1 + o(1))$.

([11]) The function $f_{\delta/2}$ is defined in the formulation of Lemma 4 in §3.6.

For this, in turn, we compose a table whose cells contain the composition quotients $\tilde{\sigma} = \rho_{\text{next}} \circ \sigma$, $\rho = \sigma^{-1}$, $\rho_{\text{next}} = \sigma_{\text{next}}^{-1}$; the germ σ is the same along the rows, and the germ σ_{next} is the same along the columns; the lower left corner is empty, since $\sigma_{\text{next}} \prec \sigma$. The table is written for germs of class $\mathscr{D}_1^{n-1}$; the upper left-hand cell with $1-n$ replaced by $-n$ corresponds to germs of class $\mathscr{D}_0^n$.

TABLE 1

σ \ σ_{next}	$\exp \circ A^{2-n} g_2$ $g_2 \in G_{\text{slow}}^{n-2^-}$	$A^{1-n} g_2 \circ h_2$, $h_2 \in \mathscr{L}^{n-1}$ $g_2 \in G_{\text{slow}}^{n-2^+} \cup G_{\text{rap}}^{n-2}$
$\exp \circ A^{2-n} g_1$ $g_1 \in G_{\text{slow}}^{n-2^-}$	① $A^{2-n} g$ $g = g_2^{-1} \circ g_1$ $g \succ \text{id}$	② $h_2^{-1} \circ \exp \circ A^{2-n} g$ $g = g_2^{-1} \circ g_1$
$A^{1-n} g_2 \circ h_1$ $h_1 \in \mathscr{L}^{n-1}$ $g_1 \in G_{\text{slow}}^{n-2^+}$ $\cup G_{\text{rap}}^{n-2}$	$\varnothing$	③ $h_2^{-1} \circ A^{1-n} g \circ h_1$ $g = g_2^{-1} \circ g_1$

PROPOSITION 3. *Suppose that* $\tilde{\sigma} = \rho_{\text{next}} \circ \sigma$ *is the same as in Table* 1. *Then* $\lim_{(\mathbf{R}^+, \infty)} \tilde{\sigma}'$ *exists, perhaps infinite, and is not less than* 1.

PROOF. The cells of the table are numbered: we consider the corresponding germs $\tilde{\sigma}'$: ① $\tilde{\sigma}' = (A^{2-n} g)'$. The limit $\mu_{n-1}(g) = \lim_{(\mathbf{R}^+, \infty)} \tilde{\sigma}'$ exists by Lemma 5.3_{n-1}b. This limit is not less than 1, because $g \succ \text{id}$.

②. Let $\rho_* = A^{2-n} g$, $\sigma_* = \rho_*^{-1}$. Then

$$\tilde{\sigma}' = h_2^{-1} \circ \exp \rho_* \cdot \exp \rho_* \cdot \rho_*' = (h_2^{-1})' \circ \exp \rho_* \cdot \frac{\exp}{\sigma_*'} \circ \rho_*.$$

But by Lemma 5.6_n, $h' = (1 + o(1))$ for $h \in \mathscr{L}^{n-1}$, and by Lemma 5.5_{n-1}, $(A^{2-n} g)' \prec \exp \xi^\varepsilon$ for arbitrary $\varepsilon > 0$ when $g \in G^{n-2}$. Therefore, $\tilde{\sigma}' \to \infty$ in Case ②.

③ $\tilde{\sigma}' = (h_2^{-1} \circ A^{1-n} g \circ h_1)'$. Considering the relation $h_j' = 1 + o(1)$, we get that

$$\lim_{(\mathbf{R}^+, \infty)} \tilde{\sigma}' = \lim_{(\mathbf{R}^+, \infty)} (A^{1-n} g)'.$$

Since $g \in G^{n-2} \subset G^{n-1}$, we get by Lemma 5.3_nc that the limit on the right-hand side exists. It cannot be less than 1, because $\widetilde{\sigma} \succ \mathrm{id}$.

PROPOSITION 4. *Suppose that* ρ *and* ρ_{next} *are the same as in Table* 1. *Then* $\rho'_{\text{next}} \succ \rho'(1+o(1))$ *on* $(\mathbf{R}^+, \infty)$.

PROOF.

$$\rho_{\text{next}} = \widetilde{\sigma} \circ \rho; \qquad \rho'_{\text{next}} = \widetilde{\sigma}' \circ \rho \cdot \rho' \succ (1+o(1))\rho'.$$

The last inequality follows from Proposition 3.

Proposition 4 proves that $\widetilde{f}_\delta$ is monotone and that $\widetilde{f}_\delta \prec f$.

F. Investigation of the estimate given by Lemma 4: proof of the preliminary estimate II. Suppose that $\sigma = \sigma_k$ and $\sigma_{\text{next}} = \sigma_m$. By Proposition 1 in §3.2, it can be assumed that the cochain F in the condition of the preliminary estimate II is defined in the domain Ω_δ and has there the type $(\sigma_m, \dots, \sigma_N)$. The first condition in Lemma 4 on ψ and F (the effectiveness of the preliminary estimate in Ω_σ) was verified in subsection B, and the second condition in subsections C and D (the properness of ψ and the lower estimate for $\widetilde{\psi}'$). By Lemma 4, on $(\mathbf{R}^+, \infty)$

$$|F| \prec \exp(-C \operatorname{Re}\psi \cdot \widetilde{f}_\delta) \circ b(\xi),$$

$b(\xi) = b_\psi(\xi) - \varepsilon$. Further,

$$\widetilde{f}_\delta = c\exp((1-\delta)\rho_{\text{next}} - \alpha\rho),$$
$$\operatorname{Re}\psi = \cos\beta \exp \alpha\rho + o(1).$$

Consequently,

$$|F| \prec \exp(-C' \exp(1-\delta)\rho_{\text{next}}) \circ b(\xi). \qquad (***)$$

Here $b(\xi) = b_\psi(\xi) - \varepsilon$. By Properties III and IV of proper mappings,

$$b(\xi) \succ c_1\xi \text{ always, and } b(\xi) = (1+o(1))\xi \qquad ({}^{\,*}_{*\,*})$$

if $(\xi \operatorname{Re}\psi' / \operatorname{Re}\psi) \to \infty$. By the previous formula for $\operatorname{Re}\psi$ and by Proposition 1 in §3.10B, the last requirement always holds when the germ σ corresponding to the mapping ψ is slow.

PROPOSITION 5. *Suppose that* f *is a monotonically increasing function* $(\mathbf{R}^+, \infty) \to (\mathbf{R}^+, \infty)$, *and* $\xi f'/f \to \mu \in (0, 1)$. *Then for any* $c \in (0, 1)$ *there exists a* $c' > 0$ *such that* $f \circ (c\xi) \succ c' f(\xi)$. *If* $\xi f'/f \to \mu = 0$, *then* $f \circ (c\xi) = (1+o(1))f(\xi)$.

PROOF. The proof is the same as in Proposition 2 of §3.3. ▶

COROLLARY. *If* $\mu > 0$ *in Proposition* 5, *then* $f \circ ((1+o(1))\xi) = (1+o(1))f(\xi)$.

PROPOSITION 6. *Suppose that* $\sigma \in \mathcal{D}_0^n \cup \mathcal{D}_0^{n-1} \cup \mathcal{D}_*^{n-1} \circ \mathcal{L}^{n-1}$, $\rho = \sigma^{-1}$. *Then* $\xi\rho'/\rho \to 0$ *on* $(\mathbf{R}^+, \infty)$ *with the exception of the single case when the germ* ρ *is almost a power germ*: $\rho = h \circ A^{1-n} g$, $\mu_{n-1}(g) = \mu \in (0, 1)$.

PROOF. Let $f = A^{2-n}g$, $g \in G^{n-2}$.

CASE 1: $\sigma \in \mathscr{D}_*^{n-1} \circ \mathscr{L}^{n-1}$. Then $\rho = h \circ A^{-1}f$, $h \in \mathscr{L}^{n-1}$, and $g \in G_{\text{slow}}^{n-2^-} \cup G_{\text{rap}}^{n-2}$.

By Lemma 5.6_n, $h' = 1 + o(1)$, and $h(\xi) = (1 + o(1))\xi$. Consequently,

$$\xi\rho'/\rho = (1 + o(1)) \cdot A^{-1}f \cdot f' \circ \ln/(1 + o(1))A^{-1}f = (1 + o(1))f' \circ \ln .$$

By Lemma 5.3_{n-1}b the limit $\lim_{(\mathbf{R}^+, \infty)} f' = \mu_{n-1}(g)$ exists, and $\mu_{n-1}(g) = 0$ because the germ ρ is not almost a power germ. This proves the proposition in Case 1.

CASE 2: $\sigma \in \mathscr{D}_0^{n-1}$, $\rho = f \circ \ln$, $g \in G_{\text{slow}}^{n-2^+}$. Then

$$\xi\rho'/\xi = (f'/f) \circ \ln \to 0,$$

by Lemma 5.5_nc.

CASE 3: $\sigma \in \mathscr{D}_0^n$ is analyzed like Case 2, except that $n - 1$ is replaced by n. ▶

* * *

It follows from Propositions 5 and 6 and relations $\binom{*}{*\,*}$ that for arbitrary $\delta > 0$

$$\rho_{\text{next}} \circ b(\xi) \succ (1 - \delta)\rho_{\text{next}}(\xi) \qquad \binom{*\,*}{*}$$

with the exception of the case when the germ ρ_{next} is almost a power germ. The inequality $\binom{*\,*}{*}$ holds also for the "almost power" germ ρ_{next} in the case when σ is a slow germ: in this case $b_\psi(\xi) = (1 + o(1))\xi$. Finally, if σ is a rapid or a sectorial germ and σ_{next} is an almost power germ (in the sense that ρ_{next} is an almost power germ), then our arguments do not work.

However, it is proved in §3.2C that we can assume without loss of generality that the sequence $(*)$ contains a germ $\sigma_* = \exp \circ A^{1-m}g$, $\mu_m(g) = 0$, $m = n$, or $m = n - 1$. Therefore, the case described above is excluded.

The preliminary estimate II is proved for such a sequence; with it, the Phragmén-Lindelöf theorem is proved.

CHAPTER IV

Superexact Asymptotic Series

In this chapter we finish the proof of the shift lemmas (see §1.10) begun in the second chapter (modulo the lemmas on the properties of germs of classes $\mathscr{D}_0^n$ and $\mathscr{D}_1^{n-1}$, which are proved in Chapter V). This implies the additive and multiplicative decomposition theorems. At the end of the chapter we prove the theorem on a lower estimate for the partial sums of an STAR-n. Together with the additive decomposition theorem and the Phragmén-Lindelöf Theorem, Chapter III, it proves the finiteness theorem (see §1.9). The end of this chapter plays a double role. It is the conclusion of the investigations of the first four chapters. On the other hand, it contains general considerations about ordered holomorphic functions on a complex domain and is basic for all the constructions in Chapter V.

§4.1. The induction hypothesis

In this chapter we prove various properties of functional cochains of class m and of mappings in the group G^m for $m = n$. The properties are proved by induction on m: it is assumed that they are valid for $m < n$ and it is proved that they are valid also for $m = n$. In the proof of the next assertion here we use the validity for $m < n$ not only of this assertion, but also of a number of other assertions that are then also proved for $m = n$. In the present subsection we formulate theorems and lemmas appearing in the induction hypothesis, along with theorems to be proved in the induction step. The basic logical interrelations between these assertions are indicated in Figure 20. Each of them is endowed with a name and a notation, used to denote it in the figure. An arrow from an assertion A to an assertion B means that A is employed in the proof of B.

Assertions appearing in the induction hypothesis for $m < n$; n should be replaced by m in the assertions cited below from §§1.8 and 1.10.

1. First Shift Lemma ($\mathrm{SL}\,1_m$; see §1.10).

2. Multiplication Lemma (ML_m). (a) $\mathrm{ML}_{m-1,m}$. *Let* $k \leq m-1$ *and* $g \in G^{m-1}$. *Then* $\mathscr{F}^k \cdot \mathscr{F}_g^m \subset \mathscr{F}_g^m$.

(b) $\mathrm{ML}_{m,m}$. *Let* $f, g \in G^{m-1}$, $f \prec g$. *Then* $\mathscr{F}_f^m \cdot \mathscr{F}_g^m \subset \mathscr{F}_g^m$.

The combination of lemmas (a) and (b) constitutes ML_m.

For $m = n$ we prove the lemmas separately, but generally we give references to their combination.

3. LEMMA ON ANALYTIC FUNCTIONS OF DECREASING COCHAINS (LA_m). *Let* $\varphi\colon (\mathbf{C}, 0) \to (\mathbf{C}, 0)$ *be a holomorphic function. Then* $\varphi \circ \mathscr{F}^m_{+g} \subset \mathscr{F}^m_{+g}$ *for arbitrary* $g \in G^{m-1}$.

COROLLARY 1. $\exp \mathscr{F}^m_{+g} - 1 \subset \mathscr{F}^m_{+g}$.

4. DIFFERENTIATION LEMMA $\mathrm{DL}\,1_m$. *Regular functional cochains of class* $\mathscr{D}^m_0$ *and* $\mathscr{D}^{m-1}_1$ *form a differential algebra.*

$\mathrm{DL}\,2_m$. *For an arbitrary fixed* $g \in G^{m-1}$ *the set* $\mathscr{F}^m_g$ *forms a differential algebra.*

5. SECOND SHIFT LEMMA ($\mathrm{SL}\,2_m$; SEE §1.10).
6. THIRD SHIFT LEMMA ($\mathrm{SL}\,3_m$; SEE §1.10).
7. FOURTH SHIFT LEMMA ($\mathrm{SL}\,4_m$; SEE §1.10).

8. FIFTH SHIFT LEMMA $(\mathrm{SL}\,5_m)$. *Let* $m \le n-1$. *Then* $\mathscr{F}^m_1 \circ J^{m-1} \subset \mathscr{F}^m_1$.

9. LEMMAS 5.2_m–5.8_m FOR $m \le n$. *These were formulated in Chapters* II *and* III *and proved there for* $n = 1$.

10. MULTIPLICATIVE DECOMPOSITION THEOREM (MDT_m; SEE §1.8).
11. ADDITIVE DECOMPOSITION THEOREM (ADT_m; SEE §1.8).

12. ORDERING THEOREM (OT_m). *The group* G^m *is ordered on* $(\mathbf{R}^+, \infty)$ *by the relation* $\succ$. *Moreover, for any* $g \in G^m$ *there exist a* $k \le m$ *and a* $\tilde{g} \in G^{k-1}$ *such that* $|g - \mathrm{id}| \in A^k_{\tilde{g}}$. *In other words, there are positive numbers* ν_1 *and* ν_2 *such that*

$$\exp(-\nu_1 \exp^{[k]} \tilde{g}) \prec |g - \mathrm{id}| \prec \exp(-\nu_2 \exp^{[k]} \tilde{g}).$$

The pair $(k, \tilde{g})$ *is determined as follows: the first nonzero term in the decomposition given by* ADT_m *for* $g - \mathrm{id}$ *belongs to* $\mathscr{F}^k_{\tilde{g}}$.

13. LEMMA ON A LOWER ESTIMATE FOR THE PARTIAL SUMS OF A STAR-m, LEPS_m. *Let* Σ *be a nonzero weakly real partial sum of a STAR-m, with all the terms in* Σ *having principal exponents greater than* ν. *Then*

$$|\Sigma| \succ \exp(\nu \exp^{[m]}) \quad \text{on } (\mathbf{R}^+, \infty).$$

In the induction scheme of proof pictured (see Figure 20), the first two rows contain theorems and lemmas appearing in the induction hypothesis, and the third, fourth, and fifth contain assertions proved in the induction step. In addition to the theorems listed above corresponding to $m = n$, rows 3–5 contain the following.

1. Properties of regular map-cochains and functional cochains proved in Chapter II and denoted in the scheme by $\mathscr{F}\mathscr{C}_{\mathrm{reg}}(\mathscr{D}^n_0)$, $\mathscr{F}\mathscr{C}_{\mathrm{reg}}(\mathscr{D}^{n-1}_1)$.

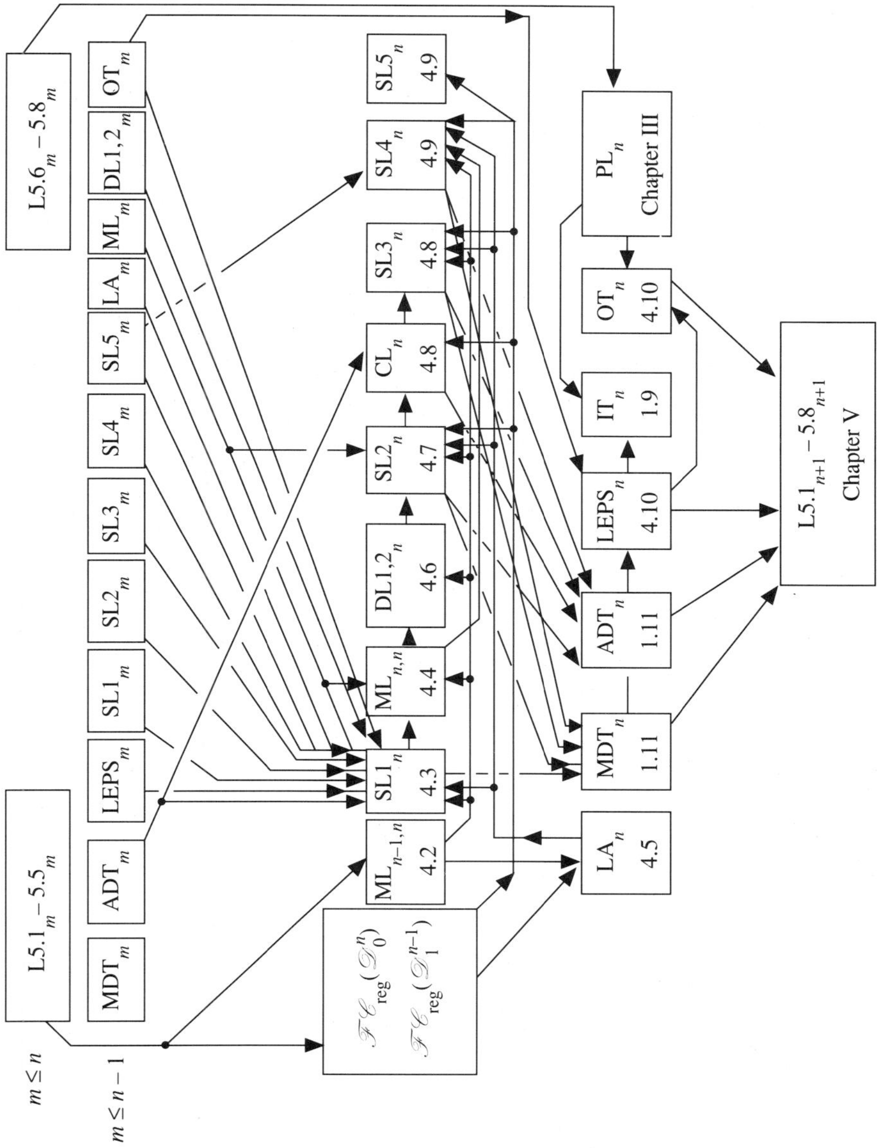

FIGURE 20. INDUCTION SCHEME OF PROOF

2. The Phragmén-Lindelöf Theorem (PL_n), formulated and proved in Chapter III.

3. Lemmas 5.2_{n+1}–5.8_{n+1} for germs of class $\mathscr{D}_0^{n+1}$ and $\mathscr{D}_1^n$, which can be expressed in terms of elements of the group G^n, to the study of which Chapter IV is devoted. These lemmas are proved in Chapter V.

4. CONJUGATION LEMMA (CL_n). *Suppose that* $m = n$ *or* $m = n-1$, $k \leq m-1$, $f \in G^{m-1}$, *and* $g \in G^k$. *Then* $\mathrm{Ad}(g)(\mathrm{id} + \mathscr{F}^m_{+f}) \subset \mathrm{id} + \mathscr{F}^m_{+fg}$.

5. Finally, the fourth row contains the Identity Theorem, the goal of the present work, denoted by IT_n, and also the Ordering Theorem, OT_n.

Besides the notation for the theorems and lemmas, the addresses of the proofs are also indicated in rows 3–5.

According to Definition 3°, §1.7, the set of germs of regular functional cochains of class 0 and type 0 is the set of almost regular germs. The sets of germs of regular functional cochains of class -1 is empty by definition. The definition is motivated by the fact that the group G^{-1} consists of affine germs, while the functional cochains indicate precisely the difference between the germs in the groups G^k $(k \geq 0)$ and affine germs. Thus, for $m = 0$ all the theorems and lemmas in the second row relate to almost regular and affine germs, and are trivial in this case.

In the first row are Lemmas 5.2_m–5.8_m for $m \leq n$. For any m these lemmas cover Lemma 5.1_m, which is therefore not included in the list. They are used for $1 \leq m \leq n$. For $m = 1$ they are proved where they are formulated.

§4.2. Multiplication Lemma $\mathrm{ML}_{n-1,n}$

Recall the convention

$$\mathscr{F}^m = \begin{cases} \mathscr{F}_1^m & \text{for } m \leq n-1, \\ \mathscr{F}_0^n & \text{for } m = n, \end{cases}$$

$$\mathscr{D}^m = \begin{cases} \mathscr{D}_1^m & \text{for } m \leq n-1, \\ \mathscr{D}_0^n & \text{for } m = n; \end{cases}$$

$\mathscr{F}^m$, $\mathscr{F}^m_{(+)g}$, etc., are deciphered similarly.

LEMMA $\mathrm{ML}_{n-1,n}$. *Suppose that* $m = n$ *or* $m = n-1$, *and* $k \leq m-1$. *Then for arbitrary* $f \in G^{k-1}$ *and* $g \in G^{m-1}$

$$\mathscr{F}_f^k \cdot \mathscr{F}^m_{(+)g} \subset \mathscr{F}^m_{(+)g}.$$

PROOF. The proof consists of two parts corresponding to the two parts of the definition of a functional cochain of class n: regularity and expandability.

A. *Regularity.* We prove regularity under the conditions of the lemma, and also in the case $k = m$, $f \prec\!\!\prec g$; this case arises in the proof of

Lemma $\mathrm{ML}_{n,n}$. It must be proved that for arbitrary cochains $F_1 \in \mathscr{F}\mathscr{C}^k$ and $F_2 \in \mathscr{F}\mathscr{C}^m_{(+)}$ and for arbitrary $f \in G^{k-1}$ such that $(k, f) \prec (m, g)$ there exists a cochain $F \in \mathscr{F}\mathscr{C}^m_{(+)}$ such that

$$F_1 \circ \exp^{[k]} \circ f \circ F_2 \circ \exp^{[m]} \circ g = F \circ \exp^{[m]} \circ g,$$

or

$$F_1 \circ \rho \cdot F_2 = F, \quad \text{where } \rho = \exp^{[k]} \circ f \circ g^{-1} \circ \ln^{[m]}.$$

We prove that $F_1 \circ \rho = \widetilde{F} \in \mathscr{F}\mathscr{C}^m$, and that $|\widetilde{F}|$ increases more slowly than any exponential in some domain of class n:

$$|\widetilde{F}| < \exp \varepsilon\xi \text{ for any } \varepsilon > 0.$$

Let us first prove the second assertion. It follows from the estimate

$$|F_1| < \exp \nu\xi$$

for $|F_1|$ on some standard domain of class k, and the inequality

$$\operatorname{Re} \rho < \varepsilon\xi \quad \text{for any } \varepsilon > 0 \tag{$*$}$$

in some standard domain of class m, to the proof of which we now proceed.

Let $h = f \circ g^{-1}$. Then $h \in G^{m-1}$,

$$\rho = \ln^{[k-m]} \circ A^{-m} h.$$

We consider two cases.

Case 1: $k \leq m - 1$. The germ $A^{-m}h$ grows more slowly than any exponential; what is more, by Lemma 5.5_mb, $m \leq n$, (§2.6)

$$|A^{-m}h| \prec \exp\sqrt{|\zeta|} \quad \text{in } (\mathbf{C}^+, \infty).$$

Then

$$\operatorname{Re} \ln A^{-m} h = \ln |A^{-m}h| \prec \sqrt{|\zeta|}.$$

In some standard domain of class m the function $\sqrt{|\zeta|}$ grows more slowly than ξ : $\sqrt{|\zeta|}/\xi \to 0$. Compositions with the logarithm only slow down growth. Accordingly, $|\rho| \prec \varepsilon\xi$ for all $\varepsilon > 0$. This proves the inequality $(*)$ in Case 1.

Case 2: $k = m$. In this case $h \prec\prec \mathrm{id}$, $\rho = A^{-m}h$. The inequality $(*)$ is obtained from Lemma 5.5_m d.

We now apply Lemma 3 in §2.3 to the composition $F_1 \circ \rho$. One condition on ρ—the estimate $\operatorname{Re} \rho < \varepsilon\xi$—has already been verified. The second condition,

$$\sigma \circ \mathscr{D}^k \subset \mathscr{D}^m \cup \{\text{the set of nonessential germs}\},$$

holds by Propositions 1 and 2 in §2.7. The third condition follows easily from Lemma 5.4_n. According to Lemma 3 in §2.3, $F_1 \circ \rho$ is a cochain of type $\mathcal{F}\mathcal{C}_{\text{reg}}(\mathcal{D}^m)$, perhaps with a trivial coboundary and growing more slowly than any exponential. Lemma 1 in §2.1 now gives us the assertion about regularity.

B. **Expandability**. We prove expandability by induction on the rank r of the factor of class m, assuming that $k \leq m-1$. Induction base: $r = 0$. In this case the first factor in the product $\mathcal{F}_k \cdot \mathcal{F}^{m,0}$ "is absorbed into the coefficients,"

$$\mathcal{F}^k \cdot \mathcal{K}^{m,0} \subset \mathcal{K}^{m,0},$$

by the definition of $\mathcal{K}^{m,0}$: $\mathcal{K}^{m,0} = \mathcal{L}(\mathcal{F}^{\tilde{m}} | 0 \leq \tilde{m} \leq n-1)$, the inequality $k \leq m-1$, and ML_{m-1}, which appears in the induction hypothesis. Accordingly,

$$\mathcal{F}^k \cdot \mathcal{F}^{m,0} \subset \mathcal{F}^{m,0}.$$

INDUCTION STEP. Suppose that

$$\mathcal{F}^k \cdot \mathcal{K}^{m,r} \subset \mathcal{K}^{m,r}, \qquad \mathcal{F}^k \cdot \mathcal{F}^{m,r} \subset \mathcal{F}^{m,r}.$$

Then the first factor on the left-hand side once more "is absorbed into the coefficients":

$$\mathcal{F}^k \cdot \mathcal{K}^{m,r+1} \subset \mathcal{K}^{m,r+1},$$

because, by definition,

$$\mathcal{K}^{m,r+1} \subset \mathcal{K}^{m,r} \cup \mathcal{F}^{m,r},$$

and because of the induction hypothesis. Consequently,

$$\mathcal{F}^k \cdot \mathcal{F}^{m,r+1} \subset \mathcal{F}^{m,r+1},$$

since the first factor "is absorbed into the coefficients" by what was proved above. This concludes the proof of Lemma $\mathrm{ML}_{n-1,n}$, because the estimate of the remainder term is trivial (an analogous argument is given in more detail in §4.4).

§4.3. First Shift Lemma, $\mathrm{SL}\,1_n$

LEMMA $\mathrm{SL}\,1_n$. *Let $m = n$ or $m = n-1$. Then*

$$\mathcal{F}_{\text{id}}^m \circ G_{\text{rap}}^{m-1} \subset \mathcal{F}_{\text{id}}.$$

PROOF. An equivalent assertion:

$$\mathcal{F}\mathcal{C}^m \circ \exp^{[m]} \circ G_{\text{rap}}^{m-1} \subset \mathcal{F}\mathcal{C}^m \circ \exp^{[m]}.$$

Regularity follows from Lemma $\mathrm{SL}\,1_{n,\text{reg}}$ in §2.4:

$$\mathcal{F}\mathcal{C}_{\text{reg}}(\mathcal{D}^m) \circ A^{-m} G_{\text{rap}}^{m-1} \subset \mathcal{F}\mathcal{C}_{\text{reg}}(\mathcal{D}^m)$$

for $m = n$ or $m = n-1$.

We prove expandability. We first use the additive decomposition theorem: ADT_m, $m \leq n-1$, which appears in the induction hypothesis, to describe the elements of the group G_{rap}^{n-1}.

A. The group G^{n-1}_{rap}. Let us begin with a remark.

REMARK. $g \in G^{n-1}_{\text{rap}}$ if and only if $g \in G^{n-1}$ and the correction of the germ $A^{1-n}g$ is bounded on $\mathbf{R}^+$.

PROPOSITION 1. *Let m be an arbitrary positive integer. If the correction of the germ $A^{-m}(\text{id}+\varphi)$ is bounded on $\mathbf{R}^+$, then*

$$A^{-m}(\text{id}+\varphi) = \text{id}+\left(\prod_0^{m-1} \ln^{[j]}\right) \varphi \circ \ln^{[m]} \cdot (1+o(1)).$$

PROOF. The proof is by induction on m. Induction base: $m = 1$.

$$A^{-1}(\text{id}+\varphi) = \exp\circ(\ln+\varphi\circ\ln) = \zeta\cdot\exp\varphi\circ\ln.$$

By assumption, the correction of the germ $A^{-1}(\text{id}+\varphi)$ is bounded on $\mathbf{R}^+$. Consequently, $\varphi \to 0$ on $(\mathbf{R}^+, \infty)$. Then

$$A^{-1}(\text{id}+\varphi) = \zeta(1+\varphi\circ\ln\cdot(1+o(1))) = \text{id}+\zeta\cdot\varphi\circ\ln(1+o(1)).$$

INDUCTION STEP. Let $A^{-k}(\text{id}+\varphi) = \text{id}+\varphi_k$. As above, we prove that if the correction φ_m is bounded, then $\varphi_k \to 0$ on $(\mathbf{R}^+, \infty)$ for $k \leq m-1$. By the induction hypothesis,

$$\varphi_k = \left(\prod_0^{k-1} \ln^{[j]}\right)\cdot\varphi\circ\ln^{[k]}\cdot(1+o(1)).$$

Then

$$\text{id}+\varphi_{k+1} = A^{-1}(\text{id}+\varphi_k) = \text{id}+\zeta\cdot\varphi_k\circ\ln(1+o(1)),$$

$$\varphi_{k+1} = \prod_0^{k}\ln^{[j]}\cdot\varphi\circ\ln^{[k+1]}\cdot(1+o(1)).$$

The last equality follows from the induction hypothesis. ▶

PROPOSITION 2. *Suppose that $1 \leq m \leq n-1$. Then*

$$G^m_{\text{rap}} \subset \text{id}+\mathscr{L}(\mathscr{F}^{m-1}_{\mathbf{R},\text{id}}, \mathscr{F}^{m-1}_{++}, \mathscr{F}^m_+),$$

and $g \in G^m_{\text{rap}} \Rightarrow g = \text{id}+\varphi$, where $\varphi\prod_1^m \exp^{[j]}$ is bounded on $(\mathbf{R}^+, \infty)$; here $\mathscr{F}^k_{++} = \{\varphi \in \mathscr{F}^k_{+\tilde{g}} | \tilde{g} \in G^{k-1^+}_{\text{slow}}\}$.

PROOF. Suppose that $g \in G^m_{\text{rap}}$, and ψ is the first nonzero term in the decomposition for $g - \text{id}$ given by ADT_m, $\psi \in \mathscr{F}^k_{\mathbf{R}+\tilde{g}}$ (Theorem ADT_m appears in the induction hypothesis for $m \leq n-1$).

By assumption, the correction of the germ $A^{-m}g$ is bounded on $\mathbf{R}^+$. Consequently, Proposition 1 implies the boundedness of the product

$$\left(\prod_0^{m-1}\ln^{[j]}\right)\cdot\varphi\circ\ln^{[m]} \quad \text{or} \quad \left(\prod_1^{m}\exp^{[j]}\right)\cdot\varphi.$$

Proposition 2 now follows from the assertion $*: k \geq m-1$; if $k = m-1$, then $\tilde{g} \in G_{\text{slow}}^{m-2^+} \cup G_{\text{rap}}^{m-2}$.

PROOF OF $*$. By the Ordering Theorem OT_{m-1}, which appears in the induction hypothesis, there is a positive ν such that

$$|g - \mathrm{id}| \succ \exp(-\nu \exp^{[k]} \tilde{g}).$$

By Proposition 1, the function

$$\exp \circ (\exp^{[m-1]} - \nu \exp^{[k]} \tilde{g})$$

is bounded on $(\mathbf{R}^+, \infty)$. Note that $\tilde{g} \prec \mu\xi$ for some $\mu > 0$. Thus, $k \geq m-1$. For $k = m-1$ the difference

$$\mathrm{id} - \nu A^{1-m} \tilde{g}$$

has an upper bound. This proves the assertion $*$ and hence Proposition 2. ▶

PROPOSITION 3. *$G_{\text{rap}}^{m} \subset H_{\text{rap}}^{m-1} \circ J^{m-1} \circ H^m$, where $m \leq n-1$, and $H_{\text{rap}}^{m-1} = H^{m-1} \cap G_{\text{rap}}^{m-1}$.*

PROOF. We first prove that $G_{\text{rap}}^{m} \subset H^{m-1} \circ J^{m-1} \circ H^m$. Assume the contrary. Let

$$g \in G_{\text{rap}}^m, \quad g = g_1 \circ j \circ h, \quad h \in H^m, \quad j \in J^{m-1},$$
$$g_1 \in G^{m-1} \backslash H^{m-1}.$$

Then, by the ADT_{m-1},

$$g_1 = a + \varphi + \psi, \qquad a \in \mathscr{A}\!f\!f, \quad \varphi \in \mathscr{L}(\mathscr{F}_{1+}^k | 0 \leq k \leq m-2),$$
$$\psi \in \mathscr{L}(\mathscr{F}_{0+}^{m-1}), \quad a + \varphi \neq \mathrm{id}.$$

According to assertion a in $\mathrm{SL}\,4_{m-1}$, which appears in the induction hypothesis, $j \in \mathrm{id} + \mathscr{L}(\mathscr{F}_{1+}^{m-1})$. By the definition of the group H^m, $h \in \mathrm{Gr}(\mathrm{id} + \mathscr{F}_{0+}^m)$. From $\mathrm{SL}\,2_k$, $k \leq m$, which appears in the induction hypothesis,

$$g \in a + \varphi + \mathscr{L}(\mathscr{F}_{1+}^{m-1}, \mathscr{F}_{0+}^m).$$

This contradicts Proposition 2.

Accordingly, it is proved that $g_1 \in H^{m-1} \cap G^{m-1}$ in the preceding decomposition for g. We prove that $g_1 \in G_{\text{rap}}^{m-1}$. Note that the germ $A^{-m} g$ has bounded correction. Further, it follows from Proposition 1 that the germ $A^{-m} h$, $h \in H^m$, has the same property. It was proved in §§2.9 and 2.11 that an arbitrary germ of class $A^{-m} J^{m-1}$ has the form $\zeta(1 + o(1))$.

Consequently, the germs $A^{-m} g$, $A^{-m} j$, and $A^{-m} h$ increase no more rapidly than a linear germ on $(\mathbf{R}^+, \infty)$. Hence, the germ $A^{-m} g_1 = A^{-m}(g \circ h^{-1} \circ j^{-1})$ also has this property. ▶

B. Properness of the group G^{m-1}.

PROPOSITION 4. 1°. *The group G^{m-1} is m-proper for $m \le n$, in the sense of Definition 4 in subsection* B *of* § 1.4.

2°. *If $g \in G^{m-1^-}_{\text{slow}}$ ($g \in G^{m-1^+}_{\text{slow}}$), $m \le n$, then the germ $A^{-m} g$ increases on $(\mathbf{R}^+, \infty)$ slower (faster) than any linear germ. If $g \in G^{m-1}_{\text{rap}}$, then the limit of $\mu = A^{-m} g/\xi$ exists as $\xi \to \infty$, $\mu \in (0, \infty)$.*

PROOF. 1°. We verify all the requirements of the definition of m-properness, §1.4 B. Requirement 1° follows from Theorem ADT_{m-1}, which appears in the induction hypothesis for $m \le n$. Let us prove requirement 2°.

$g_m \overset{\text{def}}{=} A^m(\mu\zeta) \in G^{m-1}$ for arbitrary $\mu > 0$. Indeed, the assertion is obvious for $m = 0$. For $m = 1$, $g_m = \zeta + \ln\mu \in G^0$. Let $m \ge 2$. Then $A^2(\mu\zeta) \in \mathscr{R}^0$ (see Example 2 in subsection B of §1.4). Consequently, $A^m(\mu\zeta) \in G^{m-2} \subset G^{m-1}$ (see Example 4 in §1.7).

Requirement 3° follows from the ordering theorem OT_{m-1}, which appears in the induction hypothesis for $m \le n$.

Assertion 2° of the proposition now follows from Proposition 1 in subsection C of §1.4 and the remarks after it.

COROLLARY 1. *For an arbitrary germ $g \in G^{m-1^+}_{\text{slow}}$ the germ $A^{-m} g$ increases on $(\mathbf{R}^+, \infty)$ more rapidly than any linear germ.*

COROLLARY 2.

$$G^{m-1^\pm}_{\text{slow}} \circ G^{m-1}_{\text{rap}} = G^{m-1\pm}_{\text{slow}}.$$

C. Shifts of functional cochains of class k by elements of the group G^m for $k \le m \le n-1$.

PROPOSITION 5. *Suppose that $k \le m \le n-1$ and $f \in G^{k-1}$. Then*

$$\mathscr{F}^k_f \circ G^m \subset \mathscr{F}^k_f + \mathscr{L}(\mathscr{F}^k_{+f} \circ G^{k-1^+}_{\text{slow}}) + \mathscr{L}(\mathscr{F}^p_+ | k+1 \le p \le m).$$

PROOF. It follows from the definition of G^m that

$$G^m \subset G^{k-1} \circ J^{k-1} \circ H^k \circ J^k \circ \cdots \circ J^{m-1} \circ H^m.$$

By the definition of the set $\mathscr{F}^k$,

$$\mathscr{F}^k \circ G^{k-1} = \mathscr{F}^k.$$

Further, by Lemmas SL 4, 5_k, which appears in the induction hypothesis for $k \le n-1$,

$$\mathscr{F}^k \circ J^{k-1} \subset \mathscr{F}^k.$$

In view of assertion a of Lemma SL 4_p, which appears in the induction hypothesis for $p \le n-1$, $J^{p-1} \subset \mathrm{Gr}(\mathrm{id} + \mathscr{F}^{p-1}_+)$. By definition, $H^p = \mathrm{Gr}(\mathrm{id} + \mathscr{F}^p_{0+})$. According to Lemmas SL 2, 3_p, which also appear in the induction hypothesis for $p \le n-1$,

$$\mathscr{F}^k \circ (\mathrm{id} + \mathscr{F}_+^p) \subset \mathscr{F}^k + \mathscr{F}_+^p \quad \text{for } p > k,$$

$$\mathscr{F}_f^k \circ (\mathrm{id} + \mathscr{F}_{+f\tilde{g}}^k) \subset \begin{cases} \mathscr{F}_f^k & \text{for } \tilde{g} \in G^{k-1} \cup G_{\mathrm{rap}}^{k-1}, \\ \mathscr{F}_f^k + \mathscr{F}_{+f\tilde{g}}^k & \text{for } \tilde{g} \in G_{\mathrm{slow}}^{k-1^+}. \end{cases}$$

This yields the proposition. ►

D. Shifts of the coefficients. We now proceed directly to the proof of Lemma $\mathrm{SL}\,1_n$. Consider the case $m = n$; the case $m = n-1$ is analyzed similarly. We prove that partial sums and remainder terms of a STAR-m pass under a shift of the argument by a germ of class G_{rap}^{n-1} into partial sums and remainder terms, respectively. Let us begin with the partial sums. We investigate separately shifts of the coefficients and of the exponents.

PROPOSITION 6.

$$\mathscr{K}^n \circ G_{\mathrm{rap}}^{n-1} \subset \mathscr{K}^n.$$

PROOF. By definition,

$$\mathscr{K}^n = \mathscr{L}(\mathscr{F}^{n-1}) + \mathscr{L}(\mathscr{F}_{+\mathrm{id}}^n \circ G_{\mathrm{slow}}^{n-1^-}).$$

It follows from Proposition 5 that

$$\mathscr{F}^{n-1} \circ G_{\mathrm{rap}}^{n-1} \subset \mathscr{L}(\mathscr{F}^{n-1}).$$

Further, by Corollary 2 to Proposition 4,

$$\mathscr{F}_{+\mathrm{id}}^n \circ G_{\mathrm{slow}}^{n-1^-} \circ G_{\mathrm{rap}}^{n-1} \subset \mathscr{F}_{+\mathrm{id}}^n \circ G_{\mathrm{slow}}^{n-1^-}.$$

This proves the proposition. ►

E. Shifts of the exponents. The corresponding assertion is technically most difficult in this part, and is distinguished as a separate lemma.

LEMMA 1. *Suppose that* $m \le n$. *Then*

$$E^m \circ G_{\mathrm{rap}}^{m-1} \subset E^m + \mathscr{L}(\mathscr{F}_{+\mathrm{id}}^{m-1} \circ G^{m-2^+}).$$

Recall that G^+ is the subset of elements of the ordered group G that are not less than the identity of the group.

We prove the lemma for $m = n$; the proof is analogous for $m < n$.

Namely, we shall prove the inclusion

$$E^n \circ G_{\mathrm{rap}}^{n-1} \subset E^n + \mathscr{L}(\mathscr{F}_{+\mathrm{id}}^{n-1} \circ G^{n-2^+}). \qquad (*)$$

PROOF. Suppose that $\mathbf{e} \in E^n$ and $g \in G_{\mathrm{rap}}^{n-1}$. By the regularity lemma, RL_{n-1} (proved in §2.12), $\mathbf{e} \in \mathscr{F}_{\mathrm{id}}^{n-1}$.

By Proposition 5,

$$\mathbf{e} \circ g = \tilde{\mathbf{e}} + \varphi, \quad \text{where } \tilde{\mathbf{e}} \in \mathscr{F}_{\mathrm{id}}^{n-1} \text{ and } \varphi \in \mathscr{L}(\mathscr{F}_{+\mathrm{id}}^{n-1} \circ G_{\mathrm{slow}}^{n-2^+}). \qquad (**)$$

By definition,

$$\tilde{\mathbf{e}} = \widetilde{F} \circ \exp^{[n-1]}, \quad \text{where } \widetilde{F} \in \mathscr{F}\mathscr{C}^{n-1}.$$

Then the composition $\tilde{\mathbf{e}}$ is expandable in a STAR-$(n-1)$. Let Σ be the partial sum of all terms in this series with nonnegative principal exponents. Then

$$\tilde{\mathbf{e}} = \Sigma + \psi, \qquad \psi \in \mathscr{F}\mathscr{C}_{+\mathrm{id}}^{n-1}.$$

Let us prove that $\Sigma \in E^n$. This will imply the lemma, because ψ is included in the second term on the right-hand side of the formula $(*)$.

We verify all the requirements of Definition 2^n, I in §1.7; they are recalled below.

1°. Σ is a weakly real partial sum of a STAR-$(n-1)$ with nonnegative principal exponents of its terms.

All these requirements except for weak realness follow from the construction of the sum Σ; weak realness is proved below together with requirement 4°.

2°. The limit

$$\nu(\Sigma) = \lim_{(\mathbf{R}^+, \infty)} \Sigma / \exp^{[n]}$$

exists.

Let us prove this. The analogous limit $\nu(\mathbf{e})$ exists by the definition of the set E^n. It suffices to prove the existence of the limit $\nu(\mathbf{e} \circ g)$, because $\Sigma = \mathbf{e} \circ g - (\varphi + \psi)$, where φ and ψ are decreasing cochains.

We have that

$$\frac{\mathbf{e} \circ g}{\exp^{[n]}} = \frac{\mathbf{e} \circ g}{\exp^{[n]} \circ g} \circ \frac{\exp^{[n]} \circ g}{\exp^{[n]}}.$$

But the limit

$$\lim_{(\mathbf{R}^+, \infty)} \frac{\exp^{[n]} \circ g}{\exp^{[n]}} = \lim \frac{A^{-n} g}{\zeta} = \mu \in (0, \infty)$$

exists. This follows from Proposition 4 in B. We note that $\nu(\mathbf{e}) \neq \nu(\Sigma)$ in general, since $\mu \neq 1$ (also in general).

3°. There exists a constant $\mu_\Sigma > 0$ and a standard domain of class n in which

$$|\operatorname{Re} \Sigma \circ \ln^{[n]}| < \mu_\Sigma \xi, \qquad |\Sigma \circ \ln^{[n]}| < \mu_\Sigma |\zeta|. \qquad (***)$$

Like requirement 2°, it suffices to prove this and the following requirement for the composition $\mathbf{e} \circ g$.

Requirement 3° for $\mathbf{e} \circ g$ instead of Σ follows from Lemma 5.2_n, which appears in the induction hypothesis. The lemma was formulated in §2.3; the assertions of it needed in the present proof are recalled below.

Requirement 3° for $\mathbf{e}$ instead of $\mathbf{e}\circ g$ holds by definition: there exists a standard domain Ω of class n and a constant $\mu_{\mathbf{e}} > 0$ such that in Ω

$$|\operatorname{Re}\mathbf{e}\circ \ln^{[n]}| < \mu_{\mathbf{e}}\xi, \qquad |\mathbf{e}\circ \ln^{[n]}| < \mu_{\mathbf{e}}|\zeta|.$$

We have that

$$\mathbf{e}\circ g\circ \ln^{[n]} = (\mathbf{e}\circ \ln^{[n]})\circ (A^{-n}g).$$

By Lemma 5.2_nb there exists a standard domain $\Omega_1 \in \mathbf{\Omega}$ in which

$$|\operatorname{Re} A^{-n}g| < \alpha\xi, \qquad |A^{-n}g| < \alpha|\zeta|$$

for some $\alpha > 0$. By the same lemma, there exists a standard domain $\widetilde{\Omega}$ of class n such that $(A^{-n}g)\widetilde{\Omega} \subset \Omega_1$. The intersection of any two standard domains of class n contains a standard domain of the same class (by Lemma 5.2_n). Therefore, it can be assumed without loss of generality that $\widetilde{\Omega} \subset \Omega_1$. Then the inequalities $(***)$ hold in Ω_1 with the constant $\mu_\Sigma = \alpha\mu_{\mathbf{e}}$.

We verify

Requirement 4°. $\operatorname{Im}\Sigma \to 0$ on $(\mathbf{R}^+, \infty)$.

At the same time we prove that the sum Σ is weakly real. For this we prove the weak realness of the cochain $\widetilde{F} = \tilde{\mathbf{e}}\circ \ln^{[n-1]} \in \mathcal{F}\mathcal{C}^{n-1}$; see formula $(**)$. We have that

$$\widetilde{F} = \mathbf{e}\circ g\circ \ln^{[n-1]} - \varphi\circ \ln^{[n-1]}.$$

Let us investigate the imaginary part of the first term on $(\mathbf{R}^+, \infty)$. We set $F = \mathbf{e}\circ \ln^{[n-1]}$. Then

$$\widetilde{F} = F\circ A^{1-n}g - \varphi\circ \ln^{[n-1]}.$$

The function $A^{1-n}g$ is real on $(\mathbf{R}^+, \infty)$ and has bounded correction, by the remark at the beginning of subsection A. The function $\operatorname{Im} F^u$ (F^u is the main function of the tuple F) decreases more rapidly than any exponential on $(\mathbf{R}^+, \infty)$ in view of the criterion for weak realness in §1.11; we prove that $\operatorname{Im}\widetilde{F}^u$ has this property.

Indeed, on $(\mathbf{R}^+, \infty)$ the second term decreases more rapidly than any exponential (a similar assertion was proved in §1.9 in the analysis of Case 2). Namely,

$$\varphi = \sum \varphi_j; \qquad \varphi_j = F_j\circ \exp^{[n-1]}\circ g_j, \quad g_j \in G^{n-2^+}_{\text{slow}}$$
$$F_j \in \mathcal{F}\mathcal{C}^{n-1}_+;$$

see the formula $(**)$. By the definition of rapidly decreasing cochains, for each j there exists a $\nu_j > 0$ such that

$$|F_j| < \exp(-\nu_j\xi) \quad \text{on } (\mathbf{R}^+, \infty).$$

Then

$$|\varphi_j\circ \ln^{[n-1]}| < \exp(-\nu_j \operatorname{Re}\sigma_j) \quad \text{on } (\mathbf{R}^+, \infty),$$

where $\sigma_j = A^{1-n} g_j$. The germ σ_j increases more rapidly than a linear germ; this follows from the fact that $g_j \in G_{\text{slow}}^{n-2^+}$ and Proposition 4 in subsection B. Consequently, the germ φ_j decreases on $(\mathbf{R}^+, \infty)$ more rapidly than any exponential (see Proposition 3 in §1.4). By the criterion for weak realness (§1.11, D) the cochain F is weakly real. Further, that partial sum Σ_ν of the STAR-$(n-1)$ corresponding to the weakly real composition $F \circ \exp^{[n]}$ that consists of all the terms with principal exponents not less than ν is weakly real. Indeed, on $(\mathbf{R}^+, \infty)$

$$F \circ \exp^{[n-1]} = \Sigma_\nu + o(\exp(-(\nu + \varepsilon)) \exp^{[n-1]}).$$

Further, $IF \equiv 0$. Consequently, $I\Sigma_\nu = o(\exp(-(\nu + \varepsilon) \exp \varepsilon^{[n-1]}))$, which contradicts the lemma on a lower estimate LEPS_{n-1} (see §4.1) if it is assumed that $I\Sigma_\nu \not\equiv 0$. This concludes the verification of requirement 4° (weak realness of the sum Σ), and with it the proof of Lemma 1. ▶

F. Shifts of partial sums. We proceed to the conclusion of the proof of SL 1_n, more precisely, the conclusion of the investigation of shifts of partial sums.

PROPOSITION 7. *Suppose that Σ is a partial sum of a STAR-n, and let $g \in G_{\text{rap}}^{n-1}$. Then $\Sigma \circ g$ is again a partial sum of a STAR-n.*

PROOF. It can be assumed that the sum Σ consists of a single term:

$$\Sigma = a \exp \mathbf{e}, \qquad \mathbf{e} \in E^n, \ \alpha \in \mathscr{K}^n.$$

By Proposition 6 and Lemma 1,

$$\Sigma \circ g = a \circ g \cdot \exp \mathbf{e} \circ g = b \exp(\mathbf{e}_1 + \varphi),$$
$$b \in \mathscr{K}^n, \ \mathbf{e}_1 \in E^n, \ \varphi \in \mathscr{L}(\mathscr{F}_{+\text{id}}^{n-1} \circ G^{n-2^+}).$$

According to Corollary 1 of Lemma LA_{n-1}, which appears in the induction hypothesis,

$$\exp \varphi = 1 + \psi, \qquad \psi \in \mathscr{L}(\mathscr{F}_{+\text{id}}^{n-1} \circ G^{n-2^+}).$$

By Lemma ML_{n-1}, which appears in the induction hypothesis, and by the regularity lemma,

$$b \exp \psi \in \mathscr{K}^n.$$

This proves the proposition. ▶

G. Shifts of remainder terms.

Shifts of partial sums have been investigated. It remains to investigate shifts of remainder terms. Let Ω be a standard domain of class n, and R a remainder term of a STAR-n, namely, a functional cochain such that in the domain $\ln^{[n]} \Omega$

$$R = o(|\exp(-\nu \exp^{[n]})|).$$

Let $g \in G_{\text{rap}}^{n-1}$. Then

$$R \circ g = o(|\exp(-\nu \exp^{[n]} \circ g)|) = o(\exp(-\nu \operatorname{Re} A^{-n} g \circ \exp^{[n]})).$$

By Lemma 5.2_n, there exist $\varepsilon > 0$ and standard domains Ω_1 and $\widetilde{\Omega}$ of class n such that in Ω_1

$$\operatorname{Re} A^{-n} g > \varepsilon \xi ,$$

and, moreover, $A^{-n} g \widetilde{\Omega} \subset \Omega$. It can be assumed without loss of generality that $\widetilde{\Omega} \subset \Omega_1$. Consequently, in the domain $ln^{[n]}\widetilde{\Omega}$

$$R \circ g = o(\exp(-\nu\varepsilon(\operatorname{Re} \exp^{[n]}))).$$

Thus, the shifted partial sums approximate the shifted germ with the accuracy required in the definition of expandability. Lemma SL 1_n is proved. ►

§4.4. The Multiplication Lemma, $\mathrm{ML}_{n,n}$

LEMMA $\mathrm{ML}_{n,n}$. *Suppose that* $m = n$ *or* $m = n - 1$, f, $g \in G^{m-1}$, *and* $f \preccurlyeq g$. *Then*

$$\mathscr{F}_f^m \cdot \mathscr{F}_g^m \in \mathscr{F}_g^m .$$

PROOF. As in the case of $\mathrm{ML}_{n-1,n}$, the proof consists of two parts: regularity and expandability.

Regularity. It must be proved that for arbitrary cochains F_1, $F_2 \in \mathscr{F}\mathscr{C}^m$ there exists a cochain $f \in \mathscr{F}\mathscr{C}^m$ such that

$$F_1 \circ \exp^{[m]} \circ f \cdot F_2 \circ \exp^{[m]} \circ g = F \circ \exp^{[m]} \circ g ,$$

or

$$F_1 \circ \rho \cdot F_2 = F , \quad \text{where } \rho = A^{-m} h, \ h = f \circ g^{-1}.$$

We consider two cases: $h \in G_{\text{slow}}^{m-1^-}$, $h \in G_{\text{rap}}^{m-1}$.

CASE 1: $h \in G_{\text{slow}}^{m-1}$. In this case the regularity of the germ F was proved in §4.2.

CASE 2: $h \in G_{\text{rap}}^{m-1}$. In this case $F_1 \circ \rho = \widetilde{F}$, and $\widetilde{F} \in \mathscr{F}\mathscr{C}^n$ by Lemma SL 1_n, already proved. The regularity of the product $\widetilde{F} \cdot F_2$ now follows from Lemma 1 in §2.1.

Let us proceed to the proof of expandability. As above, we let $h = f \circ g^{-1}$. Our goal is to prove that $\mathscr{F}_h^m \cdot \mathscr{F}_{\text{id}}^m \subset \mathscr{F}_{\text{id}}^m$. For this it suffices to prove that

$$\mathscr{F}_h^{m,p} \cdot \mathscr{F}_{\text{id}}^{m,q} \subset \mathscr{F}_{\text{id}}^m$$

for arbitrary p and q. The regularity of the left-hand side is already proved. We prove expandability. The proof is by induction on $s = p + q$, differently in the two cases: 1. $h \in G_{\text{slow}}^{n-1^-}$; 2. $h \in G_{\text{rap}}^{n-1}$. It can be assumed without loss of generality that the factors belong to $\mathscr{F}_{+h}^n$ and $\mathscr{F}_{+\text{id}}^n$.

CASE 1. INDUCTION BASE. Let $s=0$. Then $p=q=0$, because $p\geq 0$ and $q\geq 0$. By definition, $\mathscr{K}^{n,0}\subset\mathscr{L}(\mathscr{F}^{n-1})$. Consequently,

$$\mathscr{F}_{+h}^{n,0}\cdot\mathscr{K}^{n,0}\subset\mathscr{F}_{+h}^{n,0}\subset\mathscr{K}^{n}$$

by Lemma $\mathrm{ML}_{n-1,n}$. This proves expandability in Case 1 for $s=0$.

INDUCTION STEP. $\mathscr{F}_{+h}^{n,p}\subset\mathscr{K}^{n,p}$, since $h\in G_{\mathrm{slow}}^{n-1^{-}}$. We prove that the factor $\mathscr{F}_{+h}^{n,p}$ "is absorbed into the coefficients":

$$\mathscr{K}^{n,p}\cdot\mathscr{K}^{n,q}\subset\mathscr{K}^{n}.$$

This follows from the definition of $\mathscr{K}^{n,r}$, from Lemma $\mathrm{ML}_{n-1,n}$, ML_{n-1}, and the induction hypothesis for $s<p+q$.

CASE 2. $h\in G_{\mathrm{rap}}^{n-1}$. Then $\mathscr{F}_{h}^{n}\subset\mathscr{F}_{\mathrm{id}}^{n}$ by $\mathrm{SL}\,1_{n}$, §4.3. Consequently, it suffices to prove expandability of the elements of the product $\mathscr{F}_{\mathrm{id}}^{n}\cdot\mathscr{F}_{\mathrm{id}}^{n}$. For this it suffices to prove expandability of the elements of the product $\mathscr{F}_{\mathrm{id}}^{n,p}\cdot\mathscr{F}_{\mathrm{id}}^{n,q}$ for arbitrary p and q. Induction is again carried out on $s=n+q$.

PROPOSITION 2. *The set of partial sums of a STAR-* n *is closed under multiplication.*

PROOF. Denote by $\Sigma^{n,r}$ the set of partial sums of STAR-(n,r), and set $\Sigma^{n}=\bigcup_{r}\Sigma^{n,r}$. It is required to prove that $\Sigma^{n}\cdot\Sigma^{n}\subset\Sigma^{n}$, or

$$\Sigma^{n,p}\cdot\Sigma^{n,q}\subset\Sigma^{n}$$

for arbitrary p and q. This will be done by induction on $s=p+q$.

INDUCTION BASE. $s=0$. The set E^{n} of exponents is closed under addition—this is an immediate consequence of the definition. The set $\mathscr{K}^{n,0}$ of coefficients is closed under multiplication in view of Lemma ML_{n-1}. This proves the proposition for $s=0$.

INDUCTION STEP. $s=p+q$. The sets of coefficients of the partial sums in $\Sigma^{n,p}$ and $\Sigma^{n,q}$ are $\mathscr{K}^{n,p}$ and $\mathscr{K}^{n,q}$, respectively; $\mathscr{K}^{n,p}\subset\mathscr{K}^{n,0}\cup\mathscr{F}_{+}^{n,p-1}$ and $\mathscr{K}^{n,q}\subset\mathscr{K}^{n,0}\cup\mathscr{F}_{+}^{n,q-1}$. Products of these coefficients belong to $\mathscr{K}^{n}$ by the induction hypothesis: Lemma ML_{n} is assumed to be proved for $p+q<s$. This proves the proposition for arbitrary s.

We estimate the remainder term of the product. Suppose that the factors f_1, $f_2\in\mathscr{F}\mathscr{C}_{+\mathrm{id}}^{n}$ do not exceed 1 in the domain $\ln^{[n]}\Omega_{\mathrm{st}}$, and the partial sums Σ_1 and Σ_2 of the corresponding STAR-n approximate these factors to within $o(|\exp(-\nu\exp^{[n]})|)$. Then the product $\Sigma_1\Sigma_2$ approximates f_1f_2 to within $o(|\exp(-\nu\exp^{[n]})|)$. This concludes the proof of ML_{n} in Case 2. ►

§4.5. Analytic functions of decreasing cochains

A. LEMMA. *Suppose that* $m = n$ *or* $m = n - 1$, *and* $\varphi\colon (\mathbf{C}, 0) \to (\mathbf{C}, 0)$ *is a holomorphic function. Then* $\varphi \circ \mathscr{F}^m_{+g} \subset \mathscr{F}^m_{+g}$ *for arbitrary* $g \in G^{m-1}$.

PROOF. *Regularity.* It must be proved that for arbitrary $F \in \mathscr{F}\mathscr{C}^m_+$

$$\varphi \circ F \circ \exp^{[m]} \circ g \in \mathscr{F}\mathscr{C}^m_+ \circ \exp^{[m]} \circ g,$$

or, what is equivalent,

$$\varphi \circ F \in \mathscr{F}\mathscr{C}^m_+.$$

This is Corollary 1 in §2.2.

Expandability. It suffices to prove that $\varphi \circ \psi \in \mathscr{F}^m_{+\mathrm{id}}$ for arbitrary $\psi \in \mathscr{F}^m_{+\mathrm{id}}$. This is a consequence of the Taylor formula for φ and the multiplication lemma ML_n. Namely,

$$\varphi \circ \psi = \sum_1^\infty \frac{\varphi^{(j)}(0)}{j!}\psi^j. \qquad (*)$$

By the multiplication lemma ML_n, the partial sums of this series belong to the set $\mathscr{F}^m_{+\mathrm{id}}$, and hence are expandable in a STAR-m. The Nth partial sum Σ_N of the series $(*)$ approximates $\varphi \circ \psi$ to within $o(|\psi|^N)$, where $\psi = F \circ \exp^{[m]}$, $F \in \mathscr{F}\mathscr{C}^m_+$. But since the cochain F is rapidly decreasing, this implies that the corresponding remainder term R is equal to $o(|\exp(-N\varepsilon \exp^{[m]})|)$ for some $\varepsilon > 0$. Choosing N such that $\varepsilon N > \nu$, we get that

$$|R| \prec \exp(-\nu \exp^{[m]}).$$

B. COROLLARIES. 1. *Suppose that* $m = n$ *or* $m = n - 1$. *Then*

$$\exp \circ \mathscr{F}^m_+ \subset 1 + \mathscr{F}^m_+.$$

2. *Under the conditions of Corollary* 1

$$\ln \circ (1 + \mathscr{F}^m_+) \subset \mathscr{F}^m_+.$$

3. *Under the conditions of Corollary* 1

$$1/(1 + \mathscr{F}^m_+) \subset 1 + \mathscr{F}^m_+.$$

REMARK. In all the corollaries the inclusion can be replaced by equality, but this is not needed in what follows.

C. Inclusions. This section might appear right after the definitions of §1.7. We prove for any $k \geq 0$ the following natural inclusions by induction on k:

1°. $$\mathscr{D}_0^{k-1} \subset \mathscr{D}_0^k, \quad \mathscr{D}_1^{k-1} \subset \mathscr{D}_1^k;$$

2°. $$\Omega_{k-1} \subset \Omega_k;$$

3°. $$\mathscr{F}\mathscr{C}_0^{k-1} \subset \mathscr{F}\mathscr{C}_0^k, \quad \mathscr{F}\mathscr{C}_1^{k-1} \subset \mathscr{F}\mathscr{C}_1^k;$$

4°. $$AG^{k-1} \subset G^k.$$

INDUCTION BASE. $k = 0$.

$$\mathscr{D}_1^{-1} = \mathscr{D}_0^1 = \varnothing, \quad \Omega_{-1} = \varnothing, \quad \mathscr{F}\mathscr{C}_1^{-1} = \mathscr{F}\mathscr{C}_0^{-1} = \mathscr{F}\mathscr{C}_1^0 = \varnothing.$$

Therefore, the first three inclusions are trivial for $k = 0$. Further, $G^{-1} = \mathscr{A}\!f\!f$ and $G^0 = \mathscr{R}$. The inclusion $AG^{-1} \subset G^0$ follows immediately from the definitions.

INDUCTION STEP: THE PASSAGE FROM $k = m - 1$ TO $k = m$. We prove the inclusions 1°, beginning with the first one. It follows immediately from the definition and the induction hypothesis. Namely, if $\sigma \in \mathscr{D}_0^{m-1}$, then

$$\sigma = A^{1-m} g \circ \exp, \qquad g \in G_{\mathrm{slow}}^{m-2^-},$$

that is, $g \in G^{m-2}$, and $A^{1-m} g$ grows on $(\mathbf{R}^+, \infty)$ more slowly than any linear function. Then

$$\sigma = A^{-m} g_1 \circ \exp, \qquad g_1 = Ag.$$

By the induction hypothesis, $g_1 \in G^{m-1}$. Moreover, $A^{-m} g_1 = A^{1-m} g$. Consequently, $g_1 \in G_{\mathrm{slow}}^{m-1^-}$ and $\sigma \in \mathscr{D}_0^m$. This proves the first of the inclusions in 1°. In passing we have proved for any $k \le m - 1$:

COROLLARY TO INCLUSION 4°. *$AG_{\mathrm{slow}}^{k-1^-} \subset G_{\mathrm{slow}}^{k^-}$, $AG_{\mathrm{rap}}^{k-1} \subset G_{\mathrm{rap}}^k$, and $AG_{\mathrm{slow}}^{k-1^+} \subset G_{\mathrm{slow}}^{k^+}$.*

PROOF. The first inclusion was proved above; the proofs of the two others are completely analogous. ▶

We now prove the inclusion $\mathscr{D}_1^{m-1} \subset \mathscr{D}_1^m$. By Definition 16 in §1.6, $\mathscr{D}_1^m = \mathscr{D}_0^m \cup \mathscr{D}_*^m \circ \mathscr{L}^m$. The inclusion $\mathscr{D}_0^{m-1} \subset \mathscr{D}_0^m$ has already been proved. The inclusion $\mathscr{D}_*^{m-1} \subset \mathscr{D}_*^m$ follows immediately from Definition 16 in §1.6 and the preceding corollary:

$$\begin{aligned}\mathscr{D}_*^{m-1} &= A^{1-m}(G_{\mathrm{slow}}^{m-2^+} \cup G_{\mathrm{rap}}^{m-2}) = A^{-m}(AG_{\mathrm{slow}}^{m-2^+} \cup AG_{\mathrm{rap}}^{m-2}) \\ &\subset A^{-m}(G_{\mathrm{slow}}^{m-1^+} \cup G_{\mathrm{rap}}^{m-1}) = \mathscr{D}_*^m.\end{aligned}$$

Finally, we prove the inclusion $\mathscr{L}^{m-1} \subset \mathscr{L}^m$. According to Definition 15 in §1.6, the group $\mathscr{L}^k$ is generated by germs of the form

$$\tilde{h} = \operatorname{Ad}(A^{1-k} g \circ \ln)\tilde{f}, \qquad \tilde{f} = \operatorname{Ad}(g_1) f, \ f \in \mathscr{A}^0,$$
$$g_1 \in \mathscr{H}^0, \ \tilde{f} \in \mathscr{M}_{\mathbf{R}}, \ g \in G^{k-1}, \ \mu_k(g) = 0.$$

By definition, $\mu_k(g) = \lim_{(\mathbf{R}^+, \infty)} (A^{1-k} g)'$. Consequently, if $\tilde{g} = Ag$, $g \in G^{m-2}$, and $\mu_{m-1}(g) = 0$, then $\tilde{g} \in G^{m-1}$ and $\mu_m(\tilde{g}) = 0$. Further, if we take $k = m-1$ in the formula for $\tilde{h}$, then $\tilde{h} \in \mathscr{L}^{m-1}$. Moreover,

$$\tilde{h} = \operatorname{Ad}(A^{1-m} \tilde{g} \circ \ln)\tilde{f}, \qquad \tilde{g} \in G^{m-1}, \ \mu_m(\tilde{g}) = 0.$$

Consequently, $\tilde{h} \in \mathscr{L}^m$. This concludes the proof of the inclusion 1°.

We proceed to the proof of the inclusion 2°. Domains of class m and type 1 have the form $(A^{-m} g)(\mathbf{C}^+ \backslash K)$, where K is some disk, and $g \in G^{m-1}_{\text{rap}}$. Let Ω be an analogous domain of class $m-1$, and let $g \in G^{m-2}_{\text{rap}}$.

Then

$$A^{1-m} g = A^{-m} \tilde{g}, \qquad \tilde{g} = Ag \in G^{m-1}_{\text{rap}},$$

by the corollary to assertion 4° for $m-1$ instead of k.

Domains of class m and type 2 have the form

$$A^{-1}(\mathrm{id} + \psi)(\mathbf{C}^+ \backslash K),$$
$$\psi = \exp\varphi, \ \varphi \in [\mathscr{L}(\mathscr{F}^{m-2}_{\mathbf{R},\mathrm{id}}, \mathscr{F}^{m-2}_{++}, \mathscr{F}^{m-1}_{+}) \backslash \mathscr{L}(\mathscr{F}^{m-2}_{++}, \mathscr{F}^{m-1}_{+})] \circ \ln^{[m-1]},$$
$$\mathscr{F}^{m-2}_{++} = \{\tilde{\varphi} \in \mathscr{F}^{m-2}_{+g} | g \in G^{m-3^+}_{\text{slow}}\}, \ \operatorname{Re}\varphi \to -\infty, \ \varphi \to 0 \text{ in } (\Pi^\forall, \infty).$$

To prove the inclusion 2° for domains of type 2 it suffices to verify that

$$\mathscr{F}^{m-3}_{\mathbf{R},\mathrm{id}} \circ \ln^{[m-2]} \subset \mathscr{F}^{m-2}_{\mathbf{R},\mathrm{id}} \circ \ln^{[m-1]},$$
$$\mathscr{F}^{m-3}_{++} \circ \ln^{[m-2]} \subset \mathscr{F}^{m-2}_{++} \circ \ln^{[m-1]},$$
$$\mathscr{F}^{m-2}_{+} \circ \ln^{[m-2]} \subset \mathscr{F}^{m-1}_{+} \circ \ln^{[m-1]}.$$

The first inclusion is equivalent to the relation $\mathscr{F}^{m-3}_{\mathbf{R},\mathrm{id}} \circ \exp \subset \mathscr{F}^{m-2}_{\mathbf{R},\mathrm{id}}$, which follows from the induction hypothesis. The second and third are proved similarly; the inclusion $AG^{m-4^+}_{\text{slow}} \subset G^{m-3^+}_{\text{slow}}$ is used in the proof of the second; see the corollary to inclusion 4°.

We now prove inclusion 3°. As always, there are two parts in the proof: regularity and expandability. Regularity is the assertion that

$$\mathscr{F}\mathscr{C}_{\text{reg}}(\mathscr{D}^{m-1}_l) \subset \mathscr{F}\mathscr{C}_{\text{reg}}(\mathscr{D}^m_l), \qquad l = 0, 1.$$

It follows from inclusion 1°. We prove expandability of compositions of class $\mathscr{F}\mathscr{C}^{m-1}_l \circ \exp^{[m]}$ in STAR-m. For $l = 0$ the arguments are carried out in detail; for $l = 1$ they are repeated word-for-word. (Everywhere in the following argument we could write $\mathscr{F}\mathscr{C}^{\boldsymbol{\cdot}}_l$ instead of $\mathscr{F}\mathscr{C}^{\boldsymbol{\cdot}}_0$, taking $l = 0$ or

1.) We first estimate the "shift of remainder terms." Let R be a remainder term of a STAR-$(m-1)$,

$$R = o(|\exp(-\nu \exp^{[m-1]})|) \quad \text{in } \ln^{[m-1]} \Omega,$$

where Ω is a standard domain of class $m-1$: $\Omega \in \mathbf{\Omega}_{m-1}$. Then

$$|R \circ \exp| = o(|\exp(-\nu \exp^{[m]})|) \quad \text{in } \ln^{[m]} \Omega.$$

This implies that $R \circ \exp$ decreases like a remainder term of a STAR-m, since $\Omega \in \mathbf{\Omega}_m$ in view of inclusion 2°.

We now investigate shifts of partial sums, that is, we prove that if $\Sigma = \sum a_j \exp e_j$ is a partial sum of a STAR-$(k-1, r)$ for $k = m$, then $\Sigma \circ \exp$ is a partial sum of a STAR-(k, r). Assuming that for $k \le m-1$, this has already been proved (the assertion is trivial for $k = 0$), we prove that $E^{m-1} \circ \exp \subset E^m$, and

$$\mathscr{K}^{m-1,r} \circ \exp \subset \mathscr{K}^{m,r}, \qquad \mathscr{F}\mathscr{C}^{m-1,r} \subset \mathscr{F}\mathscr{C}^{m,r}. \qquad (**)$$

The last two inclusions can be proved by induction on r. The last one follows from the two preceding ones and the assertion proved above about a shift of remainder terms.

We begin with the inclusion $E^{m-1} \circ \exp \subset E^m$. By definition, an $\mathbf{e} \in E^{m-1}$ is a partial sum of a weakly real STAR-$(m-2)$ satisfying requirements 1°–4° of Definition 2^{m-1}.I in §1.7. By the induction hypothesis, $\mathbf{e} \circ \exp$ is a partial sum of a weakly real STAR-$(m-1)$. The principal exponents of the terms of the sums $\mathbf{e}$ and $\mathbf{e} \circ \exp$ (regarded as sums of class $m-2$ and $m-1$ respectively) coincide; this proves requirement 1°. Further,

$$\lim_{(\mathbf{R}^+, \infty)} \frac{\mathbf{e}}{\exp^{[m-1]}} = \lim \frac{\mathbf{e} \circ \exp}{\exp^{[m]}} = \nu(\mathbf{e}).$$

This proves requirement 2°. Requirement 3° follows from the equality

$$\mathbf{e} \circ \ln^{[m-1]} = (\mathbf{e} \circ \exp) \circ \ln^{[m]}$$

and inclusion 2°. Finally, the requirements that $\operatorname{Im} \mathbf{e} \to 0$ and $\operatorname{Im} \mathbf{e} \circ \exp \to 0$ on $(\mathbf{R}^+, \infty)$ are equivalent.

Let us prove the inclusion $(**)$ for $r = 0$ (induction base). By definition, $\mathscr{K}^{m,0} = \mathscr{L}(\mathscr{F}^{m-1})$. We have that for $g \in G^{m-3}$

$$\begin{aligned} \mathscr{F}_g^{m-1} \circ \exp &= \mathscr{F}\mathscr{C}^{m-2} \circ \exp^{[m-2]} \circ g \circ \exp \\ &= \mathscr{F}\mathscr{C}^{m-2} \circ \exp^{[m-1]} \circ A g. \end{aligned}$$

But $Ag = \tilde{g} \in G^{m-2}$, and $\mathscr{F}\mathscr{C}^{m-2} \subset \mathscr{F}\mathscr{C}^{m-1}$, by the induction hypothesis. Consequently,

$$\mathscr{F}_g^{m-2} \circ \exp = \mathscr{F}_{\tilde{g}}^{m-1}, \qquad \tilde{g} = Ag.$$

Hence, under composition multiplication from the right by exp the partial sums and remainder terms of a STAR-$(m-1, 0)$ pass into partial sums and remainder terms of a STAR-$(m, 0)$. This proves the second inclusion in $(**)$ for $r = 0$.

INDUCTION STEP: PASSAGE FROM r TO $r+1$. We prove that for arbitrary $g \in G_{\text{slow}}^{m-2^-}$ there exists a $\tilde{g} \in G_{\text{slow}}^{m-1^-}$ such that

$$\mathscr{F}\mathscr{C}_{0+}^{m-1,r+1} \circ \exp^{[m-1]} \circ g \circ \exp \subset \mathscr{F}\mathscr{C}_{0+}^{m,r+1} \circ \exp^{[m]} \circ \tilde{g}.$$

Indeed, setting $\tilde{g} = Ag$, we get from the preceding corollary that $\tilde{g} \in G_{\text{slow}}^{m-1^-}$. Further, $\mathscr{F}\mathscr{C}_{0+}^{m-1,r} \subset \mathscr{F}\mathscr{C}_{0+}^{m,r}$ by the induction hypothesis. Therefore, $\mathscr{K}_0^{m-1,r+1} \circ \exp \subset \mathscr{K}_0^{m,r+1}$. Arguing as in the case $r = 0$, we get the second of the inclusions in $(**)$ for $r+1$ instead of r. This proves inclusion 3°.

Let us prove inclusion 4°. We have that

$$G^m = G^{m-1} \circ J^{m-1} \circ H^m \cap \mathscr{M}_{\mathbf{R}},$$
$$J^{m-1} = \operatorname{Ad}(G^{m-1})A^{1-m}\mathscr{A}^0, \qquad H^m = \operatorname{Gr}(\operatorname{id} + \mathscr{F}_{0+}^m).$$

Consequently,

$$AG^{m-1} = AG^{m-2} \circ AJ^{m-2} \circ AH^{m-1} \cap \mathscr{M}.$$

By the induction hypothesis, $AG^{m-2} \subset G^{m-1}$. We prove that

$$AJ^{m-2} \subset J^{m-1}, \qquad AH^{m-1} \subset H^m.$$

We have

$$AJ^{m-2} = \operatorname{Ad}(AG^{m-2})A^{1-m}\mathscr{A}^0 \subset \operatorname{Ad}(G^{m-1})A^{1-m}\mathscr{A}^0 = J^{m-1}.$$

Next, by Corollary 2 to Lemma LA_m, $m \leq n-1$,

$$A(\operatorname{id} + \mathscr{F}_{+g}^{m-1}) = \operatorname{id} + \mathscr{F}_{+g}^{m-1} \circ \exp \subset \operatorname{id} + \mathscr{F}_{+\tilde{g}}^m,$$

where $\tilde{g} = Ag \in G^{m-1}$, as proved above. Consequently, $AH^{m-1} \subset H^m$. ►

§4.6. Differentiation of functional cochains

In this section it is proved that functional cochains of classes $\mathscr{F}\mathscr{C}_0^n$ and $\mathscr{F}\mathscr{C}_1^{n-1}$ form a differential algebra. The same property is enjoyed by the set $\mathscr{F}_g^m$, where $m = n$ or $m = n-1$. It is obvious that these sets are closed under addition and subtraction. Closedness with respect to multiplication was proved in §§4.2 and 4.4. It remains to prove closedness with respect to differentiation.

FIRST DIFFERENTIATION LEMMA, $\mathrm{DL}\,1_n$. *The sets* $\mathscr{F}\mathscr{C}_0^n$ *and* $\mathscr{F}\mathscr{C}_1^{n-1}$ *are closed with respect to differentiation. What is more, for every cochain* F *of class* $\mathscr{F}\mathscr{C}^m$, $m = n$ *or* $m = n-1$, *there exists a standard domain in which all derivatives of the cochain belong to this class. Moreover, there exist*

$c > 0$ and $\nu > 0$ such that $|F^{(k)}/k!| \prec c^k \exp \nu\xi$ in some standard domain of class m.

PROOF. 1. *Regularity.* The set $\mathscr{F}\mathscr{C}_{\mathrm{reg}}(\mathscr{D})$ is closed with respect to differentiation for an arbitrary set $\mathscr{D}$ of admissible germs. For every cochain in this set there exists a standard domain in which all derivatives of this cochain are defined and belong to $\mathscr{F}\mathscr{C}_{\mathrm{reg}}(\mathscr{D})$. This was proved in §2.1.

2. *Expandability.* Consider the case $m = n$; the case $m = n - 1$ is treated similarly. Let $F \in \mathscr{F}\mathscr{C}_0^n$. We prove that there exists a standard domain Ω_{st} of class n such that for every k the cochain $F^{(k)} \circ \exp^{[n]}$ is expandable in a STAR-n in the domain $\ln^{[n]} \Omega_{\mathrm{st}}$. We prove that the cochain $(F \circ \exp^{[n]})^{(k)}$ is expandable in a STAR-n in the same domain. The last two assertions are equivalent, according to the chain rule and the multiplication lemma: $\exp^{[n]}$ belongs to $\mathscr{F}_{\mathrm{id}}^{n-1}$ together with all its derivatives and their inverse functions (in the sense of pointwise multiplication). We reduce the question of expandability of the derivative to the question of differentiation of the partial sums of a STAR-n, namely, we prove that a remainder term of a STAR-n "withstands differentiation."

PROPOSITION 1. *Suppose that Ω is a standard domain of class n and that in the domain $\ln^{[n]} \Omega$ the holomorphic functions R has the form:*

$$R = o(|\exp(-\nu \exp^{[n]} \zeta)|), \ \nu > 0 \quad \text{as } \zeta \to \infty, \ \zeta \in \ln^{[n]} \Omega.$$

Then there exists a standard domain Ω_{st} of class n, $\Omega_{\mathrm{st}} \subset \Omega$, such that for any $\mu < \nu$ and any k

$$R^{(k)} o(|\exp(-\mu \exp^{[n]} \zeta)|) \quad \text{as } \zeta \to \infty, \quad \zeta \in \ln^{[n]} \Omega_{\mathrm{st}}.$$

PROOF. The proof follows from Cauchy's inequality and properties of elementary functions (multiple exponentials and logarithms); it is verified schematically. Any standard domain with $\rho(\Omega_{\mathrm{st}}, \partial\Omega) \geq 1$ works as Ω_{st}. Such a domain exists because the class Ω_n is proper (Lemma 5.2_n.d). Then the inequality

$$\rho(\zeta, \partial \ln^{[n]} \Omega) \geq 1/|\exp^{[n]} c\zeta|$$

holds in the domain $\ln^{[n]} \Omega_{\mathrm{st}}$ for sufficiently large $c > 0$. Consequently, by the Cauchy estimate,

$$R^{(k)} = o(|\exp(-\nu \exp^{[n]} \zeta + k \exp^{[n-1]} c\zeta)|) = o(|\exp(-\mu \exp^{[n]} \zeta)|)$$

for all $\mu < \nu$. ▶

PROPOSITION 2. *The set of partial sums of STAR-n is closed with respect to differentiation.*

PROOF. We prove Proposition 2 and at the same time DL 1_n by induction on the rank.

INDUCTION BASE. Let $r = 0$.

$$\Sigma = \sum_{j=1}^{N} a_j \exp \mathbf{e}_j, \quad a_j \in \mathscr{K}^{n,0}, \ \mathbf{e}_j \in E^n.$$

Then

$$\Sigma' = \sum_{j=1}^{N} (a_j' + a_j \mathbf{e}_j') \exp \mathbf{e}_j. \tag{$*$}$$

We have that $a_j' \in \mathscr{K}^{n,0}$ by the definition of $\mathscr{K}^{n,0}$ and by DL 1_{n-1} (which appears in the induction hypothesis); $\mathbf{e}_j' \in \mathscr{K}^{n,0}$ for the same reason, because $E^n \subset \mathscr{F}_{\mathrm{id}}^{n-1}$ according to the regularity lemma in §2.12. Further, $a_j \mathbf{e}_j' \in \mathscr{K}^{n,0}$ by the ML_{n-1}. Consequently, Σ' is a partial sum of a STAR-$(n, 0)$.

Proposition 2 is proved for $r = 0$. It follows from Propositions 1 and 2 that Lemma DL 1_n and DL 2_n stated below are valid for cochains of rank 0.

INDUCTION STEP. Suppose that Proposition 2 and Lemma DL 1_n have been proved for cochains of rank $s \le r - 1$. We prove them for $s = r$.

In the formula $(*)$

$$a_j \in \mathscr{K}^{n,r} \subset \mathscr{K}^{n,0} + \mathscr{L}(\mathscr{F}_+^{n,r-1}).$$

Then $a_j' \in \mathscr{K}^{n,r}$ by the induction hypothesis. The rest of the proof is just as for $r = 0$.

The last assertion of Lemma DL 1_n follows from the Cauchy estimates.

SECOND DIFFERENTIATION LEMMA DL 2_n. *For any $g \in G^{n-1}$ the set $\mathscr{F}_g^n$ is closed with respect to differentiation. What is more, for any germ $\varphi \in \mathscr{F}_g^n$ there exists a standard domain Ω_{st} of class n such that all derivatives $\varphi^{(k)}$ belong to the class $\mathscr{F}_g^n$ in the domain $g^{-1} \circ \ln^{[n]} \Omega_{\mathrm{st}}$.*

PROOF. Suppose that $\varphi = F \circ \exp^{[n]} \circ g$, $F \in \mathscr{F}\mathscr{C}^{[n]}$, $g \in G^{n-1}$, $F \in \mathscr{F}\mathscr{C}^{[n]}$, and $F^{(k)}$ are regular in Ω_{st}. The assertion of the lemma is equivalent to the following: $\varphi^{(k)} \circ g^{-1} \in \mathscr{F}_{\mathrm{id}}^n$ in the domain $\ln^{[n]} \Omega_{\mathrm{st}}$. It suffices to carry out the proof for $k = 1$, and then to use induction on k. Let $\psi = \varphi' \circ g^{-1}$. By the chain rule,

$$\psi = F' \circ \exp^{[n]} \cdot \prod_{1}^{n-1} \exp^{[j]} \cdot g' \circ g^{-1}.$$

According to DL 1_n, $F' \in \mathscr{F}\mathscr{C}^{n,2}$. Consequently, $F' \circ \exp^{[n]} \in \mathscr{F}_{\mathrm{id}}^n$. Further, $\prod_1^{n-1} \exp^{[j]} \in \mathscr{F}_{\mathrm{id}}^{n-1}$. By the ML_n, the product of the first two factors in the expression for ψ belongs to $\mathscr{F}_{\mathrm{id}}^n$. We prove that the third factor belongs to $\mathscr{L}(\mathbf{R}_+, \mathscr{F}_+^k | 0 \le k \le n-1)$. Then the DL 2_n will follow from the multiplication lemma.

Since G^{n-1} is a group, $g^{-1} \in G^{n-1}$. According to the chain rule, $g' \circ g^{-1} = 1/(g^{-1})'$. By the ADT_{n-1},

$$g^{-1} = \alpha + \sum \varphi_j, \qquad \varphi_j \in \mathscr{F}_{g_j}^{k_j}, \ (k_j, g_j) \prec\!\prec (k_{j+1}, g_{j+1}).$$

From the $\mathrm{DL}\,2_{m-1}$, which appears in the induction hypothesis, for $m \le n$

$$(g^{-1})' = \alpha + \sum \tilde{\psi}_j, \qquad \alpha \in \mathbf{R}_+, \ \tilde{\psi}_j \in \mathscr{F}_{+g_j}^{k_j}.$$

Further, by Corollary 3 to the lemma in §4.5,

$$1/(1 + \mathscr{F}_{+g}^{k}) = 1 + \mathscr{F}_{+g}^{k}.$$

From this, the multiplication lemma gives us that

$$(g^{-1})' = \alpha \prod (1 + \psi_j), \qquad \psi_j \in \mathscr{F}_{+g_j}^{k_j}.$$

Again using Corollary 3 in §4.5 and the multiplication lemma, we get from this that

$$1/(g^{-1})' \in \mathscr{L}(\mathbf{R}_+, \mathscr{F}_+^{k} | 0 \le k \le n-1). \quad \blacktriangleright$$

§4.7. The Second Shift Lemma, $\mathrm{SL}\,2_n$

LEMMA $\mathrm{SL}\,2_n$. *Let $m = n$ or $m = n-1$, $(k, f) \prec (m, g)$, and $\varphi \in \mathscr{F}_f^k$. Then*

$$\varphi \circ (\mathrm{id} + \mathscr{F}_{+g}^{m}) \subset \varphi + \mathscr{F}_{+g}^{m}.$$

PROOF. Suppose that under the conditions of the lemma $\psi \in \mathscr{F}_{+g}^{m}$. It must be proved that

$$\varphi \circ (\mathrm{id} + \psi) - \varphi = \kappa \in \mathscr{F}_{+g}^{m}.$$

The regularity—the relation $\kappa \circ g^{-1} \circ \ln^{[m]} \in \mathscr{F}\mathscr{C}_{\mathrm{reg}}^{+}(\mathscr{D}^m)$—was already proved in §2.7. We prove expandability. It is required to prove that there exists a standard domain Ω of class n such that the composition $\kappa \circ g^{-1}$ can be approximated in the domain $\ln^{[m]} \Omega$ by partial sums of one and the same STAR-n to within a remainder term

$$R = o(\exp(-\nu \exp^{[m]} \xi)) \qquad (*)$$

for any $\nu > 0$ (Definition 3^n in §1.7).

By the Taylor formula, for any N

$$\kappa = \Sigma_N + R_N, \quad \text{where } \Sigma_N = \sum_{1}^{N} \frac{1}{j!} \varphi^{(j)} \psi^j.$$

According to the differentiation lemma $\mathrm{DL}\,2_n$ and the multiplication lemma ML_n, each term of this sum belongs to $\mathscr{F}_{+g}^{m}$; hence $\Sigma_N \in \mathscr{F}_{+g}^{m}$. Consequently, $\Sigma_N \circ g^{-1} \in \mathscr{F}_{+\mathrm{id}}^{M}$. Therefore, the composition $\Sigma_N \circ g^{-1}$ is expandable in the sense of Definition 3^n in §1.7.

It remains to estimate the remainder term $R_N \circ g^{-1}$ and to prove that for each $\nu > 0$ there is an N such that $R = R_N \circ g^{-1}$ satisfies the estimate $(*)$.

The remainder term in Taylor's formula admits the following estimate:

$$|R_N(\zeta)| \leq \frac{1}{(N+1)!}|\varphi^{(N+1)}(\zeta + \theta\psi(\zeta))\psi^{N+1}(\zeta)|$$

for some $\theta \in [0, 1]$. Estimating the derivatives $\varphi^{(N+1)}$ as in Proposition 1 in §4.6, we get that

$$|R_N \circ g^{-1}| \leq |\exp \circ ((o(1) - \varepsilon(N+1))\exp^{[m]})| \leq |\exp(-\nu \exp^{[m]})|$$

in the domain $\ln^{[m]}\Omega$ for sufficiently large N. ►

§4.8. The conjugation lemma CL_n and the third shift lemma

A. Lemma CL_n. *Suppose that* $m = n$ *or* $m = n-1$, $k \leq m-1$, $f \in G^{m-1}$, *and* $g \in G^k$. *Then*

$$\mathrm{Ad}(g)(\mathrm{id} + \mathscr{F}^m_{+f}) \subset \mathrm{id} + \mathscr{F}^m_{+fg}.$$

Proof. We verify the proof for $m = n$; the proof is completely analogous for $m = n-1$.

In view of the ADT_{n-1}, which appears in the induction hypothesis,

$$g^{-1} = a + \sum \varphi_j, \qquad a \in \mathscr{Aff}, \quad \varphi_j \in \mathscr{F}^{k_j}_+, \quad k_j \leq n-1.$$

Let $\psi \in \mathscr{F}^n_{+f}$. Then

$$\varphi_j \circ (\mathrm{id} + \psi) = \varphi_j + \psi_j, \qquad \psi_j \in \mathscr{F}^n_{+f},$$

by the $\mathrm{SL}\,2_n$. Moreover, if $a' = \lambda$, then $a \circ (\mathrm{id} + \psi) = a + \lambda\psi$. Consequently,

$$\begin{aligned}\mathrm{Ad}(g)(\mathrm{id} + \psi) &= \left(a + \sum \varphi_j\right) \circ (\mathrm{id} + \psi) \circ g \\ &= \left(a + \sum \varphi_j + \lambda\psi + \sum \psi_j\right) \circ g \\ &\subset (g^{-1} + \mathscr{F}^n_{+f}) \circ g = \mathrm{id} + \mathscr{F}^n_{+fg}.\end{aligned}$$ ►

B. The Third Shift Lemma, $\mathrm{SL}\,3_n$. *Suppose that* $m = n$ *or* $m = n-1$, *and* $f \succ\succ g$ *in* G^{m-1} *or* $f \circ g^{-1} \in G^{m-1}_{\mathrm{rap}}$. *Then*

(a) $\mathscr{F}^m_{(+)f} \circ (\mathrm{id} + \mathscr{F}^m_{+g}) \subset \mathscr{F}^m_{(+)f}$.

(b) $(\mathrm{id} + \mathscr{F}^m_{+g})^{-1} = \mathrm{id} + \mathscr{F}^m_{+g}$ *for any* $g \in G^{m-1}$.

Proof. We prove the lemma for the case $m = n$; the proof is analogous in the case $m = n-1$.

(a) Let us prove the assertions about regularity and expandability.

Regularity. The regularity assertion goes as follows. Suppose that f, g are as required. Then

$$\mathscr{F}^{n^+}_{\mathrm{reg},\,g} \circ (\mathrm{id} + \mathscr{F}^{n^+}_{\mathrm{reg},\,f}) \subset \mathscr{F}^{n^+}_{\mathrm{reg},\,g}.$$

This is Lemma $\mathrm{SL}\,3_{n,\mathrm{reg}}a$, which was proved in §2.8.

Expandability. The proof of expandability consists of three parts: reduction to the case $g = \mathrm{id}$; estimation of a shifted remainder term; investigation of a shifted partial sum.

1. Reduction to the case $g = \mathrm{id}$. Let $\varphi \in \mathscr{F}_g^n$ and $\psi \in \mathscr{F}_{+f}^n$. By the $\mathrm{CL}_{n-1,n}$,

$$\mathrm{Ad}(g^{-1})(\mathrm{id}+\psi) = \mathrm{id}+\tilde{\psi}, \qquad \tilde{\psi} \in \mathscr{F}_{+h}^n, \quad h = f \circ g^{-1} \in G_{\mathrm{slow}}^{n-1} \cup G_{\mathrm{rap}}^{n-1}.$$

Let $\varphi = F \circ \exp^{[n]} \circ g$, $F \in \mathscr{F}\mathscr{C}^n$. Then

$$\varphi \circ (\mathrm{id}+\psi) \circ g^{-1} = F \circ \exp^{[n]} \circ (\mathrm{id}+\tilde{\psi}),$$
$$F \circ \exp^{[n]} \in \mathscr{F}_{\mathrm{id}}^n.$$

The reduction is complete.

2. Estimation of a shifted remainder term. We prove that under a shift of the argument by the correction $\tilde{\psi}$ a remainder term of a STAR-n keeps its order of smallness; the corresponding inequality is written at the end of the paragraph. Suppose that $\tilde{\psi} \in \mathscr{F}_+^N$. Then

$$\exp^{[n]} \circ (\mathrm{id}+\tilde{\psi}) = \exp^{[n]} + \psi_1 = \exp^{[n]} + O(1), \qquad \psi_1 \in \mathscr{F}_+^n.$$

The first equality follows from the relation $\exp^{[n]} \in \mathscr{F}^{n-1}$ and $\mathrm{SL}\,2_n$, and the second from the boundedness of cochains of class $\mathscr{F}_+^n$. Consequently, if R is a remainder term of a STAR-n and satisfies in the domain $\ln^{[n]} \Omega_{\mathrm{st}}$ the estimate

$$R = o(|\exp(-\nu \exp^{[n]} \zeta)|) \quad \text{as } \zeta \to \infty,$$

then

$$\begin{aligned} R \circ (\mathrm{id}+\tilde{\psi}) &= o(|\exp(-\nu \exp^{[n]}(\zeta + \tilde{\psi}))|) \\ &= o(|\exp(-\nu \exp^{[n]} \zeta + O(1))|) \\ &= o(|\exp(-\nu \exp^{[n]} \zeta)|). \end{aligned}$$

The smallness of the shifted remainder term is proved.

3. It remains to prove that the shift of a partial sum of STAR-n by the correction $\tilde{\psi}$ ($\tilde{\psi}$ is the same, as in the beginning of the proof) is once more a sum of this kind, plus perhaps a cochain of class $\mathscr{F}_{+\mathrm{id}}^n$. This implies the whole lemma. We first investigate a shift of an exponent. Let $\mathbf{e} \in E^n \subset \mathscr{F}_{\mathrm{id}}^{n-1}$. Then by the $\mathrm{SL}\,2_n$,

$$\tilde{\mathbf{e}} = \mathbf{e} \circ (\mathrm{id}+\tilde{\psi}) = \mathbf{e} + \psi_2, \qquad \psi_2 \in \mathscr{F}_{+h}^n.$$

According to the corollary to the completeness lemma,

$$\exp \tilde{\mathbf{e}} \in (1 + \mathscr{F}_{+h)}^n \exp \mathbf{e}.$$

Let us now investigate a shift of the coefficients.

CASE 1: $h \in G_{\text{slow}}^{n-1^-}$. We prove two assertions simultaneously by induction on r: for $h \prec\prec \mathrm{id}$ in G^{n-1}

$$\mathscr{K}^{n,r} \circ (\mathrm{id} + \mathscr{F}_{+h}^{n}) \subset \mathscr{K}^{n}, \tag{$*$}$$

$$\mathscr{F}^{n,r} \circ (\mathrm{id} + \mathscr{F}_{+h}^{n}) \subset \mathscr{F}^{n}. \tag{$**$}$$

INDUCTION BASE. Let $r = 0$. Then

$$\mathscr{K}^{n,0} \circ (\mathrm{id} + \mathscr{F}_{+h}^{n}) \subset \mathscr{K}^{n,0} + \mathscr{F}_{+h}^{n},$$

by $\mathrm{SL}\,2_n$ and the definition of $\mathscr{K}^{n,0}$. Consequently, if s is a term of a STAR-$(n, 0)$, then

$$\tilde{s} = s \circ (\mathrm{id} + \tilde{\psi}) \qquad (\tilde{\psi} \in \mathscr{F}_{+h}^{n})$$

is again a term of a STAR-$(n, 0)$ with the same exponent and with coefficient in $(\mathscr{K}^{n,0} + \mathscr{F}_{+h}^{n})(1 + \mathscr{F}_{+h}^{n}) \subset \mathscr{K}^{n}$. The last inclusion follows from the definition of $\mathscr{K}^{n}$ and the multiplication lemma ML_n. We mention that the rank of the coefficients on the left-hand side is no longer equal to zero. Namely, the shift studied in $\mathrm{SL}\,3_n$ gives rise to coefficients of nonzero rank. For $r = 0$ the assertion $(*)$ is proved. It and the preceding arguments immediately yield the assertion $(**)$ for $r = 0$.

INDUCTION STEP. Suppose that $(*)$ and $(**)$ have been proved for $r = q$; we prove them for $r = q + 1$. By the definition of $\mathscr{K}^{n,q+1}$,

$$\mathscr{K}^{n,q+1} \circ (\mathrm{id} + \mathscr{F}_{+h}^{n}) \subset \mathscr{L}(\mathscr{K}^{n,0}, \mathscr{F}_{+g}^{n,q} | g \prec\prec \mathrm{id}) \circ (\mathrm{id} + \mathscr{F}_{+h}^{n}).$$

For $g \preccurlyeq h$ we have that $\mathscr{F}_{+g}^{n,q} \circ (\mathrm{id} + \mathscr{F}_{+h}^{n}) \subset \mathscr{F}_{+g}^{n,q} + \mathscr{F}_{+h}^{n}$ by $\mathrm{SL}\,2_n$, and $\mathscr{F}_{+g}^{n,q} \circ (\mathrm{id} + \mathscr{F}_{+h}^{n}) \subset \mathscr{F}_{+g}^{n}$ for $g \succcurlyeq h$ by the induction hypothesis. This implies $(*)$ for $r = q + 1$; this implies in turn assertion $(**)$ for $r = q + 1$. This completes the induction and the proof of part (a) of the lemma in the case 1.

CASE 2. $h \in G_{\text{rap}}^{n-1^-}$. In this case

$$\mathscr{K}_0^{n}(\mathrm{id} + \mathscr{F}_{+h}^{n}) \subset \mathscr{K}^{n} + \mathscr{F}_{+h}^{n} = \mathscr{K}^{n} + \mathscr{F}_{+\mathrm{id}}^{n}$$

by $\mathrm{SL}\,2_n$, $\mathrm{SL}\,1_n$. The multiplication lemma, ML_n, completes the proof of part (a) in the case 2.

C. Inversion. We prove (b) of Lemma $\mathrm{SL}\,3_n$: suppose that $m = n$ or $m = n - 1$, and $g \in G^{m-1}$ is arbitrary. Then

$$(\mathrm{id} + \mathscr{F}_{+g}^{m})^{-1} = \mathrm{id} + \mathscr{F}_{+g}^{m}.$$

PROPOSITION 1. (a) $\mathrm{id} + \mathscr{F}_{+g}^{m} = \mathrm{Ad}(g)(\mathrm{id} + \mathscr{F}_{+\mathrm{id}}^{m})$.
(b) $\mathrm{id} + \mathscr{F}_{+\mathrm{id}}^{m} = A^{m}(\mathrm{id} + \mathscr{F}\mathscr{C}_{+}^{m})$.

PROOF. Assertion (a) is an immediate consequence of the conjugation lemma. The proof below of assertion (b) is similar to the arguments in subsection A. We prove the two inclusions

$$A^m(\mathrm{id}+\mathscr{F}\mathscr{C}_+^m) \subset \mathrm{id}+\mathscr{F}_{+\mathrm{id}}^m, \qquad ({}^{\ *}_{**})$$

$$A^{-m}(\mathrm{id}+\mathscr{F}_{+\mathrm{id}}^m) \subset \mathrm{id}+\mathscr{F}\mathscr{C}_+^m. \qquad ({}^{**}_{\ *})$$

PROOF OF $({}^{\ *}_{**})$. For any $k \le m-1$ we have the following chain of equalities and inclusions:

$$\begin{aligned} A(\mathrm{id}+\mathscr{F}\mathscr{C}_+^m \circ \exp^{[k]}) &= \ln\circ(\exp+\mathscr{F}\mathscr{C}_+^m \circ \exp^{[k+1]}) \\ &= \zeta + \ln(1+(\exp(-\zeta))\mathscr{F}\mathscr{C}_+^m \circ \exp^{[k+1]}) \\ &\subset \zeta + \ln(1+\mathscr{F}\mathscr{C}_+^m \circ \exp^{[k+1]}) \subset \mathrm{id}+\mathscr{F}\mathscr{C}_+^m \circ \exp^{[k+1]}. \end{aligned}$$

The first inclusion in the chain follows from the multiplication lemma, with the relation $\exp(-\zeta) \subset \mathscr{F}\mathscr{C}^0$ taken into account. The last inclusion follows from Corollary 2 in §4.5. We get the inclusion $({}^{\ *}_{**})$ from this by induction on k.

PROOF OF $({}^{**}_{\ *})$. For arbitrary $k \le m$ we have the chain of equalities and inclusions

$$\begin{aligned} A^{-1}(\mathrm{id}+\mathscr{F}\mathscr{C}_+^m \circ \exp^{[k]}) &= \exp\circ(\ln+\mathscr{F}\mathscr{C}_+^m \circ \exp^{[k-1]}) \\ &= \zeta\exp\circ\mathscr{F}\mathscr{C}_+^m \circ \exp^{[k-1]} \subset \zeta(1+\mathscr{F}\mathscr{C}_+^m \circ \exp^{[k-1]}) \\ &= \zeta + \zeta\cdot\mathscr{F}\mathscr{C}_+^m \circ \exp^{[k-1]} \subset \zeta+\mathscr{F}\mathscr{C}_+^m \circ \exp^{[k-1]}. \end{aligned}$$

The first inclusion in the chain follows from Corollary 1 to the completeness lemma, and the second from the multiplication lemma, with the relation $\zeta \in \mathscr{F}\mathscr{C}^0$ taken into account.

The inclusions $({}^{\ *}_{**})$ and $({}^{**}_{\ *})$ imply assertion (b) of Proposition 1. ►

We proceed to a proof of Lemma $\mathrm{SL}\,3_n$ b. By Proposition 1, it remains to prove that:

$$(\mathrm{id}+\mathscr{F}_{+\mathrm{id}}^m)^{-1} = \mathrm{id}+\mathscr{F}_{+\mathrm{id}}^m,$$

Regularity. It suffices to prove that

$$(\mathrm{id}+\mathscr{F}_{\mathrm{reg},\mathrm{id}}^{m^+})^{-1} = \mathrm{id}+\mathscr{F}_{\mathrm{reg},\mathrm{id}}^{m^+}.$$

This is assertion b of Lemma $\mathrm{SL}\,3_{n,\mathrm{reg}}$ in §2.8 for $g = \mathrm{id}$.

Expandability. Let $h \in \mathscr{F}_{+\mathrm{id}}^m$. Then the cochain $\mathrm{id}-\tilde{h}$ inverse to $\mathrm{id}+h$ satisfies the equation

$$\tilde{h} = h\circ(\mathrm{id}-\tilde{h}). \qquad (***)$$

We consider the "successive approximations" h_p of $\tilde{h}$:

$$h_0 = h, \quad h_1 = h\circ(\mathrm{id}-h_0), \ \dots, \quad h_p = h\circ(\mathrm{id}-h_{p-1}).$$

By Lemma $\mathrm{SL}\,3_n$ a, each of the approximations h_p belongs to $\mathscr{F}_{+\mathrm{id}}^m$. In particular, h_p can be approximated by a partial sum of an STAR-m, as required in the definition of expandability (Definition 3^n in §1.7).

We prove that for sufficiently large p the difference $\tilde{h} - h_p$ satisfies the estimate on the remainder term in the same definition. It will be shown by induction on p that for some $c > 0$, $\varepsilon > 0$ and some standard domain Ω

$$|\tilde{h} - h_p| \leq c^{p+2} \exp(-\varepsilon(p+2)\exp^{[m]}) \quad \text{in } \ln^{[m]} \Omega.$$

Let Ω be a domain in which the cochain $h \circ \ln^{[m]}$ is defined.

We begin with an estimate of $|\tilde{h}|$. The equation $(***)$ implies successively the assertions: in $\ln^{[m]} \Omega$

$$|\tilde{h}| = o(1), \qquad |\tilde{h}| \prec |\exp(-\varepsilon \exp^{[m]} \circ (\mathrm{id} + o(1)))|$$
$$\prec |\exp(-\varepsilon \exp^{[m]} \circ (\mathrm{id} + 1))| \stackrel{\text{def}}{=} \kappa_\varepsilon,$$
$$|\tilde{h}| \prec |\exp(-\varepsilon \exp^{[m]} \circ (\mathrm{id} + \kappa_\varepsilon))| \prec c|\exp(-\varepsilon \exp^{[m]})|.$$

Similarly, there exists a $c > 0$ such that for all $\theta \in [0, 1]$

$$|h' \circ (\mathrm{id} - \theta \tilde{h})| \prec c|\exp(-\varepsilon \exp^{[m]})|.$$

Induction base. $p = 0$.

$$\begin{aligned} |\tilde{h} - h_0| &= |\tilde{h} - h| = |h \circ (\mathrm{id} - \tilde{h}) - h| \\ &\leq |h' \circ (\mathrm{id} - \theta \tilde{h})|\, |\tilde{h}| \\ &\leq c^2 |\exp(-2\varepsilon \exp^{[m]})| \end{aligned}$$

Induction step.

$$\begin{aligned} |\tilde{h} - h_p| &= |h \circ (\mathrm{id} - \tilde{h}) - h \tilde{\circ} (\mathrm{id} - h_{p-1})| \\ &\leq c|\exp(-\varepsilon \exp^{[m]})|\, |\tilde{h} - h_{p-1}| \\ &\leq c^{p+2} |\exp(-\varepsilon(p+2) \exp^{[m]})|. \end{aligned}$$

This proves the expandability of the germ $\tilde{h}$. ▶

§4.9. The fourth and fifth shift lemmas

With account taken of the convention mentioned at the beginning of §4.2, Lemmas $\mathrm{SL}\,4_n$ and $\mathrm{SL}\,5_n$ can be formulated as follows.

Lemma a. $J^{n-1} \subset \mathrm{Gr}(\mathrm{id} + \mathscr{F}_+^{n-1})$.

Lemma $*$. *Suppose that $m = n$ or $m = n - 1$. Then for all $g \in G^{m-1}$*

$$\mathscr{F}^m_{g(+)} \circ J^{m-1} \subset \mathscr{F}^m_{g(+)}.$$

A. Proof of Lemma a. This lemma coincides with assertion a of Lemma $\mathrm{SL}\,4_n$. The argument is parallel to that used in §2.10.

Consider the generating elements of the group J^{n-1} of the form

$$h = \mathrm{Ad}(g) A^{n-1} f, \qquad g \in G^{n-1},\ f \in \mathscr{A}^0.$$

Our goal is to prove that

$$h \in \mathrm{id} + \mathscr{F}_{+}^{n-1}.$$

We prove this first for the case $g = \mathrm{id}$. For this it suffices to prove that

$$A^{n-1} f \in \mathrm{id} + \mathscr{F}_{+\,\mathrm{id}}^{n-1} \quad \text{for } f = \mathrm{id} + \varphi\,, \quad \varphi \in \mathscr{F}\mathscr{C}_{+}^{0}\,,$$

or

$$\ln^{[n-1]} \circ (\mathrm{id} + \varphi) \circ \exp^{[n-1]} - \mathrm{id} \in \mathscr{F}_{+\,\mathrm{id}}^{n-1}.$$

The regularity—the left-hand side belonging to $\mathscr{F}_{\mathrm{reg},\,\mathrm{id}}^{n-1^{+}}$—was already proved in §2.10. Let us prove expandability.

We prove by induction on m that for $m \le n-1$

$$A^{m}(\mathrm{id} + \varphi) - \mathrm{id} \in \mathscr{F}_{+\,\mathrm{id}}^{n-1} \circ \ln^{[n-1-m]} = \mathscr{F}\mathscr{C}_{+}^{n-1} \circ \exp^{[m]}. \qquad (*)$$

INDUCTION BASE. $m = 0$. The assertion follows from the fact that $\mathscr{F}\mathscr{C}_{+}^{0} \subset \mathscr{F}\mathscr{C}_{+}^{n-1}$.

INDUCTION STEP. Suppose that relation $(*)$ has been proved for m. We prove it for $m+1$.

Let

$$A^{m}(\mathrm{id} + \varphi) = \mathrm{id} + \varphi_{m}.$$

Then

$$\begin{aligned}\varphi_{m+1} &= \ln \circ (\mathrm{id} + \varphi_{m}) \circ \exp - \mathrm{id} \\ &= \ln \circ (\exp + \varphi_{m} \circ \exp) - \mathrm{id} = \ln(1 + \varphi_{m} \circ \exp / \exp).\end{aligned}$$

But $\varphi_{m} = F_{m} \circ \exp^{[m]}$, $F_{m} \in \mathscr{F}\mathscr{C}_{+}^{n-1}$ by the induction hypothesis. According to Corollary 2 in §4.5,

$$\varphi_{m+1} = F_{m+1} \circ \exp^{[m+1]}, \qquad F_{m+1} \in \mathscr{F}\mathscr{C}_{+}^{n-1}.$$

This implies $(*)$ for any $m \le n-1$, in particular, for $m = n-1$.

Let us proceed to the case of an arbitrary $g \in G^{n-1}$. We use Lemma $*$ for $n = m-1$, which is proved below and does not use the Lemma a in the case $n = m-1$. We prove that for arbitrary $g \in G^{n-1}$ and $\varphi \in \mathscr{F}_{+\,\mathrm{id}}^{n-1}$

$$\mathrm{Ad}(g)(\mathrm{id} + \varphi) \in \mathrm{id} + \mathscr{F}_{+g}^{n-1}.$$

The proof repeats word-for-word the arguments at the end of §2.10, except that $\mathscr{F}_{+g}^{n-1}$ must be written instead of $\mathscr{F}_{\mathrm{reg},\,g}^{n-1^{+}}$, and CL_{n}, $\mathrm{Sl}\,4_{n-1}$.a, and $\mathrm{SL}\,2_{n}$ instead of $\mathrm{CL}_{n,\,\mathrm{reg}}$.a, and $\mathrm{SL}\,2_{n,\,\mathrm{reg}}$. The second of these lemmas appears in the induction hypothesis, and the others were proved above.

B. Plan of proof of Lemma $*$. The proof consists of the same three parts as for Lemma $\mathrm{SL}\,3_{n}$: reduction to the case $g = \mathrm{id}$; preservation of partial sums under shifts; estimation of a shifted remainder term. The first part is trivial. Indeed, suppose that $h \in J^{m-1}$, $g \in G^{m-1}$, and $m = n$ or

$m = n-1$. Then the relations $\mathscr{F}^m_{(+)g} \circ h \subset \mathscr{F}^m_{(+)g}$ and $\mathscr{F}^m_{(+)\,\mathrm{id}} \circ \tilde{h} \subset \mathscr{F}^m_{(+)\,\mathrm{id}}$, where $\tilde{h} = \mathrm{Ad}(g)h$, are equivalent. But $\tilde{h} \in J^{m-1}$ together with h, by the definition of J^{m-1}. This concludes the reduction. The second and third parts of the proof proceed according to the following plan.

Cochains in $\mathscr{F}^m_{+\,\mathrm{id}}$ can be expanded in asymptotic series STAR-m. We verify that a shift of the argument—replacement of ζ by an $h \in J^{m-1}$—carries the partial sums of a STAR-m into partial sums of a similar series, and remainder terms into remainder terms.

We prove that under this shift of the argument each monomial in a STAR-m remains a monomial of the same class. Shifts of exponents and shifts of coefficients are investigated separately.

C. Shift of exponents.

LEMMA 2. *Suppose that* $m \leq n$. *Then*

$$E^m \circ J^{m-1} \subset E^m + \mathscr{L}(\mathscr{F}^{m-1}_{1+\mathrm{id}} \circ G^{m-2^+}). \qquad (**)$$

REMARK. This is an exact analogue of Lemma 1 in §4.3; the proofs are parallel. Lemma 2 will be proved for $m = n$; for $m < n$ the proofs are analogous and use lemmas with index $m < n$ that appear in the induction hypothesis.

PROOF. It follows from $\mathrm{SL}\,4_n\mathrm{a}$ that $J^{n-1} \subset \mathrm{Gr}(\mathrm{id} + \mathscr{F}^{n-1}_{1+})$. What is more, the generators in J^{n-1} can be taken from the set $\mathrm{Ad}(G^{n-1})A^{n-1}\mathscr{A}^0 \subset \mathrm{id} + \mathscr{F}^{n-1}_{1+}$. By Lemmas $\mathrm{SL}\,2_{n-1}$ and $\mathrm{SL}\,3_{n-1}$, which appear in the induction hypothesis, and by Lemma CL_{n-1}

$$\mathscr{L}(\mathscr{F}^{n-1}_{1+\mathrm{id}} \circ G^{n-2^+} \circ (\mathrm{id} + \mathscr{F}^{n-1}_{1+})) \subset \mathscr{L}(\mathscr{F}^{n-1}_{1+\mathrm{id}} \circ G^{n-2^+}),$$

the second term on the right-hand side of $(**)$ "does not notice the shift of the argument." Therefore, it suffices to prove the lemma with J^{n-1} replaced by $\mathrm{Ad}(G^{n-1})A^{n-1}\mathscr{A}^0$.

Accordingly, let $\mathbf{e} \in E^n$, and

$$h = \mathrm{Ad}(g)A^{n-1}f, \qquad g \in G^{n-1}, \ f \in \mathscr{A}^0.$$

We construct an expansion

$$\mathbf{e} \circ h = \Sigma + \varphi + \psi, \qquad \varphi \in \mathscr{L}(\mathscr{F}^{n-1}_{1+\mathrm{id}} \circ G^{n-2^+}_{\mathrm{slow}}),\ \psi \in \mathscr{F}^{n-1}_{1+\mathrm{id}},$$

just as done in the proof of Lemma 1 in §4.3. Namely, by Lemmas $\mathrm{SL}\,2_{n-1}$ and $\mathrm{SL}\,3_{n-1}$, which appear in the induction hypothesis,

$$\mathbf{e} \circ h = \tilde{\mathbf{e}} + \varphi, \qquad \tilde{\mathbf{e}} \in \mathscr{F}^{n-1}_{1\,\mathrm{id}},\ \varphi \in \mathscr{L}(\mathscr{F}^{n-1}_{1+\mathrm{id}} \circ G^{n-2^+}_{\mathrm{slow}}).$$

Let Σ be that partial sum of the STAR-$(n-1)$ for $\tilde{\mathbf{e}}$ that consists of all the terms of this series with nonnegative principal exponents.

Then

$$\tilde{\mathbf{e}} = \Sigma + \psi, \qquad \psi \in \mathscr{F}\mathscr{C}^{n-1}_{1+\mathrm{id}}.$$

It suffices to prove that $\Sigma \subset E^n$. In large part the proof recalls that of Lemma 1 in §4.3. We verify all the requirements of the definition of E^n for the sum Σ; they are recalled in the proof of Lemma 1.

Requirement 1° holds by construction, with the exception of weak realness, which is proved together with requirement 4°.

We precede the verification of requirements 2° and 3° by the description obtained in §2.9 for the germ $A^{-n}h$. It was shown there that three cases are possible in dependence on the value of the limit $\mu = \lim(A^{1-n}g)'$:

1°. $\mu = \infty \Rightarrow A^{-n}h = \mathrm{id} + \varphi$, $\varphi \in \mathscr{F}\mathscr{C}_{\mathrm{wr}}(\mathscr{D}^n_0)$ (Lemma 6_n in §2.9.).

2°. $\mu \in (0, \infty) \Rightarrow$ there exist a $g \in G^{n-1}_{\mathrm{rap}}$ and a $\varphi \in \mathscr{F}\mathscr{C}_{\mathrm{wr}}(\mathscr{D}^n_0)$ such that

$$A^{-n}h = g \circ (\mathrm{id} + \varphi) \qquad \text{(Lemma } 7_n \text{ in §2.9).}$$

3°. $\mu = 0 \Rightarrow A^{-n}h \in L^n$ (Definition 15 in §1.6C).

As in Lemma 1, we prove that requirements 2° and 3° on the germs Σ and $\mathbf{e} \circ h$ are equivalent. Note that

$$\mathbf{e} \circ h \circ \ln^{[n]} = (\mathbf{e} \circ \ln^{[n]}) \circ (A^{-n}h).$$

In Case 1° the correction of the germ $A^{-n}h$ tends to zero in some standard domain Ω of class n. Therefore, $A^{-n}h/\zeta \to 1$ on $(\mathbf{R}^+, \infty)$, and the inequalities

$$|\operatorname{Re}(A^{-n}h) - \mathrm{id}| < \varepsilon\xi, \qquad |A^{-n}h - \mathrm{id}| < \varepsilon|\zeta|$$

hold in Ω for arbitrary $\varepsilon > 0$. This gives us requirements 2° and 3° for the sum Σ.

In Case 2°, requirements 2° and 3° on the composition $\mathbf{e}_1 = \mathbf{e} \circ g$ have already been proved in Lemma 1 in §4.3 E. The same requirements on $\mathbf{e} \circ h = (\mathbf{e} \circ g) \circ (\mathrm{id} + \varphi)$ are proved in the same way as in Case 1°. (We use only that $\varphi = o(1)$ in Ω.)

In Case 3°, requirements 2° and 3° follow from the corollary in §2.11B and Proposition 1 in §2.11A. By this corollary, $\lim_{(\mathbf{R}^+, \infty)} A^{-n}h/\zeta = 1$. Consequently,

$$\lim_{(\mathbf{R}^+, \infty)} \frac{\mathbf{e} \circ h}{\exp^{[n]}} = \lim_{(\mathbf{R}^+, \infty)} \frac{\mathbf{e} \circ h}{\exp^{[n]} \circ h} \cdot \frac{\exp^{[n]} \circ h}{\exp^{[n]}}$$
$$= \nu(\mathbf{e}) \lim_{(\mathbf{R}^+, \infty)} \frac{A^{-n}h}{\zeta} = \nu(\mathbf{e}).$$

This proves requirement 2°. Let us prove the third requirement. Recall that in Case 3°, which we now consider, $A^{-n}h \in L^n$. It follows from Proposition 1 in §2.11 that there exists a standard domain Ω of class n in which

$$\operatorname{Re} A^{-n}h \prec \mu\xi, \qquad |A^{-n}h| = (1 + o(1))|\zeta|$$

for arbitrary $\mu > 1$. Consequently,

$$|\operatorname{Re} \mathbf{e} \circ h \circ \ln^{[n]}| = |\operatorname{Re} \mathbf{e} \circ \ln^{[n]} \circ A^{-n} h| \preccurlyeq \nu \operatorname{Re} A^{-n} h \prec \mu\nu\xi,$$
$$|\mathbf{e} \circ h \circ \ln^{[n]}| = |\mathbf{e} \circ \ln^{[n]} \circ A^{-n} h| \prec \nu |A^{-n} h| \prec (\nu\mu)|\zeta|.$$

This concludes the verification of requirement 3°.

We prove the weak realness of the sum Σ, and at the same time

Requirement 4°. $\operatorname{Im} \Sigma \to 0$ on $(\mathbf{R}^+, \infty)$.

For this it suffices to prove that

$$\operatorname{Im} \mathbf{e} \circ h \circ \ln^{[n-1]}$$

decreases on $(\mathbf{R}^+, \infty)$ more rapidly than any exponential. Arguing as in §4.3 F, we then get that the sum Σ is weakly real.

Let h^u be the main function of the tuple h, which is defined on the domain adjacent from above to the real axis, more precisely, to $(\mathbf{R}^+, \infty)$. We estimate $\operatorname{Im} h^u$. It will be proved that for some $\nu > 0$

$$|\operatorname{Im} h(\xi)| \prec \exp(-\exp^{[n]} \nu\xi) \qquad \left({}^{\ *}_{**}\right)$$

on $(\mathbf{R}^+, \infty)$.

The cochain $f \in \mathscr{A}^0$ is weakly real. By formula $\left({}^{*}_{*}\right)$ in §1.11 D,

$$|\operatorname{Im} f||_{(\mathbf{R}^+, \infty)} < \exp(-\exp c\xi)$$

for some $c > 0$; here f is the same as in the expression for h. Consequently,

$$|\operatorname{Im} f \circ \exp^{[n-1]}| < \exp(-\exp c \exp^{[n-1]} \xi),$$

and

$$f \circ \exp^{[n-1]}(\mathbf{R}^+, \infty) \subset (\Pi^\forall, \infty).$$

Therefore,

$$|\operatorname{Im} A^{n-1} f| < c_1 \exp(-\exp c \exp^{[n-1]} \xi)$$

for some $c_1 > 0$, because the function $\ln^{[n-1]}$ has bounded derivative in the domain $(\mathbf{C}^+, \infty)$. It follows from the additive decomposition theorem that the germ g in the expression for h (see the beginning of the proof) extends biholomorphically to a domain of the form $\Pi_1 = c \ln^{[n-1]} \Omega$ for some standard domain Ω of class $n-1$ and some $c > 0$. The derivative of the extended germ is bounded. Further, the germ of the domain $|\eta| < c_1 \exp(-\exp c \exp^{[n-1]} \xi)$ at infinity belongs to the germ (Π_1, ∞) for arbitrary positive c_1 and c. For any $\nu \in (0, c) \cap (0, 1)$ we get the inequality $\left({}^{\ *}_{**}\right)$.

For an arbitrary sector S, for example, $S = \{|\eta| \le \xi\}$, the domain $\ln^{[n]} S$ is bounded by curves whose distance from $\mathbf{R}^+$ decreases no more rapidly than $\exp(-c \exp^{[n-1]} \xi)$ for some $c > 0$. Therefore, the germ at infinity of the domain $\widetilde{\Omega}$ given by

$$|\eta| < \exp(-\exp^{[n]} \nu\xi)$$

belongs to the germ at infinity of the domain $\Omega_0 = \ln^{[n]}\Omega$ for an arbitrary standard domain Ω, and the margin between the domains $\widetilde{\Omega}$ and Ω_0 is greater than $1/\exp^{[n]}\xi$. By the definition of the exponent $\mathbf{e} \in E^n$ (requirement 3° of Definition 2^n in §1.7), there exists $\mu > 0$ such that in some standard domain Ω we have that $|\mathbf{e} \circ \ln^{[n]}| \prec \mu|\zeta|$, or

$$|\mathbf{e}| \prec \mu|\exp^{[n]}\zeta| \leq \mu \exp^{[n]}\xi$$

in Ω_0. In view of Cauchy's estimate,

$$|\mathbf{e}'| \prec \mu \exp \circ 2\exp^{[n-1]}\xi \qquad ({}^{\ *}_{***})$$

in the domain $\widetilde{\Omega}$. By the mean value theorem,

$$\begin{aligned}|\operatorname{Im}(\mathbf{e} \circ h - \mathbf{e})| &\prec \mu \exp(2\exp^{[n-1]}\xi - \exp^{[n]}\nu\xi) \\ &\prec \exp(-\exp^{[n]}\nu'\xi)\end{aligned}$$

on $(\mathbf{R}^+, \infty)$ for some $\nu' > 0$. But $\operatorname{Im}\mathbf{e}|_{(\mathbf{R}^+, \infty)} \to 0$ by requirement 4° of Definition 2^n in §1.7. This implies requirement 4° for $\tilde{\mathbf{e}}$.

According to the criterion for weak realness, $|\operatorname{Im}\mathbf{e} \circ \ln^{[n-1]}|$ decreases on $(\mathbf{R}^+, \infty)$ more rapidly than any exponential. Further,

$$|(\operatorname{Im}\mathbf{e} \circ h - \operatorname{Im}\mathbf{e}) \circ \ln^{[n-1]}| \prec |\exp(-\exp A^{1-n}\nu'\xi)|. \qquad ({}^{**}_{\ *})$$

By Lemma 5.5_nb, $A^{1-n}\nu'\xi \succ c\ln\xi$ for some $c > 0$. Consequently, the right-hand side of inequality $({}^{**}_{\ *})$ decreases more rapidly than any exponential and even more rapidly than any function of the form $\exp(-\xi^c)$, $c > 0$. This implies that Σ is weakly real. This concludes the proof of Lemma 2. ▶

D. Shifts of coefficients. We prove simultaneously the following two assertions by induction on r:

$$\mathscr{K}^{n,r} \circ J^{n-1} \subset \mathscr{K}^n, \qquad (***)$$

$$\mathscr{F}^{n,r}_{(+)g} \circ J^{n-1} \subset \mathscr{F}^{n,r}_{(+)g}. \qquad ({}^{*}_{*})$$

INDUCTION BASE: $r = 0$. The formula $(***)$ for $r = 0$ follows from the definition of $\mathscr{K}^{n,0}$ and Lemmas SL 2_{n-1}, SL 3_{n-1}, and SL 4_{n-1}.a. This and the preceding results of the section give us the formula $({}^{*}_{*})$, which for $r = 0$ yields Lemma $*$, or, what is the same, assertion b of Lemma SL 4_n (and Lemma SL 5_n if n is replaced by $n - 1$ in the preceding arguments). The subsequent arguments concluding the proof of this formula and these lemmas for any r are the same ([12]) as in §4.8 in the proof of formulas $(*)$ and $(**)$.

([12]) The estimate of the remainder term obtained in subsection D independently of the arguments in subsection C is used here.

This completes the proof of all the assertions of subsection B; in particular, it has been proved that under composition (from the right) with a germ $h \in J^{n-1}$ a partial sum of a STAR-n remains a partial sum of a STAR-n. We proceed to an estimate of the remainder term.

E. Shifts of a remainder term. Consider the case $m = n$; the case $m = n-1$ is analyzed similarly. We prove that under a shift of the argument by a correction $\psi \in \mathscr{F}_+^{n-1}$ a remainder term of a STAR-n "almost keeps" its order of smallness (see the estimate on $R \circ (\mathrm{id} + \psi)$ at the end of the paragraph). Let

$$R = o(|\exp(-\nu \exp^{[n]} \zeta)|).$$

We prove that there exists a standard domain $\Omega \in \Omega_n$ such that for arbitrary $\mu < \nu$

$$R \circ (\mathrm{id} + \psi) = o(|\exp(-\mu \exp^{[n]} \zeta)| \quad \text{as } \zeta \to \infty$$

in the domain $\ln^{[n]} \Omega$.

For $\psi \in \mathscr{F}_+^{n-1}$

$$\begin{aligned} \exp^{[n]} \circ (\mathrm{id} + \psi) &= \exp \circ \exp^{[n-1]} \circ (\mathrm{id} + \psi) \\ &= \exp \circ (\exp^{[n-1]} + \tilde{\psi}), \qquad \tilde{\psi} \in \mathscr{F}_+^{n-1}. \end{aligned}$$

The last equality follows from the relation $\exp^{[n-1]} \in \mathscr{F}^{n-2}$ and $\mathrm{SL}\,2_{n-1}$. By the corollary to Lemma LA_n, there exists a $\psi_1 \in \mathscr{F}_+^{n-1}$ such that

$$\exp \circ (\exp^{[n-1]} + \tilde{\psi}) = (1 + \psi_1) \exp^{[n]}.$$

Then

$$\exp(-\nu \exp^{[n]} \circ (\zeta + \psi)) = \exp \circ ((-\nu + \psi_1) \exp^{[n]}).$$

We prove that there exists a standard domain Ω of class n such that for arbitrary $\mu < \nu$

$$\exp \circ (-\nu + \psi_1) \exp^{[n]} = o(|\exp(-\mu \exp^{[n]})|) \qquad \binom{**}{**}$$

in the domain $\ln^{[n]} \Omega$. Since $\psi_1 \in \mathscr{F}_+^{n-1}$, it follows that $\psi_1 \to 0$ in $\ln^{[n]} \Omega$ for some domain Ω of class n. But if the term ψ_1 were real, the preceding equality would be obvious, at least on $\mathbf{R}^+$. The fact that ψ_1 is not real and the necessity of going out into the complex plane require further estimates. We verify them in detail for $n \geq 2$, and then briefly for $n = 1$.

Suppose that $\psi_1 = F \circ \exp^{[n-1]} \circ h$, $F \in \mathscr{F}\mathscr{C}_+^{n-1}$, $h \in G^{n-2}$. By Lemma 5.5_nc, there exists a standard domain of class n in which

$$|A^{-n} h| > |\ln \zeta|^c$$

for any $c > 0$. The domain $(\zeta + c_1 \zeta / \ln \zeta)(\mathbf{C}^+ \backslash k)$ is a standard domain of class n for arbitrary $n \geq 2$. For sufficiently large c_1 the estimate

$$|\eta| \prec \varepsilon \xi |\ln \zeta|^3$$

holds in it for any $\varepsilon > 0$ (see Example 3 in §1.6). Since the intersection of two standard domains contains a standard domain (by Lemma 5.2_nd), we can take a standard domain Ω in which the preceding two inequalities hold: the value of c will be chosen below.

We prove that in Ω we have the equality

$$\exp(-\nu + \psi_1 \circ \ln^{[n]})\zeta = o(|\exp(-\mu\zeta)|), \qquad \forall \mu < \nu, \qquad \binom{*}{{*}{*}}_{*}$$

which implies $\binom{**}{**}$.

By definition,

$$\psi_1 \circ \ln^{[n]} = F \circ \ln \circ A^{-n} h.$$

It follows from the relation $F \in \mathscr{F}\mathscr{C}_+^{n-1}$ that there exist a ν_0 such that

$$|F| \prec \exp(-\nu_0 \xi).$$

Consequently, by Lemma 5.5_nb, for arbitrary $C > 0$, in Ω

$$|F \circ \ln \circ A^{-n} h| \prec |A^{-n} h|^{-\nu_0} \prec (\ln|\zeta|)^{-\nu_0 c}.$$

Choose c such that $\nu_0 c > 3$. Then in Ω

$$|\zeta \psi_1 \circ \ln^{[n]} \prec | < |\zeta||\ln\zeta|^{-\nu_0 c} \prec |\zeta||\ln\zeta|^{-3} \prec \varepsilon\xi$$

for every $\varepsilon > 0$. This proves the inequality $\binom{*}{{*}{*}}_{*}$, and with it the estimate $\binom{**}{**}$.

For $n = 1$ the analogous arguments work with the use of Example 2 in §1.6 and the fact that $|A^{-1}h| > |\zeta|^{\varepsilon}$ for some $\varepsilon > 0$. ►

This finishes the proof of all the lemmas in §1.10. We have thereby proved the multiplicative and additive decomposition theorems: the arguments in §1.11 were conditional but now become unconditional. The Phragmén-Lindelöf theorem was proved in Chapter III. It remains only to prove the lemma in Chapter V and the theorem on a lower estimate; we now proceed to the latter.

The arguments of the next section serve simultaneously as a conclusion to Chapter I–IV and as a foundation for Chapter V: §5.1 is a translation of the real results of §4.10 to "complex language."

§4.10. The identity theorem and the ordering theorem

In this section we use induction on the class and the rank to prove the following assertions, which we then use to derive the theorems in the heading.

A. Formulations.

Lemma on a Lower Estimate for the Partial Sums of a STAR-(m, r), LEPS$_{m,r}$. *Suppose that Σ is a partial sum of a weakly real STAR-(m, r), $0 \le m \le n$, $r \ge 0$, with all the terms of Σ having the same principal exponent greater than ν. Then*:

1°. $\arg\Sigma \pmod{\pi} \to 0$ *on* $(\mathbf{R}^+, \infty)$;

2°. $|\Sigma| \succ \exp(\nu \exp^{[m]})$ *on* $(\mathbf{R}^+, \infty)$.

FIRST ORDER LEMMA, $\mathrm{OL}\,1_{m,r}$. *Suppose that* $0 \le m \le n$, $r \ge 0$, *and* $F \in \mathcal{F}\mathcal{C}_{\mathbf{R}}^{m,r}$. *Then the following hold on* $(\mathbf{R}^+, \infty)$:

1°. *if* $F \not\equiv 0$, *then* $\arg F \pmod{\pi} \to 0$;

2°. *the set* $\mathcal{F}\mathcal{C}_{\mathbf{R}}^{m,r}$ *is ordered by the relation*

$$F_1 \prec F_2 \Leftrightarrow \operatorname{Re} F_1 \prec \operatorname{Re} F_2;$$

3°. *there exists a* $\mu > 0$ *such that*

$$|\operatorname{Re} F| \succ \exp(-\mu\xi).$$

REMARK. Assertion 3° of the lemma is a consequence of §1.9—the only fact in §4.10 that is used in the proof of the identity theorem. The subsequent lemmas are needed for the inductive proof of the preceding lemma for $m = n$ and arbitrary r.

SECOND ORDER LEMMA, $\mathrm{OL}\,2_{m,r}$. *Suppose that* $(k, f) \prec (m, g)$, $f \in G^{k-1}$, $g \in G^{m-1}$, $m \le n$, $\varphi \in \mathcal{F}_{\mathbf{R}f}^{k,p}$, $\psi \in \mathcal{F}_{+g}^{m,r}$; *if* $k = m$, *then* $p \le r$. *Then* $|\operatorname{Re}\varphi| \succ |\psi|$.

THIRD ORDER LEMMA, $\mathrm{OL}\,3_{m,r}$. *The set* $\mathcal{K}_{\mathbf{R}}^{m,r}$ *is ordered by the relation* (∗). *For arbitrary* $a \in \mathcal{K}_{\mathbf{R}}^{m,r}$ *either* $a \equiv 0$ *or* $\arg a \pmod{\pi} \to 0$.

Denote by LEPS_m the union $\bigcup_r \mathrm{LEPS}_{m,r}$; $\mathrm{OL}\,1\text{–}3_m$ are defined similarly. The proofs of the lemmas formulated above are carried out by induction on the class and the rank, according to the following scheme:

$$\mathrm{LEPS}_0 \overset{①}{\Rightarrow} \mathrm{OL}\,1_0 \overset{②}{\Rightarrow} \mathrm{OL}\,2_0,$$

$$\mathrm{OL}\,2_m \overset{③}{\Rightarrow} \mathrm{OL}\,3_{m+1,0} \overset{④}{\Rightarrow} \mathrm{LEPS}_{m+1,0}, \qquad 0 \le m \le n-1,$$

$$\mathrm{LEPS}_{m,r} \overset{①'}{\Rightarrow} \mathrm{OL}\,1_{m,r} \overset{②'}{\Rightarrow} \mathrm{OL}\,2_{m,r} \overset{③'}{\Rightarrow} \mathrm{OL}\,3_{m,r+1} \overset{④'}{\Rightarrow} \mathrm{LEPS}_{m,r+1}, \qquad 1 \le m \le n,\ r \ge 0.$$

Assuming that the cochains of class zero do not depend on the rank, we can regard the implications ① and ② as the implications ①′ and ②′ for $m = 0$. The implications ④′ and ④″ are combined into one:

$$\mathrm{OL}\,3_{m,r} \overset{④''}{\Rightarrow} \mathrm{LEPS}_{m,r}, \qquad 1 \le m \le n,\ r \ge 0.$$

Since the first chain has been gone through—the corresponding implications have been proved—as has the second chain for $m = 0$, Lemma $\mathrm{LEPS}_{1,0}$ turns out to be proved. Since the third chain has been gone through for fixed m, Lemma LEPS_m and Lemmas $\mathrm{OL}\,1\text{–}3_m$ turn out to be proved by induction on r under the assumption that Lemma $\mathrm{LEPS}_{m,0}$ has been proved. Since the second chain has been gone through, Lemma $\mathrm{LEPS}_{m+1,0}$ turns out to be proved. This enables us to use induction on m. As a result, implications ①–④ and ①′–④′ prove Lemma $\mathrm{OL}\,1_m$ for $0 \le m \le n$, which was used in §1.9 in the proof of the identity theorem.

B. Induction base. The estimate in LEPS_0 serves as an induction base. Let us prove it. Recall that a STAR-$0_{\mathbf{R}}$ is a real Dulac series; the rank is not defined for such series. For a Dulac series a partial sum whose terms have the same principal exponents greater than ν is a real quasipolynomial with exponent greater than ν. Such a quasipolynomial Σ satisfies the inequality

$$|\Sigma| \succ \exp \nu\xi \quad \text{on } (\mathbf{R}^+, \infty).$$

This completes the induction base. The induction step is implemented in the following five propositions.

C. PROPOSITION 1. $\mathrm{LEPS}_{m,r} \Rightarrow \mathrm{OL}\,1_{m,r}$ *for* $0 \le m \le n$, $r \ge 0$.

REMARK. This is a combination of implications ① and ①′ in subsection A.

PROOF. 1. We consider a STAR-m Σ for the composition $F \circ \exp^{[m]}$. Two cases are possible: $\Sigma \equiv 0$ and $\Sigma \not\equiv 0$. In the first case we have by the definition of expandability in a STAR-m that

$$|F \circ \exp^{[m]}| \prec \exp(-\nu \exp^{[m]}) \quad \text{on } (\mathbf{R}^+, \infty) \text{ for every } \nu > 0,$$

or

$$|F| \prec \exp(-\nu\xi) \quad \text{on } (\mathbf{R}^+, \infty) \text{ for every } \nu > 0.$$

But $F \in \mathscr{F}\mathscr{C}^m$. By the Phragmén-Lindelöf theorem, which appears in the induction hypothesis for $m \le n-1$ and was proved in Chapter III for $m = n$, this gives us that $F \equiv 0$ on $(\mathbf{R}^+, \infty)$, a contradiction.

2. Let us consider the second case: $\Sigma \not\equiv 0$. We prove that in the STAR-m a monomial with principal exponent μ is majorized by the germ $\exp(\mu + \varepsilon)\exp^{[m]}$ on $(\mathbf{R}^+, \infty)$ for each $\varepsilon > 0$. Indeed, suppose that the monomial under consideration has the form $s = a \exp \mathbf{e}$, $\nu(\mathbf{e}) = \mu$. Then on $(\mathbf{R}^+, \infty)$

$$|s| = |a \exp(\mu + o(1)) \exp^{[m]}|$$

by requirement 2° of Definition $2''$.I in §1.7. We prove that for each $\varepsilon > 0$

$$|a| \prec |\tilde{\varphi}|, \quad \text{where } \tilde{\varphi} = \exp \varepsilon \exp^{[m]} \in \mathscr{A}_{\mathrm{id}}^m \tag{$**$}$$

(see §1.4). Indeed, since an arbitrary regular functional cochain is majorized by an exponential, we get from the definition of the set $\mathscr{K}^{m,r}$ that

$$|a| \prec \exp^{[m]} \circ \kappa \in \mathscr{A}_\kappa^m \quad \text{for some } \kappa \in \mathbf{R}, \kappa > 0.$$

The inequality $\exp^{[m]} \circ \kappa \prec \tilde{\varphi}$ is obvious. Consequently, for each $\varepsilon > 0$

$$|s| \prec \exp(\mu + \varepsilon) \exp^{[m]} \quad \text{on } (\mathbf{R}^+, \infty).$$

3. Using Lemma $\mathrm{LEPS}_{m,r}$, we prove that all the terms of a STAR-(m, r) for a weakly real germ can be assumed to be weakly real. Let $\varphi \in \mathscr{F}_{\mathbf{R},\mathrm{id}}^{m,r}$. For each $\mu \in \mathbf{R}$

$$\varphi = \Sigma_{\ge\mu} + \Sigma_{<\mu} + \widetilde{R},$$

where $\Sigma_{\geq\mu}$ and $\Sigma_{<\mu}$ are partial sums of the STAR-(m, r) for the germ φ that have principal exponents $\geq \mu$ and $< \mu$, respectively, and $\widetilde{R}$ is a remainder term with

$$\widetilde{R} = o(\exp(\mu - 1)\exp^{[m]}) \quad \text{on } (\mathbf{R}^+, \infty).$$

Let R and I be the same operators as in §1.7:

$$I\varphi = \frac{1}{2i}(\varphi(\zeta) - \overline{\varphi(\overline{\zeta})}), \qquad R\varphi = \frac{1}{2}(\varphi(\zeta) + \overline{\varphi(\overline{\zeta})}).$$

The weak realness of the germ φ is equivalent to the equalities

$$\operatorname{Re}\varphi = \varphi, \quad I\varphi = 0 \quad \text{on } (\mathbf{R}^+, \infty).$$

Then we get that on $(\mathbf{R}^+, \infty)$

$$0 = I\varphi = I\Sigma_{\geq\mu} + I\Sigma_{<\mu} + I\widetilde{R}.$$

Further, the operator I carries all germs into weakly real germs. If $I\Sigma_{\geq\mu} \not\equiv 0$, then, by Lemma $\mathrm{LEPS}_{m,r}$,

$$|\operatorname{Re} I\Sigma_{\geq\mu}| \succ \exp((\mu - \varepsilon)\exp^{[m]}) \quad \text{on } (\mathbf{R}^+, \infty)$$

for each $\varepsilon > 0$. On the other hand, by what was proved in part 2,

$$|\operatorname{Re} I\Sigma_{\geq\mu}| = |I\Sigma_{<\mu} + IR| \prec \exp(\mu - \varepsilon)\exp^{[m]}$$

on $(\mathbf{R}^+, \infty)$ for some sufficiently small $\varepsilon > 0$. The contradiction obtained proves that $I\Sigma_{\geq\mu} \equiv 0$, and the sum $\Sigma_{\geq\mu}$ is weakly real.

Further, $\Sigma_{\geq\mu} = \sum a_j \exp \mathbf{e}_j$. Since the left-hand side and the exponents $\mathbf{e}_j$ are weakly real,

$$\Sigma_{\geq\mu} = R\Sigma_{\geq\mu} = \sum (Ra_j)\exp \mathbf{e}_j.$$

The coefficients in the last sum are weakly real.

4. To prove assertion 2° we can now exploit Lemma $\mathrm{LEPS}_{m,r}$. Namely, by the definition of expandability in a STAR-(m, r),

$$F \circ \exp^{[m]} = \Sigma_\nu + \Sigma_{\leq\nu} + \widetilde{R},$$

where Σ_ν is a sum of terms with principal exponent ν, and

$$\widetilde{R} = o(\exp \circ (\nu - 1)\exp^{[m]}) \quad \text{on } (\mathbf{R}^+, \infty).$$

Then it follows from Lemma $\mathrm{LEPS}_{m,r}$ that on $(\mathbf{R}^+, \infty)$

$$|\operatorname{Re}\Sigma_\nu| \succ \exp \circ ((\nu - \varepsilon)\exp^{[m]})$$

for arbitrary $\varepsilon > 0$. But

$$|\Sigma_{\leq\nu} + \widetilde{R}| \prec \exp \circ ((\nu - \varepsilon)\exp^{[m]})$$

for sufficiently small $\varepsilon > 0$. Consequently, on $(\mathbf{R}^+, \infty)$

$$|\operatorname{Re} F \circ \exp^{[m]}| = |\operatorname{Re}\Sigma_\nu|(1 + o(1))| \succ \exp \circ ((\nu - 2\varepsilon)\exp^{[m]}).$$

This proves that the set $\mathscr{F}_{\mathbf{R},\mathrm{id}}^{m,r}$ is ordered by the relation $\mathrm{Re}(\cdot) \prec 0$—assertion 2° of Lemma $\mathrm{OL}\,1_m$ or $\mathrm{OL}\,1_{m,r}$.

5. Let us prove assertion 1° of the lemma. We have that

$$F \circ \exp^{[m]} = \Sigma_\nu(1 + o(1)).$$

The argument of the second factor tends to zero on $(\mathbf{R}^+, \infty)$, and the argument of the first tends to zero or π according to Lemma LEPS_m or $\mathrm{LEPS}_{m,r}$.

6. Finally, we prove assertion 3° of Lemma $\mathrm{OL}\,1_{m,r}$. Let $F \in \mathscr{F}\mathscr{C}_{\mathbf{R}}^{m,r}$ and $\varphi = F \circ \exp^{[m]}$. Then we get from the preceding inequality on $\mathrm{Re}\,\varphi$ that

$$|\mathrm{Re}\,F| \succ \exp(\nu - 2\varepsilon)\xi \quad \text{on } (\mathbf{R}^+, \infty). \quad \blacktriangleright$$

REMARK. This proves that the real part of a weakly real rapidly decreasing regular functional cochain of class (m, r) or m on $(\mathbf{R}^+, \infty)$ belongs to the multiplicatively Archimedean class of the germ $\exp(-\xi)$. This and assertion 1° of Lemma $\mathrm{OL}\,1_{m,r}$ give us that for arbitrary $\varphi \in \mathscr{F}_{\mathbf{R}+g}^{m,r}$, $g \in G^{m-1}$, the germs $\mathrm{Re}\,\varphi$, $|\varphi|$, and $\tilde{\varphi} = \exp(-\exp^{[m]} \circ g)$ are multiplicatively Archimedean-equivalent.

D. PROPOSITION 2. $\mathrm{OL}\,1_{m,r} \Rightarrow \mathrm{OL}\,2_{m,r}$ *for* $0 \le m \le n$, $r \ge 0$.

REMARK. This is a combination of the implications ② and ②′ in subsection A.

PROOF. Let φ and ψ be the same as in Lemma $\mathrm{OL}\,2_{m,r}$. Then, by the preceding remark, $\mathrm{Re}\,\varphi \in \mathscr{A}_f^k$ on $(\mathbf{R}^+, \infty)$. According to the assertion in part 2 of the proof of Proposition 1 and the definition of expandability in a STAR-m, we get that $|\psi| \prec \tilde{\psi} = \exp(-\nu \exp^{[m]} g) \in \mathscr{A}_g^m$ for some $\nu > 0$. By the induction hypothesis, which includes OT_{m-1}, the group G^{m-1} is m-proper in the sense of Definition 4 in §1.4 C. By Proposition 4 in §1.4 C, $|\mathrm{Re}\,\varphi| \succ |\tilde{\psi}|$. ►

E. PROPOSITION 3. $\mathrm{OL}\,1, 2_{m,r} \Rightarrow \mathrm{OL}\,3_{m,r+1}$ *for* $1 \le m \le n$ *and* $r \ge 0$, *and* $\mathrm{OL}\,1, 2_m \Rightarrow \mathrm{OL}\,3_{m+1,0}$ *for* $0 \le m \le n+1$.

REMARK. This is a combination of the implications ③ and ③′ in subsection A.

PROOF. Let $a \in \mathscr{K}_{\mathbf{R}}^{m,r} \backslash \{0\}$. It suffices to prove that the germ a is comparable with zero. The arguments are somewhat different for the two implications of the proposition. Let us prove the first. By definition,

$$a = \sum \varphi_j + \sum \psi_j, \qquad \varphi_j \in \mathscr{F}^{m-1}, \quad \psi_j \in \mathscr{F}_{+g_j}^{m,r}, \quad g_j \in G_{\mathrm{slow}}^{m-1^-}.$$

It can be assumed without loss of generality that all the terms in this sum are weakly real. Indeed, $a = Ra = \sum R\varphi_j + \sum R\psi_j$ on $(\mathbf{R}^+, \infty)$; all the terms in the last sum are weakly real. By Lemma $\mathrm{OL}\,1_{m,r}$, which appears in the induction hypothesis, the real part of each of them belongs to one of the Archimedean classes $\mathscr{A}_g^k$, $g \in G^{k-1}$. Assume first that all these terms tend to zero. In this case we choose the term φ to which the smallest value of (k, g) corresponds. By Lemma $\mathrm{OL}\,2_{m,r}$, the other terms are small in comparison with $\operatorname{Re}\varphi$. Consequently,

$$\operatorname{Re} a = (\operatorname{Re}\varphi)(1 + o(1)). \qquad ({}^{*}_{**})$$

The general case may be reduced to the previous one through the multiplication of a by $\exp(-\exp^{[m-1]} \circ \nu\zeta)$ for sufficiently large ν.

The proof that $\arg a$ tends to 0 or π is the same as in Proposition 1. The second implication can be proved similarly, except that $\varphi_j \in \mathscr{F}^m$ and $\psi_j = 0$ in the expression for a, and Lemmas $\mathrm{OL}\,1_m$ and $\mathrm{OL}\,2_m$ are used. ►

F. Proof of the Ordering Theorem, OT_n. Looking ahead, we assume that Lemma $\mathrm{OL}\,2_n$ has already been proved. To prove the first assertion of OT_n it suffices to verify that an arbitrary germ $g \in G^n$ is comparable with id: $g \succ \mathrm{id}$, or $g = \mathrm{id}$, or $g \prec \mathrm{id}$. Indeed, the relation $g_1 \prec g_2$ on $(\mathbf{R}^+, \infty)$ is equivalent to the relation $g_1 \circ g_2^{-1} \prec \mathrm{id}$ on $(\mathbf{R}^+, \infty)$. We decompose g, using Theorem ADT_n. Let φ be the first nonzero term in the sum for $g - \mathrm{id}$, ordered as in the expansion $(*)$ in §1.8. Arguing just as in the proof of Proposition 3, we get that

$$g - \mathrm{id} = (\operatorname{Re}\varphi)(1 + o(1)).$$

This and the remark at the end of subsection C give us OT_n.

G. Proof of the implication $\mathrm{OL}\,3_{m,r} \Rightarrow \mathrm{LEPS}_{m,r}$, $1 \le m \le n$, $r \ge 0$.

REMARK. This is implication ④$''$ in subsection A.

PROPOSITION 4. *Suppose that $a_j \in \mathscr{K}_{\mathbf{R}}^{m,r}$, $\mathbf{e}_j \in E^m$, $j = 1, 2$, and Lemma $\mathrm{OL}\,3_{m,r}$ is valid. Then for any $c > 0$ there exists a neighborhood of infinity on $\mathbf{R}^+$ in which the equation*

$$|a_1 \exp \mathbf{e}_1| = c|a_2 \exp \mathbf{e}_2|$$

either does not have solutions or is satisfied identically.

PROOF. We prove that if the modulus of the quotient

$$\varphi = (a_1/a_2)\exp(\mathbf{e}_1 - \mathbf{e}_2)$$

is not constant, then it varies monotonically in some neighborhood of infinity on $\mathbf{R}^+$. Indeed,

$$\varphi' = \frac{a_1' a_2 - a_1 a_2' + \mathbf{e}' a_2^2}{a_2^2} \exp \mathbf{e},$$

where $\mathbf{e} = \mathbf{e}_1 - \mathbf{e}_2$. By the definition of the set E^m in §1.7, $\operatorname{Im} \mathbf{e}|_{(\mathbf{R}^+,\infty)} \to 0$. Consequently, $\arg(\exp \mathbf{e}) \to 0$. Further,

$$a_2^2,\ a_1' a_2 - a_1 a_2' + \mathbf{e}' a_2^2 \in \mathscr{K}_{\mathbf{R}}^{m,r}.$$

According to Lemma OL $3_{m,r}$, which is assumed, the argument of each of these cochains, hence also their quotient, tends to zero modulo π. This implies that $\arg\varphi'$ $(\operatorname{mod}\pi) \to 0$. Similarly, $\arg\varphi \to 0$ $(\operatorname{mod}\pi)$. Consequently, $|\varphi|$ varies monotonically in some neighborhood of infinity. ▶

PROPOSITION 5. OL $3_{m,r} \Rightarrow \mathrm{LEPS}_{m,r}$.

PROOF. For a STAR-(m, r) let Σ be a partial sum whose terms have principal exponents equal to zero; this assumption does not restrict the generality. We prove that for any $\varepsilon > 0$ we then have on $(\mathbf{R}^+, \infty)$ the inequality

$$|\operatorname{Re}\Sigma| \succ \exp(-\varepsilon \exp^{[m]}),$$

and

$$\arg\Sigma \ (\operatorname{mod}\pi) \to 0.$$

The proof is by induction on the number N of terms in the sum Σ.

INDUCTION BASE: Let $N = 1$; $a \exp \mathbf{e}$ is a partial sum of a STAR-(m, r) and consists of a single term, with $a \in \mathscr{K}_{\mathbf{R}}^{m,r}$ and $\mathbf{e} \in E^m$. By OL $3_{m,r}$ $\arg a$ $(\operatorname{mod}\pi) \to 0$ on $(\mathbf{R}^+, \infty)$. By the definition of the set E^m in §1.7, $\operatorname{Im}\mathbf{e} \to 0$ on $(\mathbf{R}^+, \infty)$. Consequently,

$$\arg(\exp\mathbf{e}) \to 0 \quad \text{and} \quad \arg(a\exp\mathbf{e}) \ (\operatorname{mod}\pi) \to 0 \quad \text{on } (\mathbf{R}^+, \infty).$$

Therefore, on $(\mathbf{R}^+, \infty)$

$$|\operatorname{Re}(a\exp\mathbf{e})| = |a\exp\mathbf{e}|(1 + o(1))|.$$

By assumption, zero is the principal exponent of the term $a\exp\mathbf{e}$. Thus for arbitrary $\varepsilon > 0$, $|\exp\mathbf{e}| \succ \exp(-\frac{\varepsilon}{2}\exp^{[m]})$. Moreover, $|a| \in \mathscr{A}$ for some $(k, g) \prec (m, \mathrm{id})$. This follows from the relation $\left(\begin{smallmatrix} * \\ ** \end{smallmatrix}\right)$ in E and the Remark at the end of C. The Proposition 4 in §1.4C now implies that on $(\mathbf{R}^+, \infty)$, $|a| \succ \exp(-\frac{\varepsilon}{2}\exp^{[m]})$. Finally, for arbitrary $\varepsilon > 0$

$$|a\exp\mathbf{e}| \succ \exp(-\varepsilon\exp^{[m]}).$$

This concludes the induction base.

INDUCTION STEP. Consider the sum

$$\Sigma = \sum_1^N a_j \exp\mathbf{e}_j, \qquad a_j \in \mathscr{K}_{\mathbf{R}}^{m,r}, \ \mathbf{e}_j \in E^m, \ \nu(\mathbf{e}_j) = 0.$$

It follows from Proposition 4 that the modulus of the quotient of any two terms of this sum has a limit, finite or infinite. Indeed, suppose that the limit does not exist, that is, the limit superior and the limit inferior do not coincide. Then an arbitrary value between these limits is assumed in any neighborhood of infinity—a contradiction.

Suppose that the first term in the sum under consideration is nonstrictly maximal, that is, is such that for arbitrary $j \geq 2$

$$\lim_{\xi\to\infty}\left|\frac{a_j\exp\mathbf{e}_j}{a_1\exp\mathbf{e}_1}\right| \leq 1.$$

As proved above (the induction base), the arguments of the numerator and denominator in the fraction within the modulus signs tend to zero modulo π. Therefore, not only the modulus but also the fraction itself has a limit, and a finite limit. Denote it by l_j, and let $l = 1 + \sum_2^N l_j$. We consider two cases.

CASE 1: Let $l \neq 0$. Then for each $\varepsilon > 0$

$$|\Sigma| < \frac{|l|}{2}|a_1 \exp \mathbf{e}_1| \succ \exp(-\varepsilon \exp^{[m]}).$$

The last inequality was obtained in the proof of the assertion constituting the induction base. The proposition is proved in the first case.

CASE 2: $l = 0$. In this case

$$\Sigma_1(x) \overset{\text{def}}{=} (\Sigma / a_1 \exp \mathbf{e}_1)(x) = \int_\infty^x \Sigma_1'. \qquad (***)$$

Note that $a_1^2 \Sigma_1'$ is again a partial sum of a STAR-(m, r), its terms have zero principal exponents, and it consists of $N-1$ terms. By the induction hypothesis, for any $\varepsilon > 0$

$$|a_1^2 \Sigma_1'| \succ \kappa_\varepsilon \overset{\text{def}}{=} \exp(-\varepsilon \exp^{[m]}),$$

and $\arg \Sigma_1' \pmod{\pi} \to 0$. The estimate $(**)$ implies that

$$|a_1^2| \prec \exp\left(\frac{\varepsilon}{2} \exp^{[m]}\right).$$

By the Lemma OL $3_{m,r}$, $\arg a_1^2 \to 0 \pmod{\pi}$ on $(\mathbf{R}^+, \infty)$. Then $|\Sigma_1'| \succ \kappa_\varepsilon$ and $\arg \Sigma_1 \pmod{\pi} \to 0$ in view of the formula $(***)$. Further, $\Sigma = (a_1 \exp \mathbf{e}_1)\Sigma_1$.

The arguments of both factors tend to zero modulo π; consequently,

$$\arg \Sigma \pmod{\pi} \to 0.$$

Further,

$$|\Sigma_1| \succ \int_x^\infty \kappa_\varepsilon \succ \kappa_\varepsilon^2.$$

Let us prove the last inequality, using only the following properties of the function κ_ε:

$$\kappa_\varepsilon > 0, \quad \kappa_\varepsilon \to 0, \quad \kappa_\varepsilon' < 0, \quad \kappa_\varepsilon' \to 0.$$

We have that

$$\kappa_\varepsilon^2(x) = -\int_x^\infty (\kappa_\varepsilon^2)' = 2\int_x^\infty \kappa_\varepsilon |\kappa_\varepsilon'| \prec \int_x^\infty \kappa_\varepsilon,$$

because $|\kappa_\varepsilon'| \to 0$. Accordingly, in Case 2

$$|\Sigma| \succ |a_1 \exp \mathbf{e}_1 \cdot \exp(-2\varepsilon \exp^{[m]})| \succ \exp(-3\varepsilon \exp^{[m]}) \quad \text{on } (\mathbf{R}^+, \infty).$$

The second inequality uses the induction base. Lemma $\mathrm{LEPS}_{m,r}$ is proved. ►

This completes the proofs of the chains of implications given at the beginning of the section and proves the lemmas on a lower estimate of the partial sums and all three order lemmas for all m: $0 \leq m \leq n$ and all $r \geq 0$. In particular, Lemma $\mathrm{OL}\,1_n$ is proved, of which the third assertion is the corollary used in §1.9 in proving the identity theorem (see the remark at the beginning of §4.10). Thus, the identity theorem is proved, modulo Lemmas 5.1_n–5.8_n.

CHAPTER V

Ordering of Functional Cochains on a Complex Domain

In this chapter the lower estimate obtained in §4.10 for germs of regular functional cochains is extended to the complex domain. Then it is used to derive Lemmas 5.2_{n+1}–5.8_{n+1}, which are reduced in the final analysis to certain assertions about constancy of the sign. Lemma 5.1_{n+1} enters by parts in 5.2_{n+1}–5.5_{n+1} and is not discussed. We prove these lemmas in the case $n \geq 1$; the case $n = 0$ is treated above.

§5.1. The complexified order lemmas

In this section we prove the complexified Lemmas LEPS_n, $\mathrm{OL1}_n$, $\mathrm{OL2}_n$, and $\mathrm{OL3}_n$, replacing the germ of the ray $(\mathbf{R}^+, \infty)$ by a neighborhood of it on the complex line in the formulations in §4.10. The superscript $\mathbf{C}$ is added on the left to the symbols for the lemmas to indicate complexification. We proceed to the detailed formulations.

A. Formulations. Let $C > 1$ be an arbitrary constant, and Π the right half-strip $\xi \geq a$, $|\eta| \leq \pi/2$.

THE COMPLEXIFIED LEMMA ON A LOWER ESTIMATE OF THE PARTIAL SUMS OF A STAR-(m, r), ${}^{\mathbf{C}}\mathrm{LEPS}_{m,r}$. *For a* STAR-$(m, r)$, $0 \leq m \leq n$, $r \geq 0$, *let* Σ *be a weakly real partial sum whose terms all have the same principal exponent, which is greater than* ν. *Then in the germ at infinity of the domain* $C \ln^{[m+1]} \Pi$

1°. $\arg \Sigma \pmod{\pi} \to 0$;
2°. $|\operatorname{Re} \Sigma| \succ |\exp(\nu \exp^{[m]})|$.

THE FIRST COMPLEXIFIED ORDER LEMMA, ${}^{\mathbf{C}}\mathrm{OL1}_{m,r}$. *Suppose that* $F \in \mathscr{F}\mathscr{C}_{\mathbf{R}}^{m,r}$, $0 \leq m \leq n$, $r \geq 0$. *Then the following hold in the germ at infinity of the domain* ([13]) $T = A^{-m} C \circ \ln \Pi$:

1°. *if* $F \not\equiv 0$ *in* T, *then* $\arg F \pmod{\pi} \to 0$;
2°. *the set* $\mathscr{F}\mathscr{C}_{\mathbf{R}}^{m,r}$ *is ordered by the relation*

$$F_1 \prec F_2 \Leftrightarrow \operatorname{Re}(F_1 - F_2) \prec 0 \quad \text{in } T; \qquad (*)$$

([13]) "T" stands for tongue, since the curvilinear half-strip T resembles a tongue.

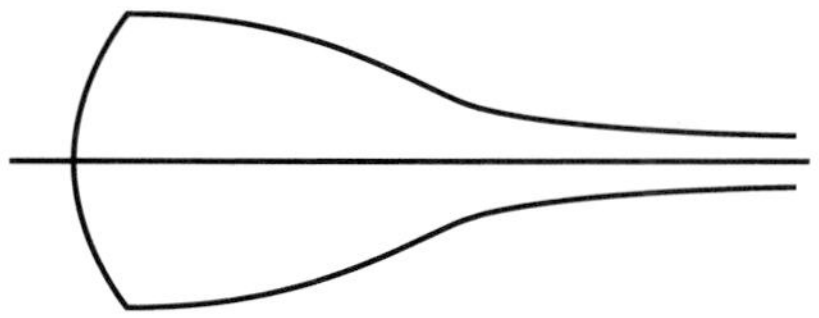

FIGURE 21

3°. *there exists a* $\mu > 0$ *such that*

$$|\operatorname{Re} F| \succ \exp(-\mu\xi) \quad in\ T.$$

THE SECOND COMPLEXIFIED ORDER LEMMA, ${}^{\mathbf{C}}\mathrm{OL2}_{m,r}$. *Suppose that* $(k, f) \prec (m, g)$, $f \in G^{k-1}$, $g \in G^{m-1}$, $m \leq n$, $\varphi \in \mathscr{F}_{\mathbf{R},f}^{k,p}$, *and* $\psi \in \mathscr{F}_{+g}^{m,r}$; $p \leq r$ *if* $k = m$. *Then*

$$|\operatorname{Re}\varphi| \succ |\psi|$$

in the germ at infinity of the domain $\Pi_1 = C\ln^{[m+1]}\Pi$, *where* Π *is a right half-strip and* C *an arbitrary positive constant. Moreover,* $\arg\varphi \pmod{\pi} \to 0$ *in the domain* Π_1 ([14]).

THE THIRD COMPLEXIFIED ORDER LEMMA, ${}^{\mathbf{C}}\mathrm{OL3}_{m,r}$. *The set* $\mathscr{K}_{\mathbf{R}}^{m,r}$ *is ordered by the relation* $(*)$ *in the domain* $\Pi_1 = C\ln^{[m+1]}\Pi$, *where* Π *is a right half-strip and* $C > 0$ *an arbitrary constant. If* $a \in \mathscr{K}_{\mathbf{R}}^{m,r}$ *and* $a \not\equiv 0$, *then* $\arg a \pmod{\pi} \to 0$ *in the domain* Π_1.

Denote by ${}^{\mathbf{C}}\mathrm{LEPS}_m$ and ${}^{\mathbf{C}}\mathrm{OL1}, 2, 3_m$ the respective unions with respect to r of the assertions in ${}^{\mathbf{C}}\mathrm{LEPS}_{m,r}$ and ${}^{\mathbf{C}}\mathrm{OL1}, 2, 3_{m,r}$. The proofs of the lemmas stated above are by induction on the class and the rank according to the same scheme as in §4.10, except that all assertions in the corresponding chain of implications are complexified: the letter **C** is added to the corresponding notation. Propositions 1–3 below are analogous to the propositions in §4.10 with the same numbers.

B. Induction base. The cochains of class 0 do not have a rank. The estimate in ${}^{\mathbf{C}}\mathrm{LEPS}_0$ serves as an induction base. We prove it.

For a real Dulac series let Σ be a partial sum whose terms have a common principal exponent $\mu > \nu$:

$$\Sigma = P(\zeta)\exp\mu\zeta.$$

It is required to prove that in the domain $C\ln\Pi$ we have the assertions:

1°. $\arg\Sigma \pmod{\pi} \to 0$;

2°. $|\Sigma| \succ \exp(\nu\xi)$.

([14]) Beginning with §5.2 C, when we refer to Lemma ${}^{\mathbf{C}}\mathrm{OL2}_{m,r}$, we understand Corollary 2 in §5.2 B, which strengthens it.

Indeed, in $\Pi_1 = C \ln \Pi$ the argument modulo π of a polynomial with real coefficients tends to zero. Moreover, in this domain the imaginary part of the exponent tends to zero, and hence so does the argument of the exponential $\exp \mu\zeta$. Consequently, assertion 1° is satisfied.

Further, in Π_1

$$|P(\zeta)\exp \mu\zeta| = (1+o(1))|P(\xi)\exp \mu\xi| \succ \exp \nu\xi,$$

which is what was required.

The induction step is implemented under the following five propositions, which are analogous to those in §4.10.

C. Proposition 1. ${}^{C}\mathrm{LEPS}_{m,r} \Rightarrow {}^{C}\mathrm{OL1}_{m,r}$ *and* ${}^{C}\mathrm{LEPS}_{m} \Rightarrow {}^{C}\mathrm{OL1}_{m}$, $0 \le m \le n$, $r \ge 0$.

PROOF. Both implications will be proved simultaneously. The rank of the cochains does not play a role in the proof. Everywhere in what follows,

$$T = A^{-m} C \circ \ln \Pi, \qquad \Pi_1 = C \ln^{[m+1]} \Pi.$$

The conclusion of ${}^{C}\mathrm{OL1}_{m,r}$ consists of three assertions. We prove first the implication $1^\circ \Rightarrow 2^\circ$.

Requirement 1° means that $\arg F(\zeta)$ is defined for all points $\zeta \in T$ with sufficiently large real part. For such ζ we have that $F(\zeta) \neq 0$. Since $\arg F(\zeta)$ $(\bmod\ \pi)$ is small for sufficiently large $\operatorname{Re}\zeta$, we get that $\operatorname{Re} F(\zeta) \neq 0$ for such ζ. Consequently, if $F = F_1 - F_2 \not\equiv 0$ in T, then one of the following two relations holds: $F \succ 0$ or $F \prec 0$. This proves assertion 2°.

PROOF OF ASSERTION 1°. Consider a STAR-$m\Sigma$ for $F \circ \exp^{[m]}$. The case $\Sigma \equiv 0$ is impossible in view of the Phragmén-Lindelöf theorem and the assumption that $F \not\equiv 0$; see the analogous argument in the proof of Proposition 1 in §4.10. Consequently, $\Sigma \not\equiv 0$. Let ν be the largest principal exponent of the terms in the series Σ. Then

$$F \circ \exp^{[m]} \cdot \exp \circ (-\nu \exp^{[m]}) = \Sigma_0 + \Sigma_+ + R,$$

where Σ_0 and Σ_+ are partial sums of the STAR-m with zero and negative principal exponents, respectively, and R is a remainder term satisfying the inequality

$$|R| \prec \exp(-\mu \operatorname{Re} \exp^{[m]})$$

in the domain $\ln^{[m]} \Omega_{\mathrm{st}}$. Here μ is a positive constant that can be chosen arbitrarily and is chosen larger than the minimal value ε of the moduli of the principal exponents of the terms in the sum Σ_+. Note that $(\Pi_1, \infty) \subset \ln^{[m]} \Omega_{\mathrm{st}}$; therefore, the preceding estimate on $|R|$ is satisfied in Π_1. The weak realness of Σ_0 was proved in §4.10 C.

It follows from ${}^{C}\mathrm{LEPS}_{m}$ or ${}^{C}\mathrm{LEPS}_{m,r}$, which are assumed in the proposition, that in $\Pi_1 = C \ln^{[m+1]} \Pi$

$$\operatorname{Re} \Sigma_0 \succ |\exp(-\tfrac{\varepsilon}{2} \exp^{[m]})|.$$

On the other hand, in the domain Π_1

$$|\Sigma_+| \prec |\exp(-\tfrac{2\varepsilon}{3}\exp^{[m]})|.$$

This follows from the upper estimates (which are actually contained in the definition) on the coefficients and from the lower estimates on the exponents of the STAR-m that are in the induction hypothesis. Let us pass to the detailed proof. It can be assumed without loss of generality that the sum Σ_+ consists of the single term $a \exp \mathbf{e}$. By definition, if $a \in \mathscr{K}_{\mathbf{R}}^{m,r} \backslash \{0\}$, then $a = \sum a_j + o(1)$, $a_j \in \mathscr{F}_{g_j}^{m-1}$, in the domain Π_1. Further, for arbitrary $g_j \in G^{m-2}$ there exists a standard domain Ω and a constant $\nu > 0$ such that

$$|a_j| \prec |\exp^{[m]} \nu\zeta|$$

in the domain $\widetilde{\Omega} = g_j^{-1} \circ \ln^{[m-1]} \Omega$. Since $(\Pi_1, \infty) \subset (\widetilde{\Omega}, \infty)$, we get that the preceding estimate is satisfied in Π_1. Further, in Π_1

$$|\exp^{[m]} \nu\zeta| \prec |\exp \circ\, \delta \exp^{[m]} \zeta|$$

for any $\delta > 0$.

According to the regularity lemma in §2.12, $\mathbf{e} \in \mathscr{F}^{m-1}$. Consequently, by ${}^{\mathrm{C}}\mathrm{OL2}_{m-1}$, which appears in the induction hypothesis, the exponent $\mathbf{e}$ is comparable with $C\exp^{[m]}$ in Π_1 for arbitrary $C > 0$. Since $\nu(\mathbf{e}) \le -\varepsilon$, we get that on Π_1

$$\operatorname{Re} \mathbf{e} \prec -\tfrac{3\varepsilon}{4} \operatorname{Re} \exp^{[m]}.$$

Consequently, in Π_1

$$|\exp \mathbf{e}| \prec |\exp(-\tfrac{3\varepsilon}{4} \operatorname{Re} \exp^{[m]})|.$$

From this, in Π_1

$$|a \exp \mathbf{e}| \prec |\exp(-\tfrac{2\varepsilon}{3}\exp^{[m]})|,$$

which is what was required.

This implies that in Π_1

$$\operatorname{Re}(\Sigma_+ + R) = o(\operatorname{Re}\Sigma_0).$$

Consequently, in $T = \exp^{[m]} \Pi_1 = A^{-m} C \circ \ln \Pi$

$$F = \exp(\nu\zeta) \cdot \Sigma_0 \circ \ln^{[m]} \cdot (1 + o(1)).$$

The argument in the first and third factors tends to zero in the domain T, whose germ at infinity belongs to an arbitrary half-strip. The argument of the second factor tends to zero modulo π in view of ${}^{\mathrm{C}}\mathrm{LEPS}_m$ or ${}^{\mathrm{C}}\mathrm{LEPS}_{m,r}$, which appears in the induction hypothesis and in the condition of the proposition. This proves assertion 1°.

We prove assertion 3°. In view of the preceding equality and the lower estimate on $\operatorname{Re}\Sigma_0$ we get that in T

$$|\operatorname{Re} F| \succ \exp((\nu - \varepsilon)\xi) \cdot (1 + o(1)).$$

Assertion 3° now follows from assertion 1° with $\mu = \nu - 2\varepsilon$.

D. PROPOSITION 2. ${}^{\mathbf{C}}\mathrm{OL1}_{m,r} \Rightarrow \mathrm{OL2}_{m,r}$, $0 \le m \le n$, $r \ge 0$.

PROOF. Suppose that the cochains φ and ψ are the same as in Lemma ${}^{\mathbf{C}}\mathrm{OL2}_{m,r}$, that is,

$$\varphi \in \mathscr{F}^{k,p}_{\mathbf{R},f}, \qquad \psi \in \mathscr{F}^{m,r}_{+g},$$

$(k, f) \prec (m, g)$, and $p \le r$ for $k = m$. By definition, there exist $\widetilde{F} \in \mathscr{F}\mathscr{C}^{k,p}_{\mathbf{R}}$ and $F \in \mathscr{F}\mathscr{C}^{m,r}_{\mathbf{R}+}$ such that

$$\varphi = \widetilde{F} \circ \exp^{[k]} \circ f, \qquad \psi = F \circ \exp^{[m]} \circ g.$$

The proposition asserts that the inequality $|\operatorname{Re}\varphi| \succ |\psi|$ holds in the domain $C \ln^{[m+1]} \Pi = \Pi_1$ for each $C > 1$. This inequality is equivalent to the inequality $|\operatorname{Re}\widetilde{F} \circ \rho| \succ |F|$ in the domain $\Pi_2 = \exp^{[m]} \circ g \circ \Pi_1$, where $\rho = \exp^{[k]} \circ f \circ g^{-1} \circ \ln^{[m]}$. We prove that in Π_2

$$\operatorname{Re}\rho \prec \varepsilon\xi \quad \text{for each } \varepsilon > 0; \tag{$**$}$$

$$|\operatorname{Re}\widetilde{F}| \succ \exp(-\mu\xi) \quad \text{for some } \mu > 0 \tag{$***$}$$

in the domain $\Pi_3 = \rho\Pi_2$. In §4.2 the inequality $(**)$ (marked as $(*)$ there) was proved in some standard domain Ω. In Lemma ${}^{\mathbf{C}}\mathrm{OL1}_{m,r}$ the inequality $(***)$ was proved for arbitrary $\widetilde{F} \in \mathscr{F}\mathscr{C}^{m,r}_{\mathbf{R}}$ and $C_1 > 0$ in the domain $T = A^{-m} C_1 \circ \ln \Pi$. Since $\mathscr{F}\mathscr{C}^k \subset \mathscr{F}\mathscr{C}^m$ (proved in §4.5 C), we get the inequality $(***)$ in (T, ∞) for the germ $\widetilde{F}$ considered. We prove that the constant C_1 can be chosen so that $(\Pi_j, \infty) \subset (T, \infty)$, $j = 2, 3$.

Let $j = 3$. By choosing C_1 we must ensure that

$$(\Pi_3, \infty) = (\exp^{[k]} \circ f \circ C \circ \ln^{[m+1]} \Pi, \infty) \subset (A^{-m} C_1 \circ \ln \Pi, \infty),$$

or, what is equivalent,

$$(\ln^{[m-k]} \circ f_1 \circ \ln^{[m+1]} \Pi, \infty) \subset (C_1 \circ \ln^{[m+1]} \Pi, \infty), \qquad f_1 = f \circ C \in G^{k-1}. \tag{${}^{\,*}_{**}$}$$

We prove this inclusion for $m = k$; the inclusion will follow from this for $m > k$, since the mapping ln only narrows the half-strip.

The mapping f_1 is real on the real axis, and its derivative is bounded in the domain $\ln^{[m+1]} \Pi$. This follows from Theorem ADT_{k-1}. Indeed, the expansion for f_1' given by this theorem contains, along with a positive constant, terms of the form $\tilde{\varphi} \in \mathscr{F}^l_+$, $l \le k - 1$, that are decreasing on $(\mathbf{R}^+, \infty)$. By Lemma ${}^{\mathbf{C}}\mathrm{OL2}_l$, which appears in the induction hypothesis, we get since $l \leq k-1$ that for some $\varepsilon > 0$ the inequality $|\tilde{\varphi}| \prec \exp(-\varepsilon\xi)$ holds in the domain $\ln^{[l+1]} \Pi \supset \ln^{[m+1]} \Pi$. Consequently, for sufficiently large C_1 the inclusion $({}^{*}_{**})$ holds. Similarly,

$$(\Pi_2, \infty) = (\exp^{[m]} \circ g \circ C \circ \ln^{[m+1]} \Pi, \infty) \subset (T, \infty).$$

This gives us that for arbitrary $\varepsilon > 0$

$$|\operatorname{Re} \widetilde{F} \circ \rho| \succ \exp(-\mu\varepsilon\xi)$$

in Π_2. On the other hand, it follows from the definition of a rapidly decreasing regular cochain that there exists a $\nu > 0$ such that $|F| \prec \exp(-\nu\xi)$ in some standard domain. Choosing $\varepsilon > 0$ such that $\varepsilon\mu < \nu$, we get the required inequality: $|\operatorname{Re} \widetilde{F} \circ \rho| \succ |F|$ in Π_2.

We prove the last assertion of Lemma ${}^{\mathbf{C}}\mathrm{OL2}_{m,r}$. It is equivalent to the assertion $\arg \widetilde{F} \circ \rho \pmod{\pi} \to 0$ in (Π_2, ∞) or $\arg \widetilde{F} \pmod{\pi} \to 0$ in (Π_3, ∞). But $(\Pi_3, \infty) \subset (T, \infty)$, and it has already been proved that $\arg \widetilde{F} \pmod{\pi}$ tends to zero on (T, ∞). ▶

E. Proposition 3. ${}^{\mathbf{C}}\mathrm{OL2}_m \Rightarrow {}^{\mathbf{C}}\mathrm{OL3}_{m+1,0}$ *and* ${}^{\mathbf{C}}\mathrm{OL2}_{m,r} \Rightarrow {}^{\mathbf{C}}\mathrm{OL3}_{m,r+1}$.

Proof. The arguments are parallel to those used in the proof of Proposition 3 in §4.10 and are carried out differently for the two implications of the proposition. We prove the first. The space of coefficients of the STAR is linear. Therefore, it suffices to prove that the germs in this space are comparable with zero. Let $a \in \mathscr{K}_{\mathbf{R}}^{m+1,0} \setminus \{0\}$. By definition, $a = \sum \varphi_j$, $\varphi_j \in \mathscr{F}_{\mathbf{R}g_j}^{m}$, $g_j \prec\prec g_{j+1}$. According to Lemma ${}^{\mathbf{C}}\mathrm{OL2}_m$, in the domain $\Pi_1 = C \ln^{[m+1]} \Pi$ we have the inequality $|\operatorname{Re} \varphi_1| \succ |\varphi_j|$ for $j \geq 2$, and even $|\operatorname{Re} \varphi_1| \succ N|\varphi_j|$ for every positive integer N. It now follows from ${}^{\mathbf{C}}\mathrm{OL1}_m$ that

$$|\operatorname{Re} a| \succ 0 \quad \text{and} \quad \arg a \pmod{\pi} \to 0$$

in Π_1.

We prove the second implication. By definition, it follows from $a \in \mathscr{K}_{\mathbf{R}}^{m,r}$ that

$$a = \sum \varphi_j + \sum \psi_j, \qquad \sum \varphi_j \in \mathscr{K}_{\mathbf{R}}^{m,0}, \ \varphi_j \in \mathscr{F}_{\mathbf{R}g_j}^{m-1},$$
$$\psi_j \in \mathscr{F}_{\mathbf{R}+h_j}^{m,r-1}, \ g_j \prec\prec g_{j+1}, \ h_j \prec\prec h_{j+1}.$$

If the first sum is nonzero, then the argument is analogous to the preceding, and

$$\operatorname{Re} a = \operatorname{Re} \varphi_1 (1 + o(1)).$$

If the first sum is equal to zero, then

$$\operatorname{Re} a = \operatorname{Re} \psi_1 (1 + o(1)).$$

In both cases the assertion that $\arg a \pmod{\pi} \to 0$ in Π_1 is deduced from the analogous assertions for φ_1 and ψ_1. These assertions are deduced, in turn, from Lemmas ${}^{\mathbf{C}}\mathrm{OL2}_{m,r}$ and ${}^{\mathbf{C}}\mathrm{OL1}_{m,r}$ in precisely the same way as above. ▶

We proceed to a proof of the last implication in the chain.

F. Implication. ${}^{\mathbf{C}}\mathrm{OL3}_{m,r} \Rightarrow {}^{\mathbf{C}}\mathrm{LEPS}_{m,r}$.

PROPOSITION 4. *Suppose that* $a_j \in \mathscr{K}_{\mathbf{R}}^{m,r}$, $a_j \not\equiv 0$, $\mathbf{e}_j \in E^m$, $j = 1, 2$, *and Lemma* ${}^{\mathbf{C}}\mathrm{OL3}_{m,r}$ *is valid. Then for the right half-strip* Π *and any* $C > 0$ *the following limit of a quotient exists and is real:*

$$\lim_{\zeta\to\infty} \varphi, \quad \text{where } \varphi = a_1 \exp \mathbf{e}_1 / a_2 \exp \mathbf{e}_2$$

in the domain $\Pi_1 = C \ln^{[m+1]} \Pi$.

PROOF. We prove two assertions. 1. $\arg\varphi \pmod{\pi} \to 0$ in the domain Π_1. 2. If $|\varphi|$ is not constant, then it varies monotonically in a suitable neighborhood of infinity along any curve $\gamma_\zeta\colon \mathbf{R}^+ \to \mathbf{C}$ that has origin at the point ζ and endpoint ∞, and has slope $\arg \dot\gamma_\zeta(\xi)$ with respect to the axis $\mathbf{R}^+$ tending to zero as $\xi \to \infty$. The family of curves γ_ζ is chosen so that $\max_\xi |\arg \dot\gamma_\zeta(\xi)| \to 0$ as $\zeta \to \infty$. These assertions yield the proposition.

Let us prove the first assertion. It follows from ${}^{\mathbf{C}}\mathrm{OL3}_{m,r}$ that $\arg a_j \pmod{\pi} \to 0$ as $\zeta \to \infty$ in Π_1. We prove that for any $\mathbf{e} \in E^m$

$$\arg \exp \mathbf{e} \to 0 \quad \text{in } (\Pi_1, \infty),$$

or, what is equivalent,

$$\operatorname{Im} \mathbf{e} \to 0 \quad \text{in } (\Pi_1, \infty).$$

By Definition $2''$ in §1.7, $\operatorname{Im}\mathbf{e} \to 0$ on $(\mathbf{R}^+, \infty)$. Further, in view of the estimate $(*\overset{*}{*}*)$ of §4.9 C, there exists a standard domain Ω of class m and a $\nu > 0$ such that

$$|\mathbf{e}'| < \exp(\nu + 1) \operatorname{Re} \exp^{[m-1]}$$

in the domain

$$\widetilde{\Omega}_m = \{\zeta \in \Omega_m | \rho(\zeta, \partial\Omega_m) > (\exp^{[m]} \xi)^{-1}\},$$

where $\Omega_m = \ln^{[m]} \Omega$. Obviously,

$$(\Pi_1, \infty) \subset (\Omega_m, \infty).$$

What is more, (Π_1, ∞) belongs to the germ at infinity of the domain

$$|\eta| \prec \exp(-\exp^{[m]} \mu\xi), \qquad \xi = \operatorname{Re}\zeta,\ \eta = \operatorname{Im}\zeta,$$

for some $\mu > 0$. Consequently, in (Π_1, ∞)

$$\begin{aligned} |\operatorname{Im} \mathbf{e}(\zeta)| &\le |\operatorname{Im} \mathbf{e}(\xi)| + |\eta| \cdot \max_{\operatorname{Re}\zeta = \xi} |\mathbf{e}'| \\ &\le o(1) + \exp \circ [-\exp^{[m]} \mu\xi + (\nu + 1) \exp^{[m-1]} \xi] \to 0 \end{aligned}$$

as $\zeta \to \infty$ in Π_1. This proves that $\arg \exp \mathbf{e} \to 0$ in Π_1, and hence $\arg \varphi \pmod{\pi} \to 0$ in Π_1. Assertion 1 is proved.

We prove assertion 2: $|\varphi|$ varies monotonically along the curve γ_ζ. For this it suffices to prove that $\arg \varphi' \pmod{\pi} \to 0$ and $\arg \dot{\gamma}_\zeta(\xi) \to 0$ as $\xi \to \infty$. The second relation is in the definition of γ_ζ; let us prove the first. We have that

$$\varphi' = \frac{a_1' a_2 - a_1 a_2' + (\mathbf{e}_1' - \mathbf{e}_2')a_2^2}{a_2^2} \exp(\mathbf{e}_1 - \mathbf{e}_2).$$

The arguments of the numerator and the denominator of the fraction tend to zero modulo π, since these expressions belong to $\mathscr{K}_{\mathbf{R}}^{m,r}$ according to the multiplication lemma and the regularity lemma. The argument of the last factor tends to zero by what was proved above, because $\mathbf{e}_1 - \mathbf{e}_2 \in E^m$. Assertion 2 is proved, and with it the proposition. ▶

PROPOSITION 5. ${}^{\mathrm{C}}\mathrm{OL3}_{m,r} \Rightarrow {}^{\mathrm{C}}\mathrm{LEPS}_{m,r}$.

PROOF. The proof is by induction on the number N of terms in the sum Σ. It can be assumed without loss of generality that Σ is a sum of terms with zero principal exponent, and proved that

$$\operatorname{Re} \Sigma \succ \exp(-\varepsilon \exp^{[m]})$$

in the domain Π_1 for arbitrary $\varepsilon > 0$.

INDUCTION BASE: $N = 1$. Let

$$\Sigma = a \exp \mathbf{e}, \qquad a \in \mathscr{K}_{\mathbf{R}}^{m,r}, \ \mathbf{e} \in E^m.$$

By Proposition 4, if we set $a_1 = a$, $\mathbf{e}_1 = \mathbf{e}$, $a_2 = 1$, $\mathbf{e}_2 = 0$, we get that

$$\arg \Sigma \pmod{\pi} \to 0$$

in Π_1. Therefore,

$$|\operatorname{Re}(a \exp \mathbf{e})| = |a \exp \mathbf{e}|(1 + o(1)).$$

Further, it follows from the two last equalities of subsection E and from ${}^{\mathrm{C}}\mathrm{OL2}_{m,r}$ that in Π_1

$$|a| \succ |\exp(-\varepsilon \exp^{[m]})|$$

for any $\varepsilon > 0$. By assumption, 0 is the principal exponent of the term $\exp \mathbf{e}$. It follows from the regularity lemma and ${}^{\mathrm{C}}\mathrm{OL1}_{m-1}$ that for any $\varepsilon > 0$

$$\operatorname{Re} \mathbf{e} \succ (-\varepsilon) \operatorname{Re} \exp^{[m]}$$

in Π_1. Consequently, in Π_1

$$|a \exp \mathbf{e}| \succ |\exp(-2\varepsilon) \exp^{[m]}|.$$

This completes the induction base.

INDUCTION STEP. Suppose that the first term in the sum Σ is a nonstrict maximum, that is, such that for any $j \geq 2$

$$l_j = \lim_{\zeta\to\infty} \left| \frac{a_j \exp \mathbf{e}_j}{a_1 \exp \mathbf{e}_1} \right| \leq 1$$

in the domain $\Pi_1 = C \ln^{[m+1]} \Pi$. Let: $\Sigma_1 = \Sigma / a_1 \exp \mathbf{e}_1$. Then $\lim_{\zeta\to\infty} \Sigma_1 = \Sigma l_j \overset{\text{def}}{=} l$, $|l| \leq N$. We consider two cases.

CASE 1: $l \neq 0$. Then for arbitrary $\varepsilon > 0$

$$|\Sigma| \succ \frac{|l|}{2} |a_1 \exp \mathbf{e}_1| \succ |\exp \circ(-\varepsilon) \exp^{[m]}|$$

in Π_1. The last inequality was obtained in the proof of the assertion constituting the induction base.

CASE 2: $l = 0$. In this case

$$\Sigma_1(\zeta) = -\int_{\gamma_\zeta} \Sigma_1'. \qquad ({}^{*\,*}_{\ *})$$

Note that $a_1^2 \Sigma_1'$ is again a partial sum of a STAR-(m, r) and has $N-1$ terms, each having zero principal exponent. By the induction hypothesis, for any $\varepsilon > 0$

$$|a_1^2 \Sigma_1'| \succ |\exp(-\tfrac{\varepsilon}{2} \exp^{[m]})|,$$

and $\arg(a_1^2 \Sigma_1') \pmod{\pi} \to 0$ in Π_1. It follows from the last relation and Lemma ${}^{\mathbf{C}}\mathrm{OL3}_{m,r}$ that $\arg \Sigma_1' \pmod{\pi} \to 0$ in Π_1. In the proof of Proposition 1 we obtained an upper estimate on the coefficients of class $\mathscr{K}_{\mathbf{R}}^{m,r}$. Applying it to $a = a_1$, we get that

$$|a| \prec |\exp^{[m]} \nu\zeta| \quad \text{in } (\Pi_1, \infty).$$

From this,

$$|\Sigma_1'| \succ |\exp(-\tfrac{\varepsilon}{2} \exp^{[m]} - 2\exp^{[m-1]} \circ \nu)| \succ |\exp(-\varepsilon \exp^{[m]})| \overset{\text{def}}{=} |\kappa_\varepsilon|$$

in (Π_1, ∞). The integration in the formula $({}^{*\,*}_{\ *})$ is along curves γ_ζ that have initial point ζ and endpoint ∞, and have slope with respect to the axis $\mathbf{R}^+$ tending to zero as $\zeta \to \infty$. This implies that $\arg \Sigma_1 \pmod{\pi} \to 0$ in Π_1. This also gives us for any $\delta > 0$ the first of the following two inequalities:

$$\Sigma_1(\zeta) \succ (1-\delta) \left| \int_{\gamma(\zeta)} |\kappa_\varepsilon| \right| \succ |\kappa_\varepsilon|^2$$

in (Π_1, ∞). We prove the last inequality, using the fact that $\kappa_\varepsilon \to 0$, $\kappa_\varepsilon' \to 0$, and $\arg \dot\gamma_\zeta(\xi) \to 0$ as $\xi \to \infty$, uniformly with respect to $\zeta \in \mathbf{R}^+$. We have that

$$|\kappa_\varepsilon^2(\zeta)| = \left| \int_{\gamma(\zeta)} (\kappa_\varepsilon^2)' \right| = 2 \left| \int_{\gamma(\zeta)} \kappa_\varepsilon \kappa_\varepsilon' \right|$$

$$\prec 2 \left| \int_{\gamma(\zeta)} |\kappa_\varepsilon| \max |\kappa_\varepsilon'| \right| \prec (1-\delta) \left| \int_{\gamma_\zeta} |\kappa_\varepsilon| \right|.$$

Thus, in Case 2

$$|\Sigma| \succ |a_1 \exp \mathbf{e}_1 \cdot \exp(-2\varepsilon \exp^{[m]})| \succ |\exp(-3\varepsilon \exp^{[m]})|$$

in the domain Π_1. The second inequality uses the induction base. The proposition is proved. ▶

This completes the proofs of the chain of implications described at the end of Section A, and proves the lemmas on a lower estimate of the partial sums and all three complexified order lemmas for all $m: 0 \le m \le n$ and all $r \ge 0$. Lemma ${}^{\mathbf{C}}\mathrm{OL1}_n$ will be especially important for us.

LEMMA ${}^{\mathbf{C}}\mathrm{OL1}_n$. *Let* $F \in \mathscr{F}\mathscr{C}^n_{\mathbf{R}}$, $F \not\equiv 0$. *Then for any* $C > 0$ *and the right half-strip* Π:

1°. $\arg F \pmod \pi) \to 0$ *in the domain* $T = A^{-n} C \circ \ln \Pi$;
2°. *in the domain* T *the set* $\mathscr{F}\mathscr{C}^n_{\mathbf{R}}$ *is ordered by the relation*

$$F_1 \prec F_2 \Leftrightarrow \mathrm{Re}(F_1 - F_2) \prec 0 \quad \text{in } T;$$

3°. *there exists a* $\mu > 0$ *such that in* T

$$|\mathrm{Re}\, F| \succ \exp(-\mu\xi).$$

Moreover, a strengthened formulation of Lemma ${}^{\mathbf{C}}\mathrm{OL2}_n$ is given in §5.2 B. These lemmas lie at the basis of all the subsequent proofs.

§5.2. The generalized exponent

This section contains preparatory material needed for all the rest of the chapter, along with a proof of Lemma 5.3_{n+1}, which asserts, in particular, that the germs of the group $A^{-(n+1)}G^n$ have a generalized exponent. We begin with the simplest properties of this group.

A. The action of the operator A^{-1} and its iterates on the germs of diffeomorphisms $(\mathbf{R}^+, \infty) \to (\mathbf{R}^+, \infty)$. This operator after iterations can carry germs with a bounded correction into rapidly increasing or rapidly decreasing germs. For example:

$$\xi + \ln 2 \overset{A^{-1}}{\mapsto} 2\xi \overset{A^{-1}}{\mapsto} \xi^2 \overset{A^{-1}}{\mapsto} \exp(\ln^2 \xi) \equiv \xi^{\ln \xi} \mapsto \cdots,$$

$$\xi - \ln 2 \overset{A^{-1}}{\mapsto} \frac{1}{2}\xi \overset{A^{-1}}{\mapsto} \xi^{1/2} \overset{A^{-1}}{\mapsto} \exp(\sqrt{\ln \xi}) \equiv \xi^{1/\sqrt{\ln \xi}} \mapsto \cdots.$$

The last germ in the first row grows more rapidly than any power, and the last in the second row grows more slowly than any power.

The investigation of the action of A^{-1} on real germs of diffeomorphisms is essentially facilitated by the following monotonicity property:

$$f \prec g \Leftrightarrow A^{-1} f \prec A^{-1} g \quad \text{on } (\mathbf{R}^+, \infty).$$

The operator A^{-1} commutes with an arbitrary composition power of the germ exp, positive or negative. This and the monotonicity indicated above imply that if $\xi + \ln C \prec f \prec \exp(\xi - \ln C)$, $C > 1$, then

$$C\xi \prec A^{-1}f \prec \exp\tfrac{\xi}{C}, \qquad \xi^C \prec A^{-2}f \prec \exp\xi^{1/C},$$
$$\xi^{\ln^{C-1}\xi} \prec A^{-3}f \prec \exp\circ\,\xi^{\ln^{-1+1/C}\xi}, \dots .$$

The rate of growth of the functions on the left-hand side increases, and that of the functions on the right-hand side decreases. It seems likely that there exists a function increasing more rapidly than any power, but more slowly than any exponential, and invariant under the action of A^{-1}. In my view it is of independent interest to determine and investigate such a function. The preceding chain of inequalities (upper estimates) proves

PROPOSITION 1. *Suppose that the germ of the diffeomorphism $f\colon(\mathbf{R}^+, \infty) \to (\mathbf{R}^+, \infty)$ grows no more rapidly than a linear germ. Then for any $\varepsilon > 0$ and any positive integer n*

$$A^{-n}f \prec \exp\xi^{\varepsilon}.$$

Stronger estimates are possible, but we do not need them. The passage to inverse mappings together with the monotonicity property

$$f \prec g \Leftrightarrow f^{-1} \succ g^{-1}$$

gives us the following proposition.

PROPOSITION 2. *Suppose that the germ of a diffeomorphism $f\colon(\mathbf{R}^+, \infty) \to (\mathbf{R}^+, \infty)$ increases more rapidly than some linear germ. Then for any $C > 0$ and any positive integer n*

$$A^{-n}f \succ (\ln\xi)^C.$$

We estimate the derivative $(A^{-n}f)'$.

PROPOSITION 3. *Suppose that the germ of a diffeomorphism $f\colon(\mathbf{R}^+, \infty) \to (\mathbf{R}^+, \infty)$ increases more rapidly than some linear germ and has a bounded derivative. Then for any $\varepsilon > 0$ and any positive integer n*

$$(A^{-n}f)' \prec \exp\xi^{\varepsilon}.$$

The proof is by induction on n. Induction base: $n = 0$. The assertion follows from the boundedness of the derivative f'.

INDUCTION STEP. Suppose that the derivative of the diffeomorphism $h\colon(\mathbf{R}^+, \infty) \to (\mathbf{R}^+, \infty)$ satisfies the estimate

$$h' \prec \exp\xi^{\varepsilon}.$$

Then

$$(A^{-1}h)' \prec \exp\xi^{\varepsilon}.$$

Indeed,

$$(A^{-1}h)' = (A^{-1}h) \cdot h' \circ \ln \cdot \xi^{-1}.$$

By Proposition 1, the assumption about h', and the inequality $\xi^\varepsilon \prec \varepsilon\xi$ for $\varepsilon < 1$, we get that

$$(A^{-1}h)' \prec (\exp \xi^\varepsilon)\xi^{\varepsilon-1} \prec \exp \xi^\varepsilon. \quad ▶$$

PROPOSITION 3 BIS. *Under the conditions of Proposition 3 the logarithmic derivative of the germ $A^{-n}f$ tends to zero for any positive integer n.*

PROOF. Suppose that $h' \prec \exp \varepsilon\xi$, $\varepsilon < 1$. Then

$$(A^{-1}h)'/A^{-1}h \to 0 \quad \text{on } (\mathbf{R}^+, \infty).$$

Indeed,

$$0 < (A^{-1}h)'/A^{-1}h = (h' \circ \ln) \cdot \xi^{-1} \prec \xi^{\varepsilon-1} \to 0.$$

Proposition 3 bis now follows from Proposition 3. ▶

The elementary arguments in this section serve as a model for analogous arguments in the complex domain.

B. The action of the operator A^{-1} on germs of conformal mappings close to linear mappings. This subsection contains preparatory material to be used later for proving that germs of the class $A^{-(n+1)}g$, $g \in G^n$, are biholomorphic in the germ of an arbitrary sector, and for getting estimates of such germs. Moreover, we give a strengthening of Lemma ${}^{\mathbf{C}}\mathrm{OL2}_n$ that is used constantly in what follows, and we prove related facts: Corollary 1–4.

Let us first investigate the germ $A^{-n}C$, $C > 1$; in the expression $A^{-n}C$, C denotes the map of multiplication by C. As above, $\Pi^\forall$ and $S^\forall$ are an arbitrary rectilinear half-strip and an arbitrary sector containing $(\mathbf{R}^+, \infty)$.

PROPOSITION 4. *For any $C > 1$ and any n the germ $A^{-n}C$ carries the germ of an arbitrary half-strip $(\Pi^\forall, \infty)$ into the germ of an arbitrary sector $(S^\forall, \infty)$.*

PROOF. It suffices to prove that $\arg A^{-n}C \to 0$ in $(\Pi^\forall, \infty)$. This is equivalent to the assertion that $\operatorname{Im} A^{1-n}C \to 0$ in $(\ln \Pi^\forall, \infty)$. The width of the curvilinear half-strip $\ln \Pi^\forall$ tends to zero exponentially, and $\operatorname{Im} A^{1-n}C = 0$ on the real axis. Therefore, it suffices to prove that $|(A^{1-n}C)'| \prec \exp \xi^\varepsilon$ in $\ln \Pi^\forall$ for arbitrary $\varepsilon > 0$. We prove a stronger assertion. Let $\Pi_m^\forall = \ln^{[n-m]} \Pi^\forall$. For example, $\Pi_0^\forall = \ln^{[n]} \Pi^\forall$, $\Pi_n^\forall = \Pi^\forall$, and $\Pi_{n-1}^\forall = \ln \Pi^\forall$. We prove by induction on m that for $0 \le m \le n-1$

$$|A^{-m}C| \prec \exp \xi^\varepsilon, \qquad |(A^{-m}C)'| \prec \exp \xi^\varepsilon$$

in $\Pi_m^\forall$ for arbitrary $\varepsilon > 0$.

INDUCTION BASE: $m = 0$. The assertions are obvious.

INDUCTION STEP: PASSAGE FROM m TO $m+1$. In the domain $\Pi^{\forall}_{m+1}$ we have the relations: $|\zeta| = O(1)\xi$,

$$\begin{aligned}|A^{-(m+1)}C| &= \exp \circ \operatorname{Re} A^{-m}C \circ \ln \prec \exp |A^{-m}C \circ \ln| \\ &\prec \exp |\exp(\ln|\zeta|)^{\varepsilon}| \prec \exp(\exp \varepsilon \ln |\zeta|) \\ &= \exp |\zeta|^{\varepsilon} = \exp \circ O(1)\xi^{\varepsilon}.\end{aligned}$$

Since ε is arbitrary, $|A^{-(m+1)}C| \prec \exp \xi^{\varepsilon}$ in $\Pi^{\forall}_{m+1}$. Further, in $\Pi^{\forall}_{m+1}$

$$\begin{aligned}|(A^{-(m+1)}C)'| &= |A^{-(m+1)}C| \cdot |(A^{-m}C)' \circ \ln| \cdot |\zeta|^{-1} \\ &\prec (\exp \xi^{\varepsilon})|\zeta^{\varepsilon-1}| \prec \exp \xi^{\varepsilon},\end{aligned}$$

by the preceding estimate and the induction hypothesis (we again use the inequality $|\exp(\ln|\zeta|)^{\varepsilon}| \prec |\zeta|^{\varepsilon})$. This finishes the induction step and proves the proposition. ►

COROLLARY 1. *$\varphi \to 0$ in $(C \ln^{[n]} \Pi, \infty)$ for arbitrary $C > 0$, $m \leq n$, and $\varphi \in \mathscr{F}^{m}_{+}$.*

PROOF. We prove the corollary for $m = n$; for $m < n$ it will follow from the relation $(C \ln^{[m]} \Pi, \infty) \supset (C \ln^{[n]} \Pi, \infty)$ for $m < n$.

Thus, let $\varphi = F \circ \exp^{[n]} \circ g$, $g \in G^{n-1}$, $F \in \mathscr{FC}^{n}_{+}$. Then $F \to 0$ in (S_{α}, ∞) for arbitrary $\alpha \in (0, \pi/2)$ (recall that $S_{\alpha} = \{\zeta \| \arg \zeta| < \alpha\})$. The convergence $\varphi \to 0$ in $(C \ln^{[n]} \Pi, \infty)$ is equivalent to the convergence $F \to 0$ in (Π', ∞), where $\Pi' = \exp^{[n]} \circ g \circ C \circ \ln^{[n]} \Pi$.

In the germ of $(\ln^{[n]} \Pi, \infty)$ the derivative g' is bounded for arbitrary $g \in G^{n-1}$; this follows from the ADT_{n-1}. The domain $\ln^{[n]} \Pi$ narrows monotonically at infinity. Therefore, $(g \ln^{[n]} \Pi, \infty) \subset (C_1 \ln^{[n]} \Pi, \infty)$ for sufficiently large C_1.

It now follows from Proposition 4 that $(\Pi', \infty) \subset (S^{\forall}, \infty)$. Consequently, $F \to 0$ in (Π', ∞). ►

REMARK. This also implies that for arbitrary $k \leq m$, $m \leq n$, and $\varphi \in \mathscr{F}^{k}_{+}$ we have the relation $\varphi \circ \ln^{[m]} \to 0$ in $(A^{-m}C\Pi, \infty)$ for arbitrary $C > 1$; in particular, $\varphi \to 0$ in $(\tilde{\Pi}, \infty)$ if $\varphi \in \mathscr{F}^{n}_{+} \circ \ln^{[n]}$, and $\tilde{\Pi} = A^{-n}C\Pi$.

COROLLARY 2 (strengthening of Lemma ${}^{\mathbf{C}}\mathrm{OL2}_{m}$ for $m \leq n$). *Suppose that $k \leq m-1$, $\varphi \in \mathscr{F}^{k}_{\mathbf{R}} \circ \ln^{[m]}$, and $\psi \in \mathscr{F}^{m}_{+} \circ \ln^{[m]}$. Then $|\operatorname{Re} \varphi| \succ |\psi|$ in the domain $(A^{-m}C\Pi, \infty)$ for arbitrary $C > 1$.*

REMARK. A weaker inequality is proved in Lemma ${}^{\mathbf{C}}\mathrm{OL2}_{m}$:

$$|\operatorname{Re} \varphi \circ \exp^{[m]}| \succ |\psi \circ \exp^{[m]}| \quad \text{in } (C \ln^{[m+1]} \Pi, \infty).$$

It follows from the preceding strengthening that the same inequality is valid in the domain $(\Pi'_1, \infty) = (C \ln^{[m]} \Pi, \infty)$. The strengthening is based on the fact that the upper estimate holds for regular functional cochains in a broader domain than the lower estimate. The strengthening is not extended to the case $k = m$.

PROOF. As in the arguments in §5.1 D, suppose that

$$\varphi = \widetilde{F} \circ \exp^{[k]} \circ f, \qquad \psi = F \circ \exp^{[m]} \circ g,$$
$$\widetilde{F} \in \mathscr{F}\mathscr{C}_{\mathbf{R}}^{k}, F \in \mathscr{F}\mathscr{C}_{+}^{m}, f \in G^{k-1}, g \in G^{m-1}.$$

It suffices to prove the inequality $|\operatorname{Re} \widetilde{F} \circ \rho| \prec |F|$, where $\rho = \exp^{[k]} \circ f \circ g^{-1} \circ \ln^{[m]}$, in the germ of the domain $(\Pi_2', \infty) = (\exp^{[m]} \circ g \circ C \circ \ln^{[m]} \Pi, \infty)$. According to the argument in the preceding proof, there exists a $C_1 > 0$ such that $(\Pi_2', \infty) \subset (A^{-m} C_1 \Pi, \infty)$. By Proposition 4, $(A^{-m} C_1 \Pi, \infty) \subset (S^{\forall}, \infty)$; hence $(\Pi_2', \infty) \subset (S^{\forall}, \infty)$. Consequently, by the inequality $(*)$ in §4.2 A, $\operatorname{Re} \rho \prec \varepsilon\xi$ in (Π_2', ∞) for all $\varepsilon > 0$. We prove that $(\Pi_3', \infty) \subset (T', \infty)$ for some $C_1' > 0$, where $\Pi_3' = \rho \Pi_2'$ and $T' = A^{1-m} C_1' \circ \ln \Pi$. Indeed,

$$\Pi_3' = \exp^{[k]} \circ f \circ C \circ \ln^{[m]} \Pi, \qquad k \le m - 1.$$

The inclusion $(\Pi_3', \infty) \subset (T', \infty)$ can be proved in the same way as the inclusion $(\Pi_3, \infty) \subset (T, \infty)$ in §5.1 D. Consequently, for arbitrary $\varepsilon > 0$

$$|\operatorname{Re} \widetilde{F} \circ \rho| \succ \exp(-\varepsilon\xi) \quad \text{in } (\Pi_2', \infty).$$

Further, since $(\Pi_2', \infty) \subset (S^{\forall}, \infty)$, we get that for some $\varepsilon > 0$

$$|F| \prec \exp(-\varepsilon\xi) \quad \text{in } (\Pi_2', \infty).$$

This implies the required inequality. ▶

An important role in what follows is played by the set

$$\mathscr{L}_*(k, n) = \mathscr{L}_*(\mathscr{F}_{\mathbf{R}, \mathrm{id}}^{k}, \mathscr{F}_{++}^{k}, \mathscr{F}_{+}^{k+1}, \ldots, \mathscr{F}_{+}^{n}), \qquad k \le n - 1,$$

where $\mathscr{F}_{++}^{k} = \{\varphi \in \mathscr{F}_{+g}^{k} | g \in G_{\mathrm{slow}}^{k-1^+}\}$. By definition, the elements of the set $\mathscr{L}_*(k, n)$ are sums of germs belonging to the classes in the parentheses, with the first term (of class $\mathscr{F}_{\mathbf{R}, \mathrm{id}}^{k}$) necessarily nonzero.

Denote by $\mathscr{L}_{\mathbf{R}}(\mathscr{F}_{+}^{k}, \ldots, \mathscr{F}_{+}^{n})$ the set of sums of germs of classes $\mathscr{F}_{+}^{k}, \ldots, \mathscr{F}_{+}^{n}$, in which a germ of class $\mathscr{F}_{+g}^{m}$ with smallest value of (m, g) (in comparison with the remaining terms of the sum) is weakly real. The corrections of the germs $g \in G^n$ with $k(g) = k \ge 0$ are precisely such germs. This follows from Theorem ADT_n. We investigate the germs in these sets.

COROLLARY 3. *Suppose that* $k \le n - 1$, *and*

$$\varphi \in \mathscr{L}_*(k, n) \cup (\mathscr{L}_{\mathbf{R}}(\mathscr{F}_{+}^{k}, \ldots, \mathscr{F}_{+}^{n}) \setminus \mathscr{L}(\mathscr{F}_{+}^{n})).$$

Then the real part of the germ φ *is of constant sign in* $(\Pi_1', \infty) = (C \ln^{[n]} \Pi, \infty)$ *for arbitrary* $C > 1$; *the limit* $l = \lim \varphi$ *exists in* (Π_1', ∞), *finite or infinite*; $\arg \varphi \to 0 \pmod{\pi}$ *in* (Π_1', ∞).

PROOF. By definition, $\varphi = \varphi_s + \varphi_r$, where $\varphi_s \in \mathscr{F}^m_{\mathbf{R},g}$, $m \le n-1$, and $\varphi_r \in \mathscr{L}(\mathscr{F}^m_{+f}, \mathscr{F}^l_+ | m+1 \le l \le n;\ f \succ\succ g \text{ in } G^{m-1})$ (s stands for slow, and r for rapid in the symbol for, respectively, rapid and slow decrease of germs). By Corollary 2, $|\operatorname{Re}\varphi_s| \succ |\varphi_r|$ in (Π_1', ∞). This proves the first assertion of the corollary. Further, $\varphi_r \to 0$. For fixed g the set $\mathscr{F}^m_{\mathbf{R},g}$ is ordered on $(C\ln^{[m+1]}\Pi, \infty)$ by the relation $\operatorname{Re}\cdot \prec 0$ for any $C > 1$, as proved in Lemma ${}^{\mathbf{C}}\mathrm{OL2}_m$. Since it contains constants, each germ in this set has a limit in $(C\ln^{[m+1]}\Pi, \infty)$. The second assertion of the corollary now follows from the inequality $m \le n-1$. The third assertion follows from the smallness of $|\varphi_r|$ in comparison with $\operatorname{Re}\varphi_s$ and Lemma ${}^{\mathbf{C}}\mathrm{OL2}_m$. ▶

An immediate consequence of Corollary 3 is

COROLLARY 3 BIS. *Let*

$$\varphi \in [\mathscr{L}_*(k, n) \cup (\mathscr{L}_{\mathbf{R}}(\mathscr{F}^k_+, \dots, \mathscr{F}^n_+) \backslash \mathscr{L}(\mathscr{F}^n_+))] \circ \ln^{[k+1]}.$$

Then $\operatorname{Re}\varphi \ne 0$ *in* $(\exp^{[k+1]} \circ C \circ \ln^{[n]}\Pi, \infty)$ *for arbitrary* $C > 1$, *and* $\lim\varphi$ *exists in this domain as* $\zeta \to \infty$, *finite or infinite.*

For any $C > 1$ let

$$\widetilde{\Pi} = A^{-n} C\Pi, \qquad \widetilde{\Pi}_k = \ln^{[n-k]}\widetilde{\Pi} = \exp^{[k]} \circ C \circ \ln^{[n]}\Pi.$$

These domains are encountered repeatedly in what follows. The dependence on C is not indicated in the notation. In these terms the conclusion of Corollary 3 bis holds in the germ of the domain $(\widetilde{\Pi}_{k+1}, \infty)$.

COROLLARY 4. *Suppose that* $\varphi \in \mathscr{F}^n_{+g}$ *and* $g \in G^{n-1}$. *Then the germ* $\psi = \varphi \circ \ln^{[n]}$ *is defined in* $(\Pi^\forall, \infty)$, *and* $|\varphi \circ \ln^{[n]}| \prec |\zeta|^{-N}$ *there for arbitrary* $N > 0$. *If in addition* $\mu_n(g) = 0$, *then the germ* ψ *is defined in* $(S^\forall, \infty)$, *and the preceding estimate holds there.*

REMARK. A related fact is Proposition 3 in §2.11.

PROOF. By the definition of the set $\mathscr{F}^n_{+g}$, there exists a germ $F \in \mathscr{F}\mathscr{C}^n_+$ such that

$$\varphi = F \circ \exp^{[n]} \circ g.$$

Then $\psi = F \circ A^{-n} g$. By Proposition 4,

$$(A^{-n} g\Pi^\forall, \infty) \subset (S^\forall, \infty).$$

Consequently, the germ ψ is defined in $(\Pi^\forall, \infty)$.

Suppose now that $\mu_n(g) = 0$. We prove that

$$(A^{-n} gS^\forall, \infty) \subset (S^\forall, \infty);$$

in particular, the germ of an arbitrary "broad" sector passes into the germ of an arbitrary "narrow" sector. We have that

$$(A^{-n} gS^\forall, \infty) = \exp(A^{1-n} g\Pi^\forall, \infty) \subset \exp(\Pi^\forall, \infty) = (S^\forall, \infty).$$

The first inclusion follows from the fact that g is real and the fact that the derivative $(A^{1-n}g)'$ tends to zero in $(\Pi^{\forall}, \infty)$. Consequently, the germ ψ is defined in $(S^{\forall}, \infty)$.

By the definition of a rapidly decreasing cochain, there exist an $\varepsilon > 0$ and a standard domain Ω such that $|F| \prec \exp(-\varepsilon\xi)$ in (Ω, ∞). It was just proved that $\arg A^{-n}g \to 0$ in $(S^{\forall}, \infty)$ for $\mu_n(g) = 0$. The same convergence in $(\Pi^{\forall}, \infty)$ for arbitrary $g \in G^{n-1}$ follows from Proposition 4. Therefore,

$$|A^{-n}g| = |\operatorname{Re} A^{-n}g|(1 + o(1))$$

in $(\Pi^{\forall}, \infty)$ for an arbitrary value of $\mu_n(g)$, and in $(S^{\forall}, \infty)$ for $\mu_n(g) = 0$. By Lemma 5.5_nb, $|A^{-n}g| \succ (\ln|\zeta|)^C$ in $(S^{\forall}, \infty)$ for any $C > 1$. Consequently, for any $N > 0$

$$|F \circ A^{-n}g| \prec \exp(-\varepsilon(\ln|\zeta|)^C) \prec \exp(-N\ln|\zeta|) = |\zeta|^{-N}. \quad ►$$

We now prove one of the main results of the section: a natural generalization of Proposition 4.

PROPOSITION 5. *Suppose that for any* $C > 1$ *the germ* h *is holomorphic in the germ of the domain* $(C\ln^{[n]}\Pi, \infty)$ *and that* $h' \to \alpha > 0$ *there. Then the germ* $A^{-n}h$ *is defined in the germ of the domain* $(\tilde{\Pi}, \infty)$ *and satisfies the estimates*

$$(\ln|\zeta|)^{C'} \prec |A^{-n}h| \prec \exp\xi^{\varepsilon}, \qquad |(A^{-n}h)'| \prec \exp\xi^{\varepsilon}$$

in $(\tilde{\Pi}, \infty)$ *for arbitrary* $\varepsilon > 0$ *and* $C' > 1$; *moreover, for any* $C' > 1$

$$(\ln|\zeta|)^{C'} \prec |A^{-(n+1)}h| \prec \exp|\zeta|^{\varepsilon},$$
$$|(A^{-(n+1)}h)'| \prec \exp|\zeta|^{\varepsilon},$$
$$(\ln A^{-(n+1)}h)' \to 0$$

in $(S^{\forall}, \infty) = \{\zeta \mid |\arg\zeta| < \alpha \in (0, \pi)\}$.

PROOF. We use induction on m to prove for any $\varepsilon > 0$ and $0 \le m \le n+1$ the inequalities

$$|A^{-m}h| \prec \exp|\zeta^{\varepsilon}|, \qquad |(A^{-m}h)'| \prec \exp|\zeta^{\varepsilon}| \tag{$*$}$$

in the domain $\widetilde{\Pi}_m \overset{\text{def}}{=} \ln^{[n-m]}\widetilde{\Pi}$. This proves all the assertions in Proposition 5 except the last—the estimate of the logarithmic derivative—and the lower estimates. Indeed, $\widetilde{\Pi}_n = \widetilde{\Pi}$, $\widetilde{\Pi}_{n+1} = \exp\widetilde{\Pi}$, and $(\widetilde{\Pi}_{n+1}, \infty) \supset (S^{\forall}, \infty)$.

For $m = 0$ (the induction base) the proof is trivial.

INDUCTION STEP (THE PASSAGE FROM m TO $m+1$).

$$|A^{-1}(A^{-m}h)| = \exp\circ\operatorname{Re} A^{-m}h \circ \ln \prec \exp|\zeta|^{\varepsilon},$$
$$|(A^{-1}(A^{-m}h))'| = |(A^{-(m+1)}h)\cdot(A^{-m}h)' \circ \ln\zeta^{-1}|$$
$$\prec (\exp|\zeta|^{\varepsilon})|\zeta|^{\varepsilon-1}.$$

By Proposition 4, $(\widetilde{\Pi}, \infty) \subset (S^{\forall}, \infty)$. Therefore, $|\zeta|$ and $\xi = \operatorname{Re}\zeta$ are comparable in $\widetilde{\Pi}$. Consequently, $|\zeta|$ can be replaced by ξ in the estimates obtained $(m \le n)$. This proves the inequality $(*)$ for $0 \le m \le n+1$.

We estimate the logarithmic derivative:

$$|\ln(A^{-(m+1)}h)'| \prec |(A^{-m}h)' \circ \ln \cdot \zeta^{-1}| \prec |\zeta|^{\varepsilon-1} \to 0$$

in (Π'_{m+1}, ∞).

Finally, we prove the lower estimate in the proposition by induction on m. Assume that

$$|A^{-m}h| \succ (\ln|\zeta|)^C \quad \text{in } \widetilde{\Pi}_m \text{ for } 0 \le m \le n;$$

this inequality is trivial for $m = 0$. Then in $\widetilde{\Pi}_{m+1}$

$$|A^{-(m+1)}h| \succ \exp(\operatorname{Re}\ln|\ln\zeta|)^C \succ (\ln|\zeta|)^C. \quad \blacktriangleright$$

It follows from assertion $*$ in subsection C below that $\operatorname{Re} A^{-n}h \succ (\ln|\zeta|)^C$ in $(\widetilde{\Pi}, \infty)$.

C. The action of the iterates of the operator A^{-1} on germs in the group G^n. Let us apply Proposition 5 to germs in the group G^n. For this we verify that the indicated germs satisfy the condition in Proposition 5, which proves

PROPOSITION 6. *Suppose that* $g \in G^n$. *Then:*

1°. *the germ* $A^{-n}g$ *extends biholomorphically from* $(\mathbf{R}^+, \infty)$ *to the germ of any half-strip* $(\Pi^{\forall}, \infty)$ *and carries it into the germ of any sector* $(S^{\forall}, \infty)$;

2°. *the germ* $A^{-(n+1)}g$ *extends biholomorphically from* $(\mathbf{R}^+, \infty)$ *to the germ of any sector* (S_α, ∞), $\alpha \in (\pi/2, \pi)$;

3°. *for any* $\varepsilon > 0$ *and any* $C > 0$

$$\begin{gathered}(\ln|\zeta|)^C \prec |A^{-n}g| \prec \exp\xi^{\varepsilon}, \\ |(A^{-n}g)'| \prec \exp\xi^{\varepsilon} \quad \textit{in } (\widetilde{\Pi}, \infty); \\ (\ln|\zeta|)^C \prec |A^{-(n+1)}g| \prec \exp|\zeta|^{\varepsilon}, \\ |(A^{-(n+1)}g)'| \prec \exp|\zeta| \quad \textit{in } (S^{\forall}, \infty), \\ (\ln(A^{-(n+1)}g))' \to 0 \quad \textit{in } (S^{\forall}, \infty).\end{gathered}$$

REMARK. These assertions constitute part of Lemmas 5.3_{n+1} and 5.5_{n+1}. Precise references will be made in the proofs of these lemmas.

PROOF. We verify the conditions of Proposition 5 for $h = g$:

1°. The mapping g is holomorphic in the domain $C\ln^{[n]}\Pi$ for arbitrary $C > 0$;

2°. $g' \to \alpha > 0$ in $(C\ln^{[n]}\Pi, \infty)$.

Let us prove assertion 1°. By the additive decomposition theorem ADT_n, which was proved in Chapter IV,

$$g = a + \sum \varphi_j, \qquad \varphi_j \in \mathscr{F}^{k_j}_{+g_j},\ 0 \le k_j \le n,\ g_j \in G^{k_j-1}, \\ (k_j, g_j) \prec (k_{j+1}, g_{j+1}). \qquad (**)$$

Recall that, by the convention in §1.10, $\mathscr{F}^k_{+g} = \mathscr{F}^k_{1+g}$ for $k \le n-1$, and $\mathscr{F}^k_{+g} = \mathscr{F}^n_{0+g}$ for $k = n$. It is obvious that the first term is holomorphic in $(C\ln^{[n]}\Pi, \infty)$; we prove that the remaining terms are holomorphic. Namely, we prove that for $k \le n$ and for all $f \in G^{k-1}$ and $F \in \mathscr{F}\mathscr{C}^k$ the composition $F^u \circ \exp^{[k]} \circ f$ (F^u is the main function in the tuple F) extends holomorphically to the domain $C\ln^{[n]}\Pi$, or, what is the same, $F^u \circ \rho$, $\rho = \exp^{[k]} \circ f \circ C\ln^{[n]}$ extends holomorphically to Π, more precisely, to the part of Π outside some compact set depending on the composition. Recall that in view of the convention in §1.10, $\mathscr{F}\mathscr{C}^k = \mathscr{F}\mathscr{C}^k_1$ for $k \le n-1$ and $\mathscr{F}\mathscr{C}^k = \mathscr{F}\mathscr{C}^n_0$ for $k = n$. The fact that the last composition is holomorphic may be proved in the following way.

For $k \le n-1$ it is sufficient to prove that $(\rho\Pi, \infty) \subset (\Pi_\varepsilon, \infty)$ for some $\varepsilon > 0$, where Π_ε is an ε-neighborhood of $\mathbb{R}^+$. We may assume, by induction, that the assertion 1° of Proposition 6 is proved for $g \in G^{n-1}$ instead of $g \in G^n$. Recall that $C \in G^k$ for any k. Then $A^{-n}(f \circ C)(\Pi^\forall, \infty) \subset (S^\forall, \infty)$. Now $\rho(\Pi^\forall, \infty) \subset \ln^{[n-k]}(S^\forall, \infty) \subset (\Pi_\varepsilon, \infty)$ for any $\varepsilon > 0$.

In the case $k = n$, F is a cochain of type $\mathscr{D}^n_0$. It is sufficient to prove the inclusion $(\rho\Pi^\forall, \infty) \subset (\sigma_0\Pi_\varepsilon, \infty)$ for any $\sigma_0 \in \mathscr{D}^n_0$, or $(\rho_0 \circ \rho\Pi^\forall, \infty) \subset (\Pi_\varepsilon, \infty)$, $\rho_0 = \sigma_0^{-1}$, which is equivalent. By Definition 16 in §1.6C $\rho_0 \circ \rho = \ln \circ A^{-n}(f \circ C \circ g)$ for some $g \in G^{n-1}$. After this the end of the proof is the same as for $k \le n-1$. This verifies condition 1°.

We verify condition 2°. By Theorem ADT_n and the differentiation lemma $\mathrm{DL2}_n$,

$$g' = \alpha + \sum \psi_j, \qquad \alpha > 0,\ \psi_j \in \mathscr{F}^{k_j}_{+g_j}.$$

Corollary 1 in subsection B implies that the germ ψ_j tends to zero in $C\ln^{[n]}\Pi$. This verifies condition 2° and by Proposition 5 proves assertions 2° and 3° of Proposition 6. Let us prove assertion 1°. To do this we prove the following assertion $*$.

Assertion $*$. *For an arbitrary $C > 1$ and an arbitrary $g \in G^n$*

$$\arg A^{-n} g \to 0$$

in the domain $\widetilde{\Pi} = A^{-n}C\Pi$ as $\zeta \to \infty$.

PROOF. The proof is according to the same scheme as the investigation of the germ $A^{-n}C$ above. Assertion $*$ follows from the fact that

$$\arg A^{-n}(g \circ C) \to 0 \quad \text{in } (\Pi, \infty).$$

Note that $\tilde{g} = g \circ C \in G^n$ and the germ $\tilde{g}$ is real. Consequently,

$$\arg(A^{-n}\tilde{g}) = \operatorname{Im} A^{1-n}\tilde{g} \circ \ln .$$

But it follows from the estimates $(*)$ in §5.2 B (the applicability of which to the germs in the group G^n was proved in this section) that

$$(A^{1-n}\tilde{g} \circ \ln)' \to 0 \quad \text{in } (\Pi, \infty).$$

From this,

$$\operatorname{Im}(A^{1-n}\tilde{g} \circ \ln) \to 0 \quad \text{in } (\Pi, \infty),$$

which proves the assertion. ►►

We now recall the formulation of Lemma 5.3_{n+1}.

LEMMA 5.3_{n+1}. a. *For each $g \in G^n$ the germ $A^{-n}g$ maps the germ of any half-strip containing $(\mathbf{R}^+, \infty)$ biholomorphically into the germ of any sector (S_α, ∞): $S_\alpha = \{|\arg \zeta| < \alpha \in (\pi/2, \pi)\}$.*

b. *For each $g \in G^n$ the limit $\mu_{n+1}(g) = \lim_{(\Pi^\forall, \infty)}(A^{-n}g)'$ exists in the domain $(\Pi^\forall, \infty)$, positive, zero, or infinite.*

The next three subsections are devoted to the proof of Lemma 5.3_{n+1}b. The first assertion of the lemma has already been proved—it is the first assertion of Proposition 6. To prove the second we introduce the characteristic of the rate of decrease of the correction of a germ $g \in G^n$.

D. The index of the Archimedean class of the correction of a germ $g \in G^n$, and the action of the operator A^{-1}. Suppose that $a \neq \mathrm{id}$ in the expansion $(**)$ for $g \in G^n$. Then we set $k(g) = -1$. Note that only nonzero terms are written in the expansion $(**)$. If $a = \mathrm{id}$ in this expansion, but $g \neq \mathrm{id}$, then we set $k(g) = k_1$. The number $k = k(g)$ characterizes the rate of decrease of the correction $g - \mathrm{id}$; namely, if $k \geq 0$, then $g - \mathrm{id}$ belongs to the Archimedean class $\mathscr{A} \in \mathscr{A}^k_{G^{k-1}}$ (see §1.4 for the definition of the Archimedean classes in the set $\mathscr{A}^k_G$). In other words, there exist positive numbers μ and ν with $\mu < \nu$ such that

$$\exp(-\exp^{[k]} \nu\xi) \prec |g(\xi) - \xi| \prec \exp(-\exp^{[k]} \mu\xi)$$

on $(\mathbf{R}^+, \infty)$.

The following notation introduced above is used often in what follows:

$$\widetilde{\Pi}_k = \ln^{[n-k]} \widetilde{\Pi} = \exp^{[k]} \circ C \circ \ln^{[n]} \Pi, \qquad C > 1,\ 0 \leq k \leq n.$$

As above, the dependence on n and C is not indicated in the notation: n and $C > 1$ are assumed to be fixed.

PROPOSITION 7. *Suppose that* $g \in G^n$, $k(g) = k \geq 0$, *and* $A^{-m} g = \mathrm{id} + \varphi_m$. *Then for arbitrary* m *with* $0 \leq m \leq k$

$$\varphi_m \in \mathscr{L}_{\mathbf{R}}(\mathscr{F}_+^k, \mathscr{F}_+^{k+1}, \ldots, \mathscr{F}_+^n) \circ \ln^{[m]}.$$

PROOF. By definition,

$$\exp^{[m]} + \varphi_m \circ \exp^{[m]} = \exp^{[m]} \circ g.$$

According to the multiplicative decomposition theorem,

$$g = g_k \circ j_k \circ h_{k+1} \circ j_{k+1} \circ \cdots \circ j_{n-1} \circ h_n, \quad \text{where } g^k \in G^k, \ j_l \in J^l, \ h_l \in H^l. \qquad (***)$$

By definition, $h_l \in \mathrm{Gr}(\mathrm{id} + \mathscr{F}_+^l)$, and by Lemma $\mathrm{SL4}_n$ a, $j_l \in \mathrm{Gr}(\mathrm{id} + \mathscr{F}_+^l)$. Setting $g_1 = g_k^{-1} \circ g$, we get that $k(g_1) \geq k$. Consequently, $k(g_k) \geq k$, because $k(g) = k$. From this, $g_k \in \mathrm{Gr}(\mathrm{id} + \mathscr{F}_+^k)$. Next, $\exp^{[m]} = \exp \circ \exp^{[m-1]} = \mathscr{F}_{\mathrm{id}}^{m-1}$. By Lemma $\mathrm{SL2}_n$ and the fact that $m - 1 < k$,

$$\exp^{[m]} \circ (\mathrm{id} + \mathscr{F}_+^k) \subset \exp^{[m]} + \mathscr{F}_+^k.$$

Similarly, using Lemma $\mathrm{SL2}_n$ several times as done in the proof of the additive decomposition theorem, we get that for $0 \leq k \leq m$

$$\exp^{[m]} \circ g \in \exp^{[m]} + \mathscr{L}(\mathscr{F}_+^k, \ldots, \mathscr{F}_+^n).$$

Consequently, $\varphi_m \in \mathscr{L}(\mathscr{F}_+^k, \ldots, \mathscr{F}_+^n) \circ \ln^{[m]}$. But the germ φ_m is real. The conclusion of the proposition is obtained from the criterion for weak realness. ►

The analogue of this proposition for φ_{k+1} consists in the following.

PROPOSITION 8. *Suppose that* $g \in G^n \setminus \mathrm{id}$, $k(g) = k \leq n-1$, *and* $A^{-m} g = \mathrm{id} + \varphi_m$. *Then*:

1°. $\varphi_{k+1} \in \mathscr{L}_{\mathbf{R}}(\mathscr{F}_{\mathrm{id}}^k, \mathscr{F}_{++}^k, \mathscr{F}_+^{k+1}, \ldots, \mathscr{F}_+^n) \circ \ln^{[k+1]}$;

2°. *the germ* $\varphi_{k+1} \circ \exp^{[k+1]}$ *can be decomposed into a sum* $\mathbf{e} + \psi$, $\mathbf{e} \in E^{k+1}$ (*the set of exponents of* STAR-$(k+1)$; *see Definition* 2^n *in* §1.7), *and* $\psi_+ \in \mathscr{L}(\mathscr{F}_{+\mathrm{id}}^k, \mathscr{F}_{++}^k, \mathscr{F}_+^{k+1}, \ldots, \mathscr{F}_+^n)$;

3°. $\varphi_{k+1}^{(l)} \to 0$ *in* $(\tilde{\Pi}_{k+1}, \infty)$ *for any positive integer* l.

PROOF. The second assertion of the proposition follows easily from the relation $\exp^{[k+1]} \in E^{k+1}$ and the lemmas on "shifts of exponents": Lemma 1 in §4.3 E and Lemma 2 in §4.9 B. Namely, by definition of φ_{k+1}

$$\exp^{[k+1]} + \varphi_{k+1} \circ \exp^{[k+1]} = \exp^{[k+1]} \circ g.$$

Suppose that the germ g is decomposed according to the formula $(***)$. Note that in this decomposition $g_k \in G_{\mathrm{rap}}^k$. Indeed, suppose that $A^{-k} g_k = \mathrm{id} + \tilde{\varphi}_k$. Then $A^{-(k+1)} g_k = \zeta \cdot \exp \circ \tilde{\varphi}_k \circ \ln$. But by Proposition 7, $\tilde{\varphi}_k \in$

$\mathscr{L}(\mathscr{F}_+^k) \circ \ln^{[k]}$. According to Corollary 1 in §5.2 B, $\tilde{\varphi}_k \to 0$ in $(\Pi^\forall, \infty)$. Consequently, $A^{-(k+1)} g_k = \zeta(1 + o(1))$ in $(\Pi^\forall, \infty)$, which implies that $g_k \in G_{\text{rap}}^k$.

Now there exist by Lemma 1 in §4.3 E an exponent $\mathbf{e} \in E^{k+1}$ and a germ $\psi_1 \in \mathscr{L}(\mathscr{F}_{+\text{id}}^k, \mathscr{F}_{++}^k)$ such that

$$\exp^{[k+1]} \circ g_k = \mathbf{e} + \psi_1.$$

Further, by Lemma 2 in §4.9 B,

$$\mathbf{e} \circ j_k = \tilde{\mathbf{e}} + \psi_2, \qquad \tilde{\mathbf{e}} \in E^{k+1}, \psi_2 \in \mathscr{L}(\mathscr{F}_{+\text{id}}^k, \mathscr{F}_{++}^k).$$

According to Lemma SL4_n a, $J^k \subset \text{Gr}(\text{id} + \mathscr{F}_+^k)$. Consequently, by Lemmas SL2_n and SL3_n,

$$\mathscr{L}(\mathscr{F}_{+\text{id}}^k, \mathscr{F}_{++}^k) \circ J^k \subset \mathscr{L}(\mathscr{F}_{+\text{id}}^k, \mathscr{F}_{++}^k).$$

Accordingly,

$$\exp^{[k+1]} \circ g_k \circ j_k = \tilde{\mathbf{e}} + \psi_3, \qquad \tilde{\mathbf{e}} \in E^{k+1}, \psi_3 \in \mathscr{L}(\mathscr{F}_{+\text{id}}^k, \mathscr{F}_{++}^k).$$

Further, it follows from Lemma SL2_n that

$$(\tilde{\mathbf{e}} + \psi_3) \circ (h_{k+1} \circ \cdots \circ h_n) = \tilde{\mathbf{e}} + \psi_+, \qquad \psi_+ \in \mathscr{L}(\mathscr{F}_{+\text{id}}^k, \dots, \mathscr{F}_+^n),$$

which proves the second assertion of the proposition.

Let us prove the first. It follows from the regularity lemma in §2.12 that $E^{k+1} \subset \mathscr{F}_{\mathbf{R},\text{id}}^k$. Then by assertion 2°,

$$\varphi_{k+1} \in \mathscr{L}(\mathscr{F}_{\text{id}}^k, \mathscr{F}_{++}^k, \mathscr{F}_+^{k+1}, \dots, \mathscr{F}_+^m) \circ \ln^{[k+1]}.$$

Assertion 1° now follows from the fact that φ_{k+1} is real and from the criterion for weak realness.

Let us prove the third assertion. We have that

$$\varphi'_{k+1} = (\exp \circ \varphi_k \circ \ln)(1 + \varphi'_k \circ \ln) - 1.$$

The set $\mathscr{L}(\mathscr{F}_+^k, \dots, \mathscr{F}_+^n) \circ \ln^{[k+1]}$ contains:

the first parenthesis, diminished by 1, on the right-hand side of the preceding equality (by Proposition 7 and Corollary 1 to Lemma LA_n in §4.5);

the second parenthesis, diminished by 1 (by Proposition 7 and the differentiation lemma);

the whole right-hand side (by the multiplication lemma).

According to Corollary 1 in §5.2 B, $\varphi'_{k+1} \to 0$ in $(\tilde{\Pi}_{k+1}, \infty)$, along with all the derivatives. ▶

The next proposition is basic in this subsection. Recall that $\mathscr{L}_*(k+1, n) = \mathscr{L}_*(\mathscr{F}_{\mathbf{R},\text{id}}^{k+1}, \mathscr{F}_{++}^{k+1}, \mathscr{F}_+^{k+2}, \dots, \mathscr{F}_+^n)$; see the definition in §5.2 B for details.

PROPOSITION 9. *Under the conditions of Proposition 8 the following hold for* $0 \le k \le n-2$:

1°. $\varphi_{k+2} \in \mathscr{L}_*(k+1, n) \circ \ln^{[k+2]}$;

2°. $|\operatorname{Re}\varphi_{k+2}| \to \infty$ *in* $(\widetilde{\Pi}_{k+2}, \infty)$;

3°. $\arg\varphi'_{k+2} \to 0$ *in* $(\widetilde{\Pi}_{k+2}, \infty)$ *for* $g \succ \mathrm{id}$,
$\arg\varphi'_{k+2} \to \pi$ *in* $(\widetilde{\Pi}_{k+2}, \infty)$ *for* $g \prec \mathrm{id}$.

PROOF. 1°. By the definition of the germs φ_{k+1} and φ_{k+2},

$$\varphi_{k+2} = \zeta \cdot (\exp\circ\varphi_{k+1} \circ \ln - 1). \qquad ({}^{*}_{**})$$

Further, by Proposition 8,

$$\varphi_{k+2} \circ \exp^{[k+2]} = \exp^{[k+2]} \cdot (\exp\circ(\tilde{\mathbf{e}} + \psi_+) - 1).$$

According to Corollary 1 to Lemma LA_n in §4.5,

$$\exp\psi_+ = 1 + \tilde{\psi}_+, \qquad \tilde{\psi}_+ \in \mathscr{L}(\mathscr{F}^k_{+\mathrm{id}}, \mathscr{F}^k_{++}, \dots, \mathscr{F}^n_{+}).$$

By the multiplication lemma,

$$\exp^{[k+2]} \cdot (\exp\circ\tilde{\mathbf{e}}) \cdot (1 + \tilde{\psi}_+) \in \mathscr{L}(\mathscr{F}^{k+1}_{\mathrm{id}}, \mathscr{F}^{k+1}_{++}, \dots, \mathscr{F}^n_{+}).$$

This proves that

$$\varphi_{k+2} \in \mathscr{L}(\mathscr{F}^{k+1}_{\mathrm{id}}, \mathscr{F}^{k+1}_{++}, \dots, \mathscr{F}^n_{+}) \circ \ln^{[k+2]}.$$

It must still be proved that in the expansion for φ_{k+2} the term of class $\mathscr{F}^{k+1}_{\mathrm{id}}$ is nonzero. This follows immediately from assertion 2° of the proposition. The weak realness of this term follows in the standard way from the criterion of §1.11 D. Thus, assertion 1° follows from what was proved above and assertion 2°, to which we now proceed.

2°. The second assertion follows from Proposition 8, the formula $({}^{*}_{**})$, and a lower estimate on φ_k that can be deduced from Lemmas ${}^{\mathrm{C}}\mathrm{OL1}_n$ and ${}^{\mathrm{C}}\mathrm{OL2}_n$. In more detail, assertion 1° of Proposition 8, assertion 1° of ${}^{\mathrm{C}}\mathrm{OL1}_n$ and Lemma ${}^{\mathrm{C}}\mathrm{OL2}_n$ give us that the limit $\lambda = \lim_{(\widetilde{\Pi}_{k+1}, \infty)} \varphi_{k+1}$ exists. Note that $(\widetilde{\Pi}_{k+1}, \infty) \subset (\ln\widetilde{\Pi}, \infty)$, because $k \le n-2$. Therefore, $\operatorname{Im}\varphi_{k+1} \to 0$ in $(\widetilde{\Pi}_{k+1}, \infty)$, and $\arg\exp\circ\varphi_{k+1}\circ\ln \to 0$ in $(\widetilde{\Pi}_{k+2}, \infty)$. We consider three cases: $|\lambda| = \infty$, $|\lambda| \in (0, \infty)$, and $\lambda = 0$.

CASE 1: $|\lambda| = \infty$. In this case if $\lambda = +\infty$, then in $(\widetilde{\Pi}_{k+2}, \infty)$

$$|\exp\circ\varphi_{k+1}\circ\ln| \to \infty, \qquad \arg\exp\circ\varphi_{k+1}\circ\ln \to 0.$$

All the more so, $\operatorname{Re}\varphi_{k+2} \to \infty$ in $(\widetilde{\Pi}_{k+2}, \infty)$ by the formula $({}^{*}_{**})$. If $\lambda = -\infty$, then $\varphi_{k+2} = \zeta(o(1) - 1)$, and $\operatorname{Re}\varphi_{k+2} \to -\infty$ in $(\widetilde{\Pi}_{k+2}, \infty)$.

CASE 2: $|\lambda| \in (0, \infty)$. In this case $l = \exp\lambda - 1 \ne 0$, and

$$\varphi_{k+2} = \zeta \cdot (l + o(1)) \quad \text{in } (\widetilde{\Pi}_{k+2}, \infty);$$

assertion 2° is again obvious.

CASE 3: $\lambda = 0$. In this case

$$\varphi_{k+2} = \zeta \cdot \varphi_{k+1} \circ \ln \cdot (1 + o(1)) = \zeta \cdot \ln \zeta \cdot \varphi_k \circ \ln^{[2]} \cdot (1 + o(1))$$

in $(\widetilde{\Pi}_{k+2}, \infty)$. By Proposition 7, there exists a decomposition $\varphi_k = \varphi_s + \varphi_r$, where $\varphi_s \in \mathscr{F}^k_{\mathbf{R}, g} \circ \ln^{[k]}$ for some $g \in G^{k-1}$, and $\varphi_r = o(\operatorname{Re} \varphi_s)$ in $(\widetilde{\Pi}_k, \infty)$. Then $\varphi_k = \varphi_s(1 + o(1))$ in $(\widetilde{\Pi}_k, \infty)$. By the definition of the set $\mathscr{F}^k_{\mathbf{R}, g}$, $\varphi_s = F \circ \exp^{[k]} \circ g$, $F \in \mathscr{FC}^k_{\mathbf{R}+}$. According to Lemma ${}^{\mathbf{C}}\mathrm{OL}1_n$, there exists an $\varepsilon > 0$ such that for any $C_1 > 1$

$$|\operatorname{Re} F| \succ \exp(-\varepsilon \xi) \quad \text{in } A^{-k} C_1 \circ \ln \Pi.$$

From this,

$$|\operatorname{Re} \varphi_s| \succ \exp(-\varepsilon \operatorname{Re} A^{-k} g) \quad \text{in } A^{-k}(g^{-1} \circ C_1) \circ \ln \Pi,$$
$$|\operatorname{Re} \varphi_s \circ \ln| \succ 1/|A^{-(k+1)} g|^{\varepsilon} \quad \text{in } A^{-(k+1)}(g^{-1} \circ C_1)\Pi.$$

We remark that

$$(A^{-(k+1)}(g^{-1} \circ C_1)\Pi, \infty) \supset (\widetilde{\Pi}_{k+1}, \infty) = ((A^{-(k+1)} C) \circ \ln^{[n-(k+1)]} \Pi, \infty)$$

because $n - (k+1) \geq 1$. Consequently, the preceding inequality holds in $(\widetilde{\Pi}_{k+1}, \infty)$. By Lemma 5.5_n b, $|A^{-(k+1)} g| \prec \exp(-\delta \xi)$ in $(\widetilde{\Pi}_{k+1}, \infty)$ for any $\delta > 0$. Consequently,

$$|\varphi_k \circ \ln^{[2]}| \succeq |(\operatorname{Re} \varphi_s \circ \ln^{[2]}) \cdot (1 + o(1))| \succ |\zeta|^{-\delta}$$

in $(\widetilde{\Pi}_{k+2}, \infty)$ for any $\delta > 0$. Therefore, $|\varphi_{k+2}| \succ |\zeta|^{1-\delta}$ in $(\widetilde{\Pi}_{k+2}, \infty)$ for any $\delta > 0$. This proves assertion 2° in Case 3.

3°. The third assertion follows immediately from the first. Indeed,

$$\varphi_{k+2} = \tilde{\varphi}_s + \tilde{\varphi}_r, \qquad \tilde{\varphi}_s \in \mathscr{F}^{k+1}_{\mathbf{R}, \mathrm{id}} \circ \ln^{[k+2]}, \ \tilde{\varphi}_r \to 0 \text{ in } (\widetilde{\Pi}_{k+2}, \infty).$$

By assertion 2°, $|\varphi_{k+2}| \to \infty$, and consequently, $\varphi_s' \neq 0$. Therefore, $\varphi_{k+2}' = \varphi_s'(1 + o(1))$. According to Lemma ${}^{\mathbf{C}}\mathrm{OL}1_n$, $\arg \varphi_s' \to 0 \pmod{\pi}$ in $(\widetilde{\Pi}_{k+2}, \infty)$. ▶

The propositions proved lie at the basis of the arguments in §§5.4, 5.5, and 5.7.

PROPOSITION 10. *Suppose that* $g \in G^1$ *and* $k(g) = -1$. *Then* $\operatorname{Re} \varphi_1 \to \infty$ *and* $\arg \varphi_1' \to 0 \pmod{\pi}$ *in* $(\widetilde{\Pi}, \infty)$, *and* $(A^{-1} g)' \to \mu \neq 1$ *in* $(\widetilde{\Pi}, \infty)$; $\mu > 1$ *or* $\mu = +\infty$ *for* $g \succ \mathrm{id}$.

PROOF. According to Theorem ADT_1, $g = \alpha\zeta + \beta + o(1)$. By assumption, $\alpha\zeta + \beta \succ \mathrm{id}$ when $g \succ \mathrm{id}$, that is, either $\alpha > 1$, or $\alpha = 1$ and $\beta > 0$. Then $A^{-1} g = \zeta^{\alpha}(\mu + o(1))$, where $\mu = \exp \beta$, and $(A^{-1} g)' = \zeta^{\alpha - 1}(\mu + o(1))$ in $(A^{-1} C\Pi, \infty)$. All the assertions in the proposition follow in an obvious way from these formulas. ▶

E. Existence of a generalized exponent for $k(g) \geq n-2$**.** In this and the next subsections we prove Lemma 5.3_{n+1} b. We consider three cases:

$$k(g) \geq n-1; \qquad k(g) = n-2; \qquad k(g) \leq n-3.$$

It will be proved that in the first case $\mu_{n+1}(g) = 1$, and in the third case $\mu_{n+1}(g) = \infty$ if $g \succ \mathrm{id}$ and $\mu_{n+1}(g) = 0$ if $g \prec \mathrm{id}$, while $\mu_{n+1}(g)$ can be arbitrary in the second case. Let $A^{-m}g = \mathrm{id} + \varphi_m$.

CASE 1: $k(g) \geq n-1$. In this case $k(g) \geq 0$, because $n \geq 1$. According to Proposition 7 in §5.2 D,

$$\varphi_{n-1} \in \mathscr{L}(\mathscr{F}_+^{n-1}, \mathscr{F}_+^n) \circ \ln^{[n-1]}.$$

Further,

$$(A^{-n}g)' = (\exp \circ \varphi_{n-1} \circ \ln) \cdot (1 + \varphi'_{n-1} \circ \ln).$$

Both factors tend to 1 in $(\widetilde{\Pi}, \infty)$, which proves Lemma 5.3_{n+1} b in Case 1.

CASE 2: $k(g) = n-2$. Let $n = 1$. Then $k(g) = -1$, and Proposition 10 is applicable and immediately yields Lemma 5.3_{n+1} b.

Let $n \geq 2$. Then $k(g) \geq 0$, and Proposition 9 is applicable. It follows from assertion 1° of this proposition that

$$\varphi_n = \varphi_s + \varphi_r, \qquad \varphi_s \in \mathscr{F}^{n-1}_{\mathbf{R}, \mathrm{id}}, \ \varphi_r \in (\mathscr{F}_{++}^{n-1}, \mathscr{F}_+^n) \circ \ln^{[n]}.$$

Further, $\varphi'_s \neq 0$: assumption of the contrary leads to a contradiction of assertion 2° in Proposition 9. The existence of the limit of φ'_n in $(\widetilde{\Pi}, \infty)$ now follows from Corollary 3 in §5.2 B. In this case Lemma 5.3_{n+1} b follows from the equality $(A^{-n}g)' = 1 + \varphi'_n$.

Case 3 $(k(g) \leq n-3)$ is analyzed in the next subsection.

F. The action of the operator A^{-1} **on germs with slowly decreasing correction.** In this subsection we finish the proof of Lemma 5.3_{n+1} b.

PROPOSITION 11. *Suppose that* $g \in G^n$, $m \leq n$, $k(g) = k \leq m-3$, $g \succ \mathrm{id}$. *Then*:

1°. $(A^{-m}g)' \to \infty$ *in* $(\widetilde{\Pi}_m, \infty)$, *in particular,* $\mu_{n+1}(g) = \infty$;

2°. $\arg(A^{-m}g)' \to 0$ *in* $(\widetilde{\Pi}_m, \infty)$;

3°. $(A^{-n}g^{-1})' \to 0$ *in* $(\Pi^\forall, \infty)$.

REMARK. Assertions 1° and 3° are used in the proof of Lemma 5.3_{n+1} b, and assertion 2° is used in the proof of assertion 1° and the convexity lemma in §5.8.

PROOF. The proof of Proposition 11 is broken up into four steps.

Step 1.

PROPOSITION 12. *Let* $g \in G^n$, $m \leq n-1$, *and* $A^{-m}g = \mathrm{id} + \varphi_m$. *Then* $\mathrm{Im}\, \varphi_m \to 0$ *in* $(\widetilde{\Pi}_m, \infty)$.

PROOF. We carry out the proof for $m = n-1$; for smaller m it is analogous. The relation

$$\operatorname{Im}(A^{1-n}g) \to 0 \quad \text{in } (\widetilde{\Pi}_{n-1}, \infty), \qquad \widetilde{\Pi}_{n-1} \stackrel{\text{def}}{=} \ln \circ A^{-n} C\Pi,$$

is equivalent to the relation

$$\operatorname{Im}(A^{1-n}(g \circ C)) \to 0 \quad \text{in } (\ln \Pi, \infty),$$

which, in turn, follows from the inequalities $(*)$ in §5.2 B. In more detail, $g_1 = g \circ C \in G^n$. By the inequalities $(*)$ in §5.2 B, for arbitrary $\varepsilon > 0$

$$|(A^{1-n}g_1)'| \prec \exp \varepsilon\xi \quad \text{in } (\ln \widetilde{\Pi}, \infty) \supset (\ln \Pi, \infty).$$

The curvilinear half-strip $\ln \Pi$ narrows exponentially: it belongs to the half-strip $|\eta| < 2\exp(-\xi)$. Consequently, since g_1 is real,

$$|\operatorname{Im} A^{1-n} g_1| \prec \exp(\varepsilon - 1)\xi \quad \text{in } (\ln \Pi, \infty). \quad \blacktriangleright$$

Step 2. We prove that for any $m\colon k+2 \le m \le n$,

$$\operatorname{Re} \varphi_m \to \infty, \ \arg \varphi_m \to 0 \quad \text{in } (\widetilde{\Pi}_m, \infty).$$

The proof is by induction on m. As an induction base for $k \ge 0$ we use assertion 2° of Proposition 9, in which the modulus sign can be omitted, since $g \succ \mathrm{id}$. For $k = -1$ we use Proposition 10.

INDUCTION STEP: PASSAGE FROM $m-1$ TO $m \ge k+3$. We have that

$$\varphi_m = \zeta(\exp \circ \varphi_{m-1} \circ \ln - 1).$$

This and the induction hypothesis give us that $|\varphi_m| \to \infty$ in $(\widetilde{\Pi}_m, \infty)$. Further, it follows from Proposition 12 that $\arg \varphi_m \to 0$ in $(\widetilde{\Pi}_m, \infty)$. This completes the induction.

Step 3. We prove that for any $m\colon k+3 \le m \le n$

$$\arg \varphi'_m \to 0 \text{ and } \operatorname{Re} \varphi'_m \to \infty \quad \text{in } (\widetilde{\Pi}_m, \infty).$$

The first relation is proved by induction on m for $m \ge k+2$; the second is then derived from this for $m \ge k+3$. As an induction base we use assertion 3° of Proposition 9 for $g \succ \mathrm{id}$ when $k(g) \ge 0$ and Proposition 10 when $k(g) = -1$. We have:

$$\varphi'_m = (\exp \circ \varphi_{m-1} \circ \ln)(1 + \varphi'_{m-1} \circ \ln) - 1.$$

The first factor tends in modulus to infinity in $(\widetilde{\Pi}_m, \infty)$, since $\operatorname{Re} \varphi_{m-1} \to +\infty$ in $(\widetilde{\Pi}_{m-1}, \infty)$. The second factor is nonzero, since $\arg \varphi'_{m-1} \to 0$ in $(\widetilde{\Pi}_{m-1}, \infty)$. Consequently, $|\varphi'_m| \to \infty$. Therefore, the term -1 does not affect the limit of the argument: in $(\widetilde{\Pi}_m, \infty)$

$$\arg \varphi'_m = \operatorname{Im} \varphi_{m-1} \circ \ln + \arg(1 + \varphi'_{m-1} \circ \ln) + o(1).$$

The convergence $\arg \varphi'_m \to 0$ in $(\widetilde{\Pi}_m, \infty)$ now follows from Proposition 12 and the induction hypothesis. Together with the relation $|\varphi'_m| \to \infty$ this gives us that $\operatorname{Re} \varphi'_m \to +\infty$ in $(\widetilde{\Pi}_m, \infty)$. This concludes the proof of the first two assertions in Proposition 11.

Step 4. We prove assertion 3°. Let $\rho = A^{-n} g^{-1}$, $\sigma = \rho^{-1}$. Then

$$\rho' = 1/\sigma' \circ \rho \to 0 \quad \text{in } (\sigma\widetilde{\Pi}, \infty) \supset (\Pi^\forall, \infty).$$

By assertion 1°, $\sigma' \circ \rho \to \infty$ in $(\Pi^\forall, \infty)$; hence $\rho' \to 0$ in $(\Pi^\forall, \infty)$. ►

Lemma 5.5_{n+1}b follows immediately from assertions 1° and 3° of Proposition 11 for $0 \le k(g) \le n-3$. For $k(g) \ge n-2$ the lemma was proved in subsection E. ►

G. Corollaries.

COROLLARY 1. *Let $g \in G^n$ and $A^{-m} g = \mathrm{id} + \varphi_m$. The germ g belongs to the group G^n_{rap} if and only if the germ φ_n is bounded on $(\mathbf{R}^+, \infty)$. The generalized exponent $\mu_{n+1}(g)$ is different from 0 and ∞ if and only if the germ φ_{n-1} is bounded on $(\mathbf{R}^+, \infty)$.*

PROOF. The proof follows immediately from the definitions. Let $g \in G^n_{\mathrm{rap}}$. Then $A^{-(n+1)} g = \zeta \cdot f(\zeta)$, where the function f is bounded and bounded away from zero by a positive constant on $(\mathbf{R}^+, \infty)$. Indeed, by the definition of G^n_{rap}, the germ $A^{-(n+1)} g$ tends to infinity no more rapidly and no more slowly than some linear germ. Then

$$\varphi_n = A^{-n} g - \mathrm{id} = \ln \circ f \circ \exp$$

is a function that is bounded on $(\mathbf{R}^+, \infty)$.

The converse assertion can be proved analogously.

Let $g \in G^n$, $\mu = \mu_{n+1}(g) \in (0, \infty)$. Then

$$A^{-n} g = \zeta \cdot (\mu + o(1)),$$
$$\varphi_{n-1} = A^{1-n} g - \mathrm{id} = \ln(\mu + o(1))$$

is a function bounded on $(\mathbf{R}^+, \infty)$. The converse assertion can be proved analogously. ►

In what follows we often encounter the "set $\mathscr{F}^{n-1}_{++} \cup \mathscr{F}^n_+$ of remainder terms." Denote it for brevity by $\widetilde{\mathscr{F}}^n_+$. Similarly, the often encountered set $\mathscr{L}_*(n-1, n)$ is denoted by $\mathscr{L}_*$ for brevity.

COROLLARY 2. *Suppose that $g \in G^n_{\mathrm{rap}}$. Then*

1°. $k(g) \ge n-1$;

2°. $A^{-n} g = \mathrm{id} + \varphi_n$, $\varphi_n \in \mathscr{L}_* \circ \ln^{[n]}$ *for* $k(g) = n-1$,
$\varphi_n \in \mathscr{L}_{\mathbf{R}}(\mathscr{F}^n_+) \circ \ln^{[n]}$ *for* $k(g) = n$;

3°. $\varphi_n = \lambda + \tilde{\varphi}$, $\tilde{\varphi} \to 0$ *in* $(\Pi^\forall, \infty)$, $\lambda \in \mathbf{R}$;

4°. $\varphi_n' \to 0$ *in* $(\Pi^\forall, \infty)$;
5°. $\operatorname{Im}\varphi_n \to 0$ *in* $(\Pi^\forall, \infty)$.

PROOF. 1°. Assume the opposite: $k(g) \leq n-2$. Then $|\varphi_n| \to \infty$ on $(\mathbf{R}^+, \infty)$ by Propositions 9 and 11, and this contradicts Corollary 1.

2°. The second and fourth assertions follow from the first assertion and Propositions 8 and 7 in subsection D. The fifth follows from the fourth. We prove the third.

3°. By assertion 2° and Corollary 3 in subsection B, the limit $\lambda = \lim_{(\Pi^\forall, \infty)} \varphi_n$ exists, finite or infinite. By Corollary 1, the germ φ_n is bounded; consequently, $\lambda \in \mathbf{R}$. ►

We terminate this subsection with two remarks and corollaries.

REMARK 1. Suppose that $\varphi \in \mathscr{L}(\mathscr{F}_+^m)$, $m \leq n$. Then

$$A^{-m}(\mathrm{id}+\varphi) = \mathrm{id}+o(1), \qquad (A^{-m}(\mathrm{id}+\varphi))' = 1+o(1)$$

in $(\Pi^\forall, \infty)$.

Indeed, Proposition 7 of D implies that

$$A^{-m}(\mathrm{id}+\varphi) = \mathrm{id}+\varphi_m, \qquad \varphi_m \in \mathscr{L}(\mathscr{F}_+^m) \circ \ln^{[m]}.$$

But the germs of class $\mathscr{F}_+^m \circ \ln^{[m]}$ tend to zero in $(\Pi^\forall, \infty)$ together with all their derivatives. This implies Remark 1.

REMARK 2. Suppose that $\varphi \in \mathscr{L}(\mathscr{F}_+^{m-1})$, $m \leq n$. Then

$$A^{-m}(\mathrm{id}+\varphi) = \zeta(1+o(1)), \qquad (A^{-m}(\mathrm{id}+\varphi))' = 1+o(1)$$

in $(\mathbf{C}^+, \infty)$.

Indeed, let $A^{1-m}(\mathrm{id}+\varphi) = \mathrm{id}+\varphi_{m-1}$. Then

$$A^{-m}(\mathrm{id}+\varphi) = \zeta \cdot \exp \circ \varphi_{m-1} \circ \ln,$$
$$(A^{-m}(\mathrm{id}+\varphi))' = (\exp \circ \varphi_{m-1} \circ \ln) \cdot (1+\varphi_{m-1}' \circ \ln).$$

Remark 2 now follows from Remark 1.

COROLLARY 3. *Suppose that* $m \leq n$. *Then*

$$A^{-m}H^m \subset \{\zeta + o(1)\}, \qquad (A^{-m}H^m)' \subset \{1+o(1)\}$$

in $(\Pi^\forall, \infty)$, *and*

$$A^{-m}J^{m-1} \subset \{\zeta(1+o(1))\}, \qquad (A^{-m}J^{m-1})' \subset \{1+o(1)\}$$

in $(\mathbf{C}^+, \infty)$.

PROOF. This is an immediate consequence of Remarks 1 and 2 and the relations

$$H^m \subset \mathrm{id}+\mathscr{L}(\mathscr{F}_{0+}^m), \qquad J^{m-1} \subset \mathrm{id}+\mathscr{L}(\mathscr{F}_{1+}^{m-1}).$$

The first relation follows from Lemmas SL2, $\mathrm{SL3}_m$, and the second is a corollary of part a in Lemma $\mathrm{SL4}_m$. ►

COROLLARY 4. *Let* $g \in G_{\text{slow}}^{n^+}$, *and* $A^{-n}g = \text{id} + \varphi_n$. *Then* $\operatorname{Re}\varphi_n \to +\infty$ *in* $(\Pi^{\forall}, \infty)$.

PROOF. By the definition of the semigroup $G_{\text{slow}}^{n^+}$, we have that $\operatorname{Re}\varphi_n = \varphi_n \to +\infty$ on $(\mathbf{R}^+, \infty)$. The analogous assertion is to be proved for any horizontal half-strip. We consider three cases: $k(g) = n-1$; $k(g) = n-2$; $k(g) \le n-3$. In the first two cases

$$\varphi_n \in \mathscr{L}_*(n-1, n) \circ \ln^{[n]}.$$

In the first (second) case it follows from Proposition 8 (respectively, 9). The notation $\mathscr{L}_*$ is introduced before Corollary 3 in subsection 5.2 B. The germ $\varphi_n - C$ also has the above property for any $C \in \mathbf{R}$, because the constants belong to $\mathscr{F}_{\text{id}}^{n-1} \circ \ln^{[n]}$, and $\varphi_n \to +\infty$ on $(\mathbf{R}^+, \infty)$. In Cases 1 and 2, Corollary 4 now follows from Corollary 3 in 5.2 B, applied to the germ $\varphi = (\varphi_n - C) \circ \exp^{[n]}$.

In Case 3, Corollary 4 is an immediate consequence of assertions 1° and 2° of Proposition 11.

§5.3. Ordering, holomorphicity, convexity

In this section we obtain auxiliary results on ordered algebras along with distortion theorems, strengthenings of which appear in the assertions of Lemmas 5.7_{n+1} and 5.8_{n+1}.

A. Exponential extensions of ordered algebras.

PROPOSITION 1. *Let* $\Pi^{\forall}$ *be an arbitrary half-strip* $|\eta| \le a$, $\xi \ge 1$, *and let* $\mathscr{K}$ *be a differential algebra of germs of holomorphic functions that is ordered in* $(\Pi^{\forall}, \infty)$ *by the relation* $\operatorname{Re} \cdot \prec 0$ *and is such that* $\arg a \to 0 \pmod{\pi}$ *in* $(\Pi^{\forall}, \infty)$ *for each* $a \in \mathscr{K} \setminus \{0\}$. *Let* E *be a subset of* $\mathscr{K}$ *such that* $\operatorname{Im}\mathbf{e} \to 0$ *in* $(\Pi^{\forall}, \infty)$ *for each* $\mathbf{e} \in E$. *Then the differential algebra generated by the sets* $\mathscr{K}$ *and* $\exp E$ *is ordered by the relation* $\operatorname{Re} \cdot \prec 0$ *in* $(\Pi^{\forall}, \infty)$.

PROOF. The differential algebra generated by $\mathscr{K}$ and $\exp E$ contains all possible sums of the form

$$\Sigma = \sum a_j \exp \mathbf{e}_j, \qquad a_j \in \mathscr{K}, \ \mathbf{e} \in E. \tag{$*$}$$

On the other hand, since $E \subset \mathscr{K}$, these sums form a differential algebra. This is the differential algebra generated by the sets $\mathscr{K}$ and $\exp E$. We prove that this algebra is ordered, namely, we prove that the real part of any germ of the form $(*)$ is of definite sign. To do this it suffices to prove that

$$\arg \Sigma \pmod{\pi} \to 0 \quad \text{in the domain } (\Pi^{\forall}, \infty). \tag{$**$}$$

The proof is analogous to that in §5.1 and is carried out by induction on the number N of terms in the sum Σ.

INDUCTION BASE: $N = 1$, $\Sigma = a \exp \mathbf{e}$. In this case

$$\arg \Sigma = \arg a + \operatorname{Im} \mathbf{e} \to 0 \pmod{\pi} \quad \text{in the domain } (\Pi^{\forall}, \infty).$$

Consequently, the germ $\operatorname{Re} \Sigma$ is of definite sign in $(\Pi^{\forall}, \infty)$.

INDUCTION STEP. Suppose that the assertion $(**)$ has been proved for nonzero exponential sums $(*)$ consisting of no more than $N - 1$ terms. Suppose that $\Sigma \not\equiv 0$, and let

$$\Sigma_1 = \Sigma / a_1 \exp \mathbf{e}_1, \qquad \widetilde{\Sigma} = a_1^2 \Sigma_1'.$$

Then $\widetilde{\Sigma}$ is an exponential sum of no more than $N - 1$ terms. If $\widetilde{\Sigma} \equiv 0$, as happens, for example, in the case when $N = 1$ or when the terms in the sum Σ are proportional even though represented differently in the form $a \exp \mathbf{e}$ $(\mathbf{e} \in E)$, then $\Sigma = c a_1 \exp \mathbf{e}_1$. For such Σ assertion $(**)$ was proved in the "induction base."

Suppose now that $\widetilde{\Sigma} \neq 0$. Then by the induction hypothesis,

$$\arg \widetilde{\Sigma} \pmod{\pi} \to 0 \quad \text{in the domain } (\Pi^{\forall}, \infty).$$

Consequently,

$$\arg \Sigma_1' \pmod{\pi} \to 0 \quad \text{in } (\Pi^{\forall}, \infty).$$

Below we prove

PROPOSITION 2. *Under the conditions of Proposition 1 the following limit of a quotient exists and is real:*

$$\lim_{(\Pi^{\forall}, \infty)} \varphi, \quad \textit{where } \varphi = a_2 \exp \mathbf{e}_2 / a_1 \exp \mathbf{e}_1.$$

It can be assumed without loss of generality that the first term is "the largest" in the sum Σ:

$$\lim_{(\Pi^{\forall}, \infty)} |a_j \exp \mathbf{e}_j / a_1 \exp \mathbf{e}_1| \leq 1.$$

Then we get from Proposition 2 that the following limit exists and is finite:

$$l = \lim_{(\Pi^{\forall}, \infty)} \Sigma_1.$$

We consider two cases: $l \neq 0$ and $l = 0$.

CASE 1: $l \neq 0$. Then

$$\arg \Sigma_1 \to \arg l = 0 \pmod{\pi}.$$

Consequently,

$$\arg \Sigma = \arg(a_1 \exp \mathbf{e}_1) + \arg \Sigma_1 \to 0 \pmod{\pi},$$

and the proposition is proved.

CASE 2: $l = 0$. Then

$$\Sigma_1(\zeta) = \int_{\infty}^{\zeta} \Sigma_1', \qquad \arg \Sigma_1' \to 0 \pmod{\pi}.$$

Consequently, $\arg \Sigma_1 \to 0 \pmod{\pi}$. This implies the assertion $(**)$, and Proposition 1 is proved. ▶

PROOF OF PROPOSITION 2. We prove two assertions: 1. $\arg \varphi \pmod{\pi} \to 0$ in the domain $(\Pi^{\forall}, \infty)$. 2. If $|\varphi|$ is not constant, then it varies monotonically along an arbitrary horizontal ray $\gamma(\zeta)$ with initial point ζ. The proposition is obtained from this.

The first assertion is obvious:

$$\arg \varphi = \arg a_2 - \arg a_1 + \operatorname{Im}(\mathbf{e}_2 - \mathbf{e}_1) \to 0 \pmod{\pi}.$$

Let us prove assertion 2. For this it suffices to prove that $\arg \varphi' \to 0 \pmod{\pi}$ in $(\Pi^{\forall}, \infty)$. We have that

$$\varphi' = \frac{a_2' a_1 - a_2 a_1' + (\mathbf{e}_2' - \mathbf{e}_1') a_1^2}{a_2^2} \exp(\mathbf{e}_2 - \mathbf{e}_1).$$

The arguments of the numerator and denominator of the fraction tend to zero modulo π, because these expressions belong to $\mathscr{K}$. The argument of the last factor tends to zero, because $\mathbf{e}_2 \in E$, $\mathbf{e}_1 \in E$, and $\operatorname{Im}(\mathbf{e}_2 - \mathbf{e}_1) \to 0$. This proves assertion 2. ▶

B. REMARK. The algebra $\mathscr{K} = \mathscr{L}(\mathscr{F}_{\mathbf{R}}^{n-1}) \circ \ln^{[n]}$ is ordered in the germ of a right half-strip (Π, ∞), as well as in the germ of an arbitrary horizontal half-strip $(\Pi^{\forall}, \infty)$. Further, $\arg a \to 0 \pmod{\pi}$ for any $a \in \mathscr{K} \setminus \{0\}$ in view of Lemma ${}^{\mathbf{C}}\mathrm{SL}2_{n-1}$. Take E to be the subset of germs

$$\varphi \in \mathscr{K}: \operatorname{Re} \varphi \to -\infty, \varphi' \to 0 \quad \text{in } (\Pi^{\forall}, \infty),$$

that are real on $(\mathbf{R}^+, \infty)$. For such germs $\operatorname{Im} \varphi \to 0$ in $(\Pi^{\forall}, \infty)$ since φ is real and $\varphi' \to 0$ in $(\Pi^{\forall}, \infty)$. Therefore, the sets $\mathscr{K}$ and E satisfy all the conditions of Proposition 1. Consequently, for such $\mathscr{K}$ and E the differential algebra generated by the sets $\mathscr{K}$ and $\exp E$ is ordered in $(\Pi^{\forall}, \infty)$ by the relation $\operatorname{Re} \cdot \prec 0$.

The results in the next two subsections are used in §5.7.

C. The connection between ordering, inclusion, and convexity.

PROPOSITION 3. *Let $\mathscr{K}$ be a differential algebra ordered by the relation* $\operatorname{Re} \cdot \prec 0$ *and consisting of germs of holomorphic functions on* (Π, ∞) *that are real on* $(\mathbf{R}^+, \infty)$; *here* Π *is a right half-strip. Let* $\varphi \in \mathscr{K}$ *be such that* $\varphi \succ 0$. *In this case*:

if $\varphi \to 0$, *then* $\varphi' \prec 0$, $\varphi^{(2n)} \succ 0$, $\varphi^{(2n-1)} \prec 0$, *and*

$$\operatorname{sgn} \operatorname{Im} \varphi = -\operatorname{sgn} \eta \quad \text{in } (\Pi, \infty);$$

if $\varphi \to \infty$, *then* $\varphi' \succ 0$ *and*

$$\operatorname{sgn} \operatorname{Im} \varphi = \operatorname{sgn} \eta \quad \text{in } (\Pi, \infty).$$

PROOF. The inequalities on the derivatives are obvious for germs of real functions of constant sign on $(\mathbf{R}^+, \infty)$, and they extend to (Π, ∞) by the ordering of the algebra. Further, the inequality $\varphi' \prec 0$ means by definition that $\operatorname{Re}\varphi' \prec 0$ in (Π, ∞). By the Cauchy-Riemann condition,

$$\operatorname{Re}\varphi' = \frac{\partial}{\partial\eta}\operatorname{Im}\varphi.$$

Consequently, the conditions

$$\operatorname{Im}\varphi|_{(\mathbf{R}^+, \infty)} = 0, \qquad \frac{\partial}{\partial\eta}\operatorname{Im}\varphi \prec 0$$

give us that

$$\eta\operatorname{Im}\varphi \prec 0 \quad \text{off } (\mathbf{R}^+, \infty).$$

The second assertion of the proposition is proved similarly. ►

PROPOSITION 4. *Suppose that* $\mathscr{K}$ *is the same differential algebra as in Proposition* 3, *and* $\arg\psi \to 0 \pmod{\pi}$ *for each* $\psi \in \mathscr{K}$ *in* (Π, ∞). *Suppose that* $\mathrm{id} + \varphi \in \mathscr{K}$, $\varphi' \to 0$, *and* φ *is bounded. Then:*

1°. *the function* φ *tends to a finite limit;*

2°. *for each horizontal ray* $L \subset \Pi\backslash\mathbf{R}^+$ *the curve* $(\mathrm{id}+\varphi)L$ *approaches its inverse image asymptotically and is the graph of a monotone and convex function* $\eta(\xi)$.

PROOF. 1°. It follows from the conditions $\varphi' \to 0$ in (Π, ∞) and $\operatorname{Im}\varphi|_{(\mathbf{R}^+, \infty)} = 0$ that $\operatorname{Im}\varphi \to 0$ in (Π, ∞). Therefore, it suffices to prove that $\operatorname{Re}\varphi$ has a limit. But this follows from the fact that φ is comparable with any constant. Namely,

$$\lim_{(\Pi, \infty)} \operatorname{Re}\varphi = \sup\{l \mid \varphi \succ l\}.$$

This supremum is finite, since φ is bounded. Assertion 1° is proved.

2°. By what was proved above, $\operatorname{Im}\varphi \to 0$. Consequently, the curve $(\mathrm{id}+\varphi)L$ approaches L asymptotically. Further, $(\mathrm{id}+\varphi)' \to 1$. Consequently, $\arg(\mathrm{id}+\varphi)' \to 0$. Therefore, the tangent vector to the curve $(\mathrm{id}+\varphi)L$ tends to the horizontal, and the curve itself is the graph of a function $\eta(\xi)$ for sufficiently large ξ.

Let us prove that the germ of the function $\arg(\mathrm{id}+\varphi)'$ is of definite sign on (L, ∞). We have that

$$\arg(\mathrm{id}+\varphi)' = \operatorname{Im}\ln(\mathrm{id}+\varphi)'.$$

By the Cauchy-Riemann condition,

$$\operatorname{Re}(\ln(\mathrm{id}+\varphi)')' = \frac{\partial}{\partial\eta}\operatorname{Im}\ln(\mathrm{id}+\varphi)'.$$

It suffices to prove that the left-hand side is of definite sign; this will imply that the function $\operatorname{Im}\ln(\mathrm{id}+\varphi)'$ is of definite sign on (L, ∞), because this function is equal to zero on the ray $(\mathbf{R}^+, \infty)$. We have that

$$(\ln(\mathrm{id}+\varphi)')' = \frac{\varphi''}{1+\varphi'}.$$

The numerator and denominator of the last fraction belong to $\mathscr{K}$, since $\mathscr{K}$ is a differential algebra. The arguments of the numerator and denominator tend to zero modulo π by a condition of the proposition. Consequently, the real part of the fraction is of definite sign in (Π, ∞). This proves that the argument $\arg(\mathrm{id}+\varphi)'$ is of definite sign on (L, ∞), and hence the germ of the function with the curve $(\mathrm{id}+\varphi)L$ as graph is monotone at infinity.

We prove convexity of this curve. For this it suffices to prove that the germ of the function $\psi = \arg(\varphi''/(1+\varphi'))$ is of constant sign on (L, ∞). We have that

$$\psi = \mathrm{Im}(\ln \varphi'' - \ln(1+\varphi')).$$

Let us prove that the function

$$\mathrm{Re}[\ln \varphi'' - \ln(1+\varphi')]' = \frac{\partial}{\partial \eta} \mathrm{Im}(\ln \varphi'' - \ln(1+\varphi'))$$

has constant sign in (Π, ∞). The last equality uses the Cauchy-Riemann condition. The fact that the left-hand side has constant sign implies that the germ of the function ψ has constant sign on (L, ∞). Further,

$$\begin{aligned}[\ln \varphi'' - \ln(1+\varphi')]' &= \frac{\varphi'''}{\varphi''} - \frac{\varphi''}{1+\varphi'} \\ &= [\varphi'''(1+\varphi') - \varphi''^2]/\varphi''(1+\varphi'). \qquad (***)\end{aligned}$$

The numerator and denominator of the last fraction belong to $\mathscr{K}$, because $\mathscr{K}$ is a differential algebra. Arguing as in the proof of monotonicity of the function with graph the curve $(\mathrm{id}+\varphi)L$, we get the convexity of this function. ►

REMARK. The monotonicity of the function with graph $(\mathrm{id}+\varphi)L$ follows from its convexity and the convergence to a constant. This monotonicity is proved separately in order to demonstrate an analogue of the proof of convexity on simpler material.

COROLLARY 1. *Suppose that* $\mathscr{K}$ *is the same algebra as in Proposition* 3, $\varphi \in \mathscr{K}$, $\varphi \to 0$, *and* $\varphi \not\equiv 0$. *Then*:

$$\begin{aligned}(\mathrm{id}+\varphi)(\Pi, \infty) &\subset (\Pi, \infty) \quad \textit{for } \varphi \succ 0; \\ (\mathrm{id}+\varphi)(\Pi, \infty) &\supset (\Pi, \infty) \quad \textit{for } \varphi \prec 0.\end{aligned}$$

PROOF. The first inclusion follows from the relation

$$\mathrm{sgn}\,\mathrm{Im}\,\varphi = -\,\mathrm{sgn}\,\eta \quad \text{for } \varphi \succ 0,$$

which appears in the conclusion of Proposition 3. The second inclusion follows from the following relation, which is equivalent to the previous one:

$$\mathrm{sgn}\,\mathrm{Im}\,\varphi = \mathrm{sgn}\,\eta \quad \text{for } \varphi \prec 0. \quad ►$$

D. Small perturbations, definiteness of sign, and convexity. We prove analogues of Propositions 3 and 4 for the set $\mathscr{L}_*$, which is not a differential algebra, but differs from one by "small perturbations." Recall that the sets $\mathscr{L}_*$ and $\widetilde{\mathscr{F}}_+^n$ were defined in §5.2 B and G.

PROPOSITION 5. *Suppose that* $\varphi \in \mathscr{L}_* \circ \ln^{[n]}$, $\varphi = \varphi_s + \varphi_r$, *where* $\varphi_s \in \mathscr{F}_{\mathbf{R},\mathrm{id}}^{n-1} \circ \ln^{[n]} = \mathscr{F}\mathscr{C}_{\mathbf{R}}^{n-1} \circ \ln$ *and* $\varphi_r \in \widetilde{\mathscr{F}}_+^n \circ \ln^{[n]}$. *Assume that none of the derivatives* $\varphi_s^{(m)}$ *is identically zero. Finally, suppose that* $\operatorname{Re}\varphi_s \to -\infty$ *and* $\varphi_s' \to 0$ *in* $(\Pi^\forall, \infty)$. *Then* $\varphi^{(m)} \to 0$, $\varphi^{(2m-1)} \prec 0$, *and* $\varphi^{(2m)} \succ 0$ *in* $(\Pi^\forall, \infty)$.

PROOF. For the germ φ_s instead of φ this follows from the ordering of the algebra $\mathscr{F}\mathscr{C}_{\mathbf{R}}^{n-1} \circ \ln$ in $(\exp T, \infty) = (\widetilde{\Pi}, \infty) \supset (\Pi^\forall, \infty)$ (Lemma ${}^{\mathrm{C}}\mathrm{OL1}_{n-1}$) and Proposition 3. For the sum $\varphi^{(m)} = \varphi_s^{(m)} + \varphi_r^{(m)}$ this follows from Corollary 3 in §5.2 B and the condition $\varphi_s^{(m)} \not\equiv 0$. ▶

PROPOSITION 6. *Suppose that* $\varphi \in \mathscr{L}_* \circ \ln^{[n]}$, $\varphi \to \infty$, *and* $\varphi' \to 0$ *in* $(\Pi^\forall, \infty)$. *Then* $(\mathrm{id} + \varphi)(\Pi, \infty)$ *is the germ of a half-strip of type* W.

PROOF. As in the proof of Proposition 4, it suffices to see that $\psi' \succ 0$ or $\psi' \prec 0$ in (Π, ∞), where

$$\psi = \ln \varphi'' - \ln(1 + \varphi').$$

To prove that ψ' is of definite sign we decompose φ into a sum

$$\varphi = \varphi_s + \varphi_r, \qquad \varphi_s \in \mathscr{L}(\mathscr{F}_{\mathbf{R},\mathrm{id}}^{n-1} \circ \ln^{[n]}),\ \varphi_r \in \mathscr{L}(\widetilde{\mathscr{F}}_+^n) \circ \ln^{[n]}.$$

Let us prove that in the expression $(***)$ for ψ' the numerator and denominator belong to $\mathscr{L}_* \circ \ln^{[n]}$. The multiplication lemma and the differentiation lemma give us that $\mathscr{L}(\widetilde{\mathscr{F}}_+^n) \circ \ln^{[n]}$ is an ideal in the differential algebra $\mathscr{L}(\mathscr{F}^{n-1}, \mathscr{F}_+^n) \circ \ln^{[n]}$. Therefore,

$$\psi' = \frac{\varphi_s'''(1 + \varphi_s') - \varphi_s''^2 + \psi_r}{\varphi_s''(1 + \varphi_s') + \tilde{\psi}_r},$$

where $\psi_r, \tilde{\psi}_r \in \mathscr{L}(\widetilde{\mathscr{F}}_+^n) \circ \ln^{[n]}$. The numerator and denominator in the expression for ψ' belong to $\mathscr{L}_*$ if and only if:

$$\begin{cases} \varphi_s'''(1 + \varphi_s') - \varphi_s''^2 \not\equiv 0, \\ \varphi_s'' \not\equiv 0. \end{cases}$$

Assume that the first of the inequalities in this system is false. Then

$$\ln \varphi_s'' - \ln(1 + \varphi_s') \equiv C, \qquad \varphi_s' = \exp(\zeta \exp C + C_1)$$

for some real C and C_1. This contradicts the condition that $\varphi' \to 0$ in (Π, ∞).

Assume that the second inequality in the system is false, that is, $\varphi_s'' = 0$. Then $\varphi_s = C\zeta + C_1$. Consequently, either $\varphi_s \equiv 0$ or $\varphi' \nrightarrow 0$, or $|\varphi| \nrightarrow \infty$ in (Π, ∞). All these assertions contradict the condition.

The fact that $\operatorname{Re}\psi'$ is of definite sign in (Π, ∞) now follows from Corollary 3 bis in §5.2 B: $\arg\tilde{\psi} \to 0 \pmod{\pi}$ in $(\tilde{\Pi}, \infty)$ for any $\tilde{\psi} \in \mathscr{L}_*$, where $\tilde{\Pi} = A^{-n}C\Pi$, $C > 1$ arbitrary. ►

§5.4. Standard domains of class $n+1$ and admissible germs of diffeomorphisms comparable with linear germs

Lemma 5.2_{n+1} and Lemma 5.6_{n+1} are proved in this section. We recall the corresponding formulations, beginning with the definitions.

A. Formulations.

Definition 1 in §1.5. A standard domain is a domain that is symmetric with respect to the real axis, belongs to the right half-plane, and admits a real conformal mapping onto the right half-plane that has derivative equal to $1 + o(1)$, and extends to the δ-neighborhood of the part of the domain outside a compact set for some $\delta > 0$.

Definition 1^{n+1}, §1.7. A standard domain of class $n+1$ and type 1 is defined to be a domain of the form

$$A^{-(n+1)}g(\mathbf{C}^+\backslash K), \quad \text{where } g \in G^n_{\text{rap}},$$

and K is a disk such that the mapping $A^{-(n+1)}g$ is biholomorphic in $\mathbf{C}^+\backslash K$.

A standard domain of class $n+1$ and type 2 is defined to be a domain of the form

$$A^{-1}(\mathrm{id} + \psi)(\mathbf{C}^+\backslash K),$$

where K is a disk such that the preceding mapping is biholomorphic in $\mathbf{C}^+\backslash K$ and ψ has the form

$$\psi = \exp\varphi, \qquad \varphi \in \mathscr{L}_* \circ \ln^{[n]}, \ \operatorname{Re}\varphi \to -\infty, \ \varphi' \to 0 \qquad (*)$$

in the domain $(\Pi^\forall, \infty)$, where $\Pi^\forall$ is an arbitrary half-strip of the form $\xi \geq 0$, $|\eta| \leq a$; ψ is real on $(\mathbf{R}^+, \infty)$. The set of all such germs is denoted by Ψ_{n+1}.

We recall the definition

$$\mathscr{L}_* = \mathscr{L}_*(\mathscr{F}^{n-1}_{\mathbf{R},\mathrm{id}}, \mathscr{F}^{n-1}_{++}, \mathscr{F}^n_+)$$

(see §5.2 B). A standard domain of class $n+1$ is a domain of one of the two types defined above.

We recall the definition of germs of admissible diffeomorphisms of class $n+1$. It is a modification of the definition of admissible germs of arbitrary class $\mathbf{\Omega}$ for the case when $\mathbf{\Omega} = \Omega_{n+1}$ (the set of standard domains of class $n+1$; this class is defined in §1.7).

Definition (§§1.5 and 1.7). A germ of a diffeomorphism $\sigma_{\mathbf{R}}\colon (\mathbf{R}^+, \infty) \to (\mathbf{R}^+, \infty)$ is said to be admissible of class $n+1$ (or $\mathbf{\Omega}_{n+1}$-admissible) if:

1°. the inverse germ ρ admits a biholomorphic extension to some standard domain of class $n+1$, and for an arbitrary standard domain Ω of this class there exists a standard domain $\widetilde{\Omega}$ of the same class such that ρ maps $\widetilde{\Omega}$ biholomorphically into Ω, and, moreover,

2°. the derivative ρ' is bounded in $\widetilde{\Omega}$,

3°. there exists a $\mu > 0$ such that $\operatorname{Re}\rho < \mu\xi$ in $\widetilde{\Omega}$,

4°. for each $\nu > 0$ the inequality $\exp \operatorname{Re}\rho \succ \nu\xi$ holds in $\widetilde{\Omega}$.

LEMMA 5.2_{n+1}. a. *A standard domain of class $n+1$ is standard in the sense of the definition in* §1.5.

b. *The germs in the sets $\mathscr{D}^n_{\mathrm{rap}}$ and $\mathscr{D}^n_{\mathrm{rap}} \circ \mathscr{L}^n$ are admissible of class $n+1$.*

c. *For an arbitrary germ $\sigma \in A^{-(n+1)}G^n_{\mathrm{rap}}$ there exists a standard domain Ω of class $n+1$ and positive constants ε and μ such that for $\rho = \sigma^{-1}$*

$$\operatorname{Re}\rho > \varepsilon\xi, \qquad |\sigma| < \mu|\zeta| \quad \textit{in } \Omega.$$

d. *The set of domains of class $n+1$ is proper, that is, has the following properties.*

PROPERTY 1. *Each standard domain of class $n+1$ has a standard subdomain of the same class whose distance to the boundary of the first domain is $\geq$ a previously specified positive C.*

PROPERTY 2. *The intersection of any two standard domains of class $n+1$ contains a standard domain of the same class.*

B. A criterion for domains to be standard, and the property of being standard for domains of class $n+1$. The following proposition gives a criterion for being standard.

PROPOSITION 1. *A domain Ω is standard if and only if there exists a conformal mapping $\mathrm{id} + \varphi\colon \ln\Omega \to \Pi$ that is real on $(\mathbf{R}^+, \infty)$ and has correction decreasing together with the derivative:*

$$\varphi \to 0,\ \varphi' \to 0 \quad \textit{in } (\ln\Omega, \infty);$$

moreover, there exist a $\delta > 0$ and a compact set such that the mapping $A^{-1}(\mathrm{id} + \varphi)$ extends to the δ-neighborhood of the part of Ω outside this compact set.

PROOF. The domain $\ln\Omega$ is symmetric with respect to $\mathbf{R}^+$ by the symmetry principle and the fact that φ is real. Further, the mapping

$$\Psi \overset{\mathrm{def}}{=} A^{-1}(\mathrm{id} + \varphi) = \zeta \cdot \exp \circ \varphi \circ \ln = \zeta \cdot (1 + O(1))$$

carries Ω into the domain $\mathbf{C}^+ \backslash K$, where K is the disk $|\zeta| \leq \omega$ for some $\omega > 0$. Consequently, the mapping $\widetilde{\Psi} = \Psi - \omega^2/\Psi$ carries Ω into $\mathbf{C}^+$ and is real on $\mathbf{R}^+$, and $\widetilde{\Psi} = \Psi + o(1) = \zeta(1 + o(1))$. Moreover,

$$\Psi' = (\exp \circ \psi \circ \ln)(1 + \varphi' \circ \ln) = 1 + o(1)$$

in (Ω, ∞). Finally, the function Ψ extends to the δ-neighborhood of the part of Ω outside some compact set by a condition of the proposition; consequently, $\widetilde{\Psi}$ has the same property. This proves that the conditions of the proposition suffice for the domain to be standard.

The necessity is proved similarly and is not used in what follows.

PROPOSITION 2. *Standard domains of class* $n+1$ *are standard in the sense of Definition* 1 *in* §1.5.

PROOF. We prove this first for domains of type 1. Let $g \in G^n_{\text{rap}}$ and $\Omega = A^{-(n+1)}g(\mathbf{C}^+\backslash K)$ (see Definition 1^{n+1} in subsection A). Let $A^{-n}g^{-1} = \operatorname{id}+\varphi$. Then

$$\operatorname{id}+\varphi\colon \ln\Omega \to \ln(\mathbf{C}^+\backslash K) = \Pi.$$

Further, by Corollary 2 in §5.2 G, $\varphi \in (\mathscr{L}_* \cup \mathscr{L}_{\mathbf{R}}(\mathscr{F}^n_+)) \circ \ln^{[n]}$, and, further, $\varphi' \to 0$ in $(\Pi^\forall, \infty)$, $\varphi = \lambda + \tilde{\varphi}$, $\lambda \in \mathbf{R}$, and $\tilde{\varphi} \to 0$ in $(\Pi^\forall, \infty)$. Consequently, the mapping $\operatorname{id}+\tilde{\varphi}$, as before, carries $\ln\Omega$ into Π and satisfies all the requirements of Proposition 1.

We now prove Proposition 2 for standard domains of class $n+1$ and type 2. Suppose that $\psi \in \Psi_{n+1}$, $\Omega = A^{-1}(\operatorname{id}+\psi)(\mathbf{C}^+\backslash K)$, and the germ ψ is real. Then $\operatorname{id}+\psi$ carries (Π, ∞) into $(\ln\Omega, \infty)$, and $(\operatorname{id}+\psi)^{-1}\colon (\ln\Omega, \infty) \to (\Pi, \infty)$. The germ $\operatorname{id}+\psi$ is defined in $(\Pi^\forall, \infty)$, and $\psi \to 0$ and $\psi' = \exp\varphi \cdot \varphi' \to 0$ in $(\Pi^\forall, \infty)$ (φ is the same germ as in Definition 1^{n+1} of subsection A). Consequently, $(\operatorname{id}+\psi)^{-1} = \operatorname{id}-\tilde{\psi}$, and the correction $\tilde{\psi}$ has the same properties as ψ; that is, it satisfies all the requirements imposed in Proposition 1 on the correction φ.

This proves assertion a of Lemma 5.2_{n+1}.

C. Boundedness of the derivatives and Property 1 of standard domains. Let us begin with a proposition.

PROPOSITION 3. *The mappings in the definition of standard domains of class* $n+1$ *have bounded derivatives in the domain* $\mathbf{C}^+\backslash K$ *for a sufficiently large disk* K.

PROOF. We use the following general formula. Let

$$A^{-1}f \stackrel{\text{def}}{=} A^{-1}(\operatorname{id}+\psi) = \zeta \cdot \exp\psi \circ \ln .$$

Then

$$(A^{-1}f)' = (\exp\circ\psi \circ \ln)(1 + \psi' \circ \ln). \tag{$**$}$$

Let us consider two cases, corresponding to two types of standard domains.

CASE 1: $\operatorname{id}+\psi = A^{-n}g$. According to assertion 2° of Corollary 2 in §5.2 G, $\psi = \lambda + \tilde{\varphi}$, $\tilde{\varphi} \to 0$ in $(\Pi^\forall, \infty)$. Assertion 4° of the same corollary gives us that $\tilde{\varphi}' \to 0$ in $(\Pi^\forall, \infty)$. Consequently, $\lim(A^{-1}(\operatorname{id}+\psi))' = \exp\lambda$ as $\zeta \to \infty$ in the domain $\exp(\Pi^\forall, \infty)$. This proves the proposition in Case 1.

CASE 2: $\psi \in \Psi_{n+1}$. Then it follows from formulas $(*)$ and $(**)$ that $(A^{-1}(\mathrm{id}+\psi))' \to 1$ in $(S^\forall, \infty)$ as $\zeta \to \infty$, because $\psi \to 0$ and $\psi' \to 0$ in $(\Pi^\forall, \infty)$. ▶

COROLLARY. *The mappings inverse to the germs of Proposition 3 have bounded derivatives.*

PROOF. This follows from the fact that the original mappings have derivative tending to a finite nonzero limit at infinity and from the theorem on the derivative of an inverse mapping. ▶

We now prove Property 1 of standard domains of class $n+1$ (see subsection A). Suppose that $\Omega = \sigma(\mathbf{C}^+ \backslash K)$ is a standard domain of class $n+1$ and type 1. We first prove that the set $\mathscr{D}^{n+1}_{\mathrm{rap}} = A^{-(n+1)} G^n_{\mathrm{rap}}$ is closed with respect to multiplication from the right by a shift: $\sigma \circ (\mathrm{id}+c) \in \mathscr{D}^{n+1}_{\mathrm{rap}}$ if $\sigma \in \mathscr{D}^{n+1}_{\mathrm{rap}}$. Indeed, $A^{n+1}(\mathrm{id}+c) \in G^n_{\mathrm{rap}}$; this was proved at the end of §1.3. But G^n_{rap} is a group. Consequently, $(A^{-(n+1)} g) \circ (\mathrm{id}+c) \in \mathscr{D}^{n+1}_{\mathrm{rap}}$.

It follows from Proposition 3 that the derivative σ' is bounded from below. Let $|(\sigma^{-1})'| < M$. Then the domain

$$\Omega_1 = \sigma \circ (\mathrm{id}+CM)(\mathbf{C}^+ \backslash K)$$

is standard of class $n+1$ and type 1, and the distance between its points and the boundary of Ω is at least $C/2$ for sufficiently large C. Property 1 is proved for domains of type 1.

We prove Property 1 for domains of class $n+1$ and type 2. Let

$$\Omega = A^{-1}(\mathrm{id}+\psi)(\mathbf{C}^+ \backslash K), \qquad \tilde{\Omega} = A^{-1}(\mathrm{id}+2\psi)(\mathbf{C}^+ \backslash \tilde{K}).$$

We prove that $\rho(\zeta, \partial\Omega) \to \infty$ as $\zeta \to \infty$, $\zeta \in \tilde{\Omega}$. Indeed, let $\Pi = \ln(\mathbb{C}^+ \backslash K)$ and $\Pi' = \ln(\mathbb{C}^+ \backslash \tilde{K})$. Then

$$\Omega = \exp(\mathrm{id}+\psi)\Pi, \qquad \tilde{\Omega} = \exp(\mathrm{id}+2\psi)\Pi'.$$

We prove first that

$$\rho(\zeta, \partial(\mathrm{id}+\psi)\Pi) \succ |\mathrm{Im}\, \psi(\xi + i\tfrac{\pi}{2})|/2 \qquad ({}^{*}_{*}{*})$$

We will investigate the distance to the "upper" part of the boundary, which belongs to the upper halfplane; the "lower" part is investigated similarly. We have: $\psi' = \varphi' \exp\varphi$, $\varphi = \varphi_s + \varphi_r$, $\varphi_s \in \mathscr{F}^{n-1}_{\mathbb{R}\,\mathrm{id}} \circ \ln^{[n]} = \mathscr{F} C^{n-1}_{\mathbb{R}} \circ \ln$, $\varphi_r \in \mathscr{L}(\widetilde{\mathscr{F}}^n) \circ \ln^{[n]}$. By Corollary 2 in §5.2B, $\varphi' = \varphi_s'(1 + O(1))$ in $(\Pi^\forall, \infty)$. Moreover, $\varphi_s' \prec 0$ and $\mathrm{sgn}\,\mathrm{Im}\,\varphi_s = -\mathrm{sgn}\,\eta$ in (Π, ∞) by the Remark in §5.3B and Proposition 3 in §5.3C. Consequently $\mathrm{sgn}\,\mathrm{Im}\,\psi = -\mathrm{sgn}\,\eta$. Thus $(\ln\tilde{\Omega}, \infty) \subset (\ln\Omega, \infty)$. We will prove the stronger assertion $({}^{*}_{*}{*})$.

By the Proposition 1 in §5.7B, whose proof uses only the material of §§5.2, 5.3, $|\mathrm{Im}\,\varphi| \succ |\eta\,\mathrm{Re}\,\varphi'|$ in $(\Pi^\forall, \infty)$. Thus

$$|\mathrm{Im}\,\psi| = |\sin \mathrm{Im}\,\varphi|(\exp\mathrm{Re}\,\varphi) \succ \frac{|\mathrm{Im}\,\varphi|}{2} \exp\mathrm{Re}\,\varphi.$$

Let $\tilde{\psi} = \psi(\xi + i\frac{\pi}{2})$. The assertion 1 in section D below gives $\tilde{\psi} \circ (\mathrm{id} + o(1)) = \tilde{\psi}(1 + o(1))$. The slope of the boundary lines of $(\mathrm{id} + \psi)\Pi$ tends to zero, as $\psi' \to 0$ in $(\Pi^\forall, \infty)$. Thus the "distance" may be replaced by "the distance along the vertical line". Now for $\operatorname{Im}\zeta > 0$

$$\rho(\zeta, \partial(\mathrm{id} + \psi)\Pi) = (1 + o(1))|2 \operatorname{Im} \tilde{\psi} \circ (\mathrm{id} + 2 \operatorname{Re} \tilde{\psi}(\xi))^{-1} - \operatorname{Im} \tilde{\psi} \circ (\mathrm{id} + \operatorname{Re} \tilde{\psi}(\xi))^{-1}|$$
$$\succ \frac{1}{2} \operatorname{Im} \tilde{\psi}(\xi), \qquad \xi = \operatorname{Re} \zeta.$$

This proves $(\,{}^*_*\,*)$. Let $\exp \zeta \in \widetilde{\Omega}$. Then

$$\rho(\exp \zeta, \partial \exp(\mathrm{id} + \psi)\Pi) \succ \frac{1}{3} \exp \xi |\operatorname{Im} \tilde{\psi}|$$
$$\succ \exp(\xi + \operatorname{Re} \varphi(\xi + i\frac{\pi}{2}))|\operatorname{Im} \varphi(\xi + i\frac{\pi}{2})|/6.$$

The first factor exceeds $\exp \xi/2$ as $\varphi' \to 0$ in $(\Pi^\forall, \infty)$. We will give the lower estimate for $\operatorname{Im} \varphi$ using only the realness of φ and the inclusion $\varphi \in \mathscr{F}_* \circ \ln^{[n]}$. Let φ_s be as above. Then $\varphi' = \varphi_s' \cdot (1 + o(1))$ in $(\Pi^\forall, \infty)$. Moreover $\varphi_s' = F \circ \ln$ for some $F \in \mathscr{F}\mathscr{C}_{\mathbb{R}}^{n-1}$. By ${}^{\mathrm{C}}\mathrm{OL1}_{n-1}$, $|\operatorname{Re} f| \succ \exp(-\varepsilon\xi)$ in (T, ∞), $T = A^{1-n} C \circ \ln \Pi$ for some $\varepsilon > 0$ and any $C > 0$. Thus $|\operatorname{Re} F \circ \ln| \succ \xi^{-\varepsilon}$, $|\operatorname{Re} \varphi'| \succ \xi^{-\varepsilon}$ in $(\exp T, \infty) = (A^{-n} C\Pi, \infty) \supset (\Pi^\forall, \infty)$. By the inequality $|\operatorname{Im} \varphi| \succ |\eta \operatorname{Re} \varphi'|$ we obtain that on the line $\eta = \frac{\pi}{2}$, $|\operatorname{Im} \varphi| \succ \xi^{-\varepsilon}$. Thus

$$\rho(\exp \zeta, \partial\Omega) \succ (\exp \frac{\xi}{2})\xi^{-\varepsilon} \to \infty$$

as desired.

D. Compositions of mappings whose corrections have real parts of definite sign. Below we must compare corrections of mappings, their compositions, and the derivatives of the compositions under the condition that only the principal terms of the corrections belong to an ordered differential algebra.

ASSERTION 1. *Let φ be the germ of a holomorphic function belonging for some $\varepsilon > 0$ to a differential algebra that is ordered in $(\Pi^\varepsilon, \infty)$ by the relation $\operatorname{Re}(\cdot) \prec 0$, and let ψ be the germ of a holomorphic function tending to zero in $(\Pi^\varepsilon, \infty)$. Then*

$$\operatorname{Re} \varphi \circ (\mathrm{id} + \psi) = (\operatorname{Re} \varphi)(1 + o(1)),$$
$$\operatorname{Re} \varphi' \circ (\mathrm{id} + \psi) = (\operatorname{Re} \varphi')(1 + o(1))$$

in (Π, ∞). Recall that Π^ε is the ε-neighborhood of Π.

PROOF. Since the germ φ belongs to an ordered differential algebra, both the germs $\operatorname{Re} \varphi$ and $\operatorname{Re} \varphi'$ are of constant sign in $(\Pi^\varepsilon, \infty)$. Assertion 1 now follows from the Harnack inequality for harmonic functions of constant sign. Namely, for $|\psi| < \delta < \varepsilon$

$$\frac{\delta - |\psi|}{\delta + |\psi|}|\operatorname{Re}\varphi| \le |\operatorname{Re}\varphi \circ (\mathrm{id} + \psi)| \le \frac{\delta + |\psi|}{\delta - |\psi|}|\operatorname{Re}\varphi|$$

in (Π, ∞). This and the convergence of the germ ψ to zero imply Assertion 1 for $\operatorname{Re}\varphi$. The assertion is proved similarly for $\operatorname{Re}\varphi'$; it is only necessary to replace φ by φ'. ►

ASSERTION 2. *Suppose that the germ of the function φ is the same as in Assertion 1, and that $\varphi \to 0$ and $\varphi' \to 0$ in $(\Pi^\varepsilon, \infty)$. Then the correction $-\tilde{\varphi}$ of the mapping inverse to $\mathrm{id} + \varphi$ and its derivative satisfy the relations*

$$\tilde{\varphi} = \varphi \cdot (1 + o(1)), \quad \tilde{\varphi}' = \varphi' \cdot (1 + o(1)) \quad \textit{in } (\Pi, \infty).$$

PROOF. The correction $\tilde{\varphi}$ satisfies the functional equation

$$\tilde{\varphi} = \varphi \circ (\mathrm{id} - \tilde{\varphi}).$$

By assumption, $\varphi \to 0$. Consequently, $\tilde{\varphi} \to 0$ in $(\Pi^{\varepsilon/2}, \infty)$. The first equality in Assertion 2 now follows from Assertion 1. Further,

$$\tilde{\varphi}' = \varphi' \circ (\mathrm{id} - \tilde{\varphi}) \cdot (1 - \tilde{\varphi}').$$

Since φ' belongs to an ordered differential ring and $\varphi \to 0$ in $(\Pi^\varepsilon, \infty)$, we get that $\varphi' \to 0$ in $(\Pi^\varepsilon, \infty)$. Consequently, $\tilde{\varphi}' \to 0$ in $(\Pi^{\varepsilon/2}, \infty)$. Hence, $1 - \tilde{\varphi}' = 1 + o(1)$. On the other hand, applying Assertion 1 to the composition $\varphi' \circ (\mathrm{id} - \tilde{\varphi})$, we obtain the second equality in Assertion 2. ►

E. Domains of class $n+1$ and ordered differential algebras. We investigate the set S_{n+1} (S for standard) of corrections of the germs used to define standard domains of class $n+1$. These germs are defined in (Π, ∞) but do not belong to a differential algebra that is ordered there. We consider an ordered differential algebra that contains "good approximations" of the germs of class S_{n+1}. The definition of domains of class $n+1$ and type 1 uses germs in the set $A^{-n}G^n_{\mathrm{rap}} - \mathrm{id}$. Let

$$\Phi_{n+1} = \{\varphi \in A^{-n}G^n_{\mathrm{rap}} - \mathrm{id} \,|\, \varphi \to 0 \text{ in } (\Pi, \infty)\}.$$

PROPOSITION 4. *For an arbitrary standard domain Ω of class $n+1$ and type 1 there exists a germ $\varphi \in \Phi_{n+1}$ such that*

$$(\ln \Omega, \infty) = (\mathrm{id} + \varphi)(\Pi, \infty).$$

PROOF. By the definition in subsection A, there exists a germ $\mathrm{id} + \varphi_1 \in A^{-n}G^n_{\mathrm{rap}}$ such that

$$(\ln \Omega, \infty) = (\mathrm{id} + \varphi_1)(\Pi, \infty).$$

According to Corollary 2 in §5.2 G, $\varphi_1 = \lambda + \tilde{\varphi}$, $\tilde{\varphi} \to 0$ in $(\Pi^\forall, \infty)$, $\lambda \in \mathbf{R}$. Further, as proved at the end of §1.3, the germ $g_0 = A^n(\mathrm{id} + \lambda)$ belongs to G^n_{rap}. Consequently,

$$A^{-n}(g \circ g_0^{-1}) = \mathrm{id} + \varphi, \qquad \varphi = \tilde{\varphi} \circ (\mathrm{id} - \lambda) \in \Phi_{n+1},$$

because $g \circ g_0^{-1} \in G^n_{\mathrm{rap}}$ and $\varphi \to 0$ in $(\Pi^\forall, \infty)$.

Thus, every standard domain of class $n+1$ and type 1 has the form $A^{-1}(\mathrm{id}+\varphi)(\mathbf{C}^{+}\backslash K)$, where $\varphi \in \Phi_{n+1}$, and K is some disk (depending on the domain). Let

$$S_{n+1} = \Phi_{n+1} \cup \Psi_{n+1},$$

with Ψ_{n+1} the set defined in subsection A.

REMARK. If $\psi \in S_{n+1}$, then $\psi \to 0$ and $\psi' \to 0$ in $(\Pi^{\forall}, \infty)$, by the definitions of the sets Ψ_{n+1} and Φ_{n+1}.

We now construct a differential algebra "approximating" to S_{n+1}. Let

$$\mathscr{K} = \mathscr{F}^{n-1}_{\mathbf{R},\mathrm{id}} \circ \ln^{[n]} = \mathscr{F}\mathscr{C}^{n-1}_{\mathbf{R}} \circ \ln;$$

$$E = \{\tilde\varphi \in \mathscr{K} \mid \operatorname{Re}\tilde\varphi \to -\infty,\ \tilde\varphi' \to 0 \text{ in } (\widetilde\Pi, \infty)\}, \qquad \widetilde\Psi_{n+1} = \exp E.$$

Denote by $\widetilde S_{n+1}$ the differential algebra generated by the sets $\mathscr{K}$ and $\widetilde\Psi_{n+1}$. We prove that it is ordered in $(\Pi^{\forall}, \infty)$ by the relation $\operatorname{Re}\cdot \prec 0$; to do this we check all the conditions in Proposition 1 of §5.3. The algebra $\mathscr{K}$ is ordered in $(\Pi^{\forall}, \infty)$, and even in $(\widetilde\Pi, \infty)$; this follows from Lemma ${}^{\mathbf{C}}\mathrm{OL1}_n$. By the same lemma, $\arg a \to 0 \pmod{\pi}$ in $(\Pi^{\forall}, \infty)$ for each $a \in \mathscr{K}\backslash\{0\}$. Further, $\operatorname{Im}\mathbf{e} \to 0$ on $(\mathbf{R}^{+}, \infty)$ for any $\tilde\varphi = \mathbf{e} \in E$, because the germ $\mathbf{e}$ is weakly real in view of the criterion for weak realness. The convergence to zero of the derivative $\mathbf{e}'$ now implies that $\operatorname{Im}\mathbf{e} \to 0$ in $(\Pi^{\forall}, \infty)$. This verifies all the conditions in Proposition 1 of §5.3 A. By this proposition, $\widetilde S_{n+1}$ is a differential algebra ordered in $(\Pi^{\forall}, \infty)$ by the relation $\operatorname{Re}\cdot \prec 0$.

The fact that the algebra $\widetilde S_{n+1}$ is ordered is the main use of Proposition 1 in §5.3.

Sometimes we use the notation $\mathscr{O} = \mathscr{L}(\widetilde{\mathscr{F}}^{n}_{+}) \circ \ln^{[n]}$ for brevity.

PROPOSITION 5. *For each germ* $\varphi \in S^{n+1}\backslash\mathscr{O}$ *there exists a germ* $\varphi_s \in \widetilde S^{n+1}$ *such that* $\varphi = \varphi_s(1+o(1))$ *and* $\varphi' = \varphi_s'(1+o(1))$ *in* $(\Pi^{\forall}, \infty)$. *Moreover, the germ* $\psi_0 = \zeta^{-1}$ *belongs to* $\widetilde S_{n+1}$, *and* $|\varphi'| = o(|\operatorname{Re}\psi_0'|)$ *in* $(\Pi^{\forall}, \infty)$ *for any germ* $\varphi \in \mathscr{O}$.

PROOF. Suppose first that $\varphi \in \Phi_{n+1}\backslash\mathscr{O}$. Then, by Corollary 2 in §5.2 G,

$$\varphi \in \mathscr{L}_{*}(\mathscr{F}^{n-1}_{\mathbf{R},\mathrm{id}}, \widetilde{\mathscr{F}}^{n}_{+}) \circ \ln^{[n]}.$$

Consequently, $\varphi = \varphi_s + \varphi_r$, $\varphi_s \in \mathscr{K}$, $\varphi_r \in \mathscr{O}$. According to Corollary 1 in §5.2 B, $\varphi_r \to 0$ in $(\Pi^{\forall}, \infty)$; hence, $\varphi_s \to 0$ there. Moreover, $\varphi_s \neq 0$, because $\varphi \notin \mathscr{O}$. Consequently, $\varphi_s' \in \mathscr{K}\backslash\{0\}$. Further, $\varphi_r' \in \mathscr{O}$. By Corollary 2 in §5.2 B, $\varphi_r' = o(\operatorname{Re}\varphi_s')$ in $(\Pi^{\forall}, \infty)$. This proves the proposition for $\varphi \in \Phi_{n+1}\backslash\mathscr{O}$.

Suppose now that $\varphi \in \mathscr{O}$. Note that $\psi_0 = \zeta^{-1} \in \mathscr{K}$, because $\zeta^{-1} = \exp(-\exp^{[n-1]}) \circ \ln^{[n]} \in \mathscr{F}^{n-1}_{\mathbf{R},\mathrm{id}} \circ \ln^{[n]}$. Consequently, $\psi_0' \in \mathscr{K}\backslash\{0\}$. By Corollary 2 in §5.2 B, $|\varphi'| = o(\operatorname{Re}\psi_0')$ in $(\Pi^{\forall}, \infty)$ for each $\varphi \in \mathscr{O}$.

Finally, suppose that $\varphi \in \Psi_{n+1}$. Then

$$\varphi = \exp\psi, \qquad \psi \in \mathscr{L}_*,\ \psi = \psi_s + \psi_r,\ \psi_s \in \mathscr{K},\ \psi_r \in \mathscr{O};$$
$$\psi_r \to 0, \quad \psi_r' \to 0, \quad \psi_s \to -\infty, \quad \psi_s' \to 0 \text{ in } (\Pi^\forall, \infty).$$

Let $\varphi_s = \exp\psi_s \in \widetilde{\Psi}_{n+1}$. Then

$$\varphi = \varphi_s \cdot \exp\psi_r,$$
$$\varphi_s' = \psi_s' \cdot \exp\psi_s,$$
$$\varphi' = (\psi_s' + \psi_r') \cdot \exp(\psi_s + \psi_r).$$

But $\psi_r \to 0$ and $\psi_s' + \psi_r' = \psi_s' \cdot (1 + o(1))$ in $(\Pi^\forall, \infty)$ by Corollary 2 in §5.2 B, since $\psi_s' \neq 0$. Further, $\psi_r = o(1)$ in $(\Pi^\forall, \infty)$. Consequently, $\varphi' = \varphi_s'(1 + o(1))$ in $(\Pi^\forall, \infty)$. ►

Proposition 5 shows that an arbitrary germ $\varphi \in \mathscr{O}$ is "negligibly small" in comparison with the germ $\zeta^{-1} \in \widetilde{S}_{n+1}$, and any germ $\varphi \in S_{n+1}\backslash\mathscr{O}$ admits distinction of a "principal part" $\varphi_s \in \widetilde{S}_{n+1}$.

PROPOSITION 6. *For any arbitrary germ $\varphi \in S_{n+1}\backslash\mathscr{O}$ and an arbitrary positive integer N there exist germs $\psi \in S_{n+1}\backslash\mathscr{O}$ and $\psi_s \in \widetilde{S}_{n+1}$ such that*

$$\psi = \psi_s(1 + o(1)), \qquad \operatorname{Re}\psi_s \succ N|\operatorname{Re}\varphi_s| \quad \textit{in } (\Pi^\forall, \infty),$$

where φ_s is the germ given by Proposition 5 for φ.

PROOF. Suppose first that $\varphi \in \Phi_{n+1}\backslash\mathscr{O}$, $\varphi \succ 0$. This implies by definition that

$$\mathrm{id} + \varphi = A^{-n} g, \qquad g \in G^n_{\mathrm{rap}},\ \varphi \to 0 \text{ in } (\Pi^\forall, \infty).$$

Let

$$\mathrm{id} + \psi = A^{-n} g^{[N+1]}.$$

Then $\psi \in \Phi_{n+1}$, and

$$\psi = \varphi + \varphi \circ (\mathrm{id} + \varphi) + \cdots + \varphi \circ (\mathrm{id} + \varphi)^{[N]}.$$

It follows from Proposition 5 that the germ $\operatorname{Re}\varphi$ is of definite sign in $(\Pi^\forall, \infty)$, since the germ $\operatorname{Re}\varphi_s$ has this property, as $\varphi_s \in \widetilde{S}_{n+1}$. Therefore, assertion 1 in subsection D is applicable, and it gives us that

$$\psi = (N + 1 + o(1))\varphi \quad \text{in } (\Pi^\forall, \infty).$$

Let $\psi_s \in \widetilde{S}_{n+1}$ be the germ given by Proposition 5 for ψ. Then $\psi = \psi_s(1 + o(1))$ and $\psi' = \psi_s'(1 + o(1))$.

This gives us the proposition in the case $\varphi \in \Phi_{n+1}\backslash\mathscr{O}$, $\varphi \succ 0$.

Suppose now that $\varphi \in \Phi_{n+1}\backslash\mathscr{O}$, $\varphi \prec 0$. Then let

$$\mathrm{id} + \psi = (\mathrm{id} + \varphi)^{[-(N+1)]}.$$

The germ ψ is the desired one; this can be proved just as in the previous case, except that assertion 2 of subsection D is used along with assertion 1.

Assume now that $\varphi \in \Psi_{n+1}$. Then let $\psi = (N+1)\varphi$. We get that $\psi \in \Psi_{n+1}$. Indeed, for arbitrary $C > 0$ the germs ψ and $C\psi$ satisfy all the requirements of the definition of the set Ψ_{n+1} simultaneously (see subsection A). ▶

PROPOSITION 7. *The set* $\mathrm{id} + S_{n+1}$ *is invariant under conjugation by shifts: for arbitrary* $\varphi \in S_{n+1}$ *and* $c \in \mathbf{R}$

$$\mathrm{Ad}(\mathrm{id}+c)(\mathrm{id}+\varphi) \in \mathrm{id} + S_{n+1}.$$

PROOF. Suppose first that $\varphi \in \Phi_{n+1}$, $\mathrm{id}+\varphi = A^{-n}g$, $g \in G^n_{\mathrm{rap}}$. Let $g_0 = A^n(\mathrm{id}+c)$; then $g_0 \in G^{n-1}_{\mathrm{rap}}$, as shown at the end of §1.3. Further, $G^{n-1} \subset G^n$ and $A^{-(n+1)}g_0 = (\exp c)\zeta$. Consequently, $g_0 \in G^n_{\mathrm{rap}}$. Then

$$\mathrm{Ad}(\mathrm{id}+c)(\mathrm{id}+\varphi) = A^{-n}g_1, \qquad g_1 = \mathrm{Ad}(g_0)g \in G^n_{\mathrm{rap}},$$

which proves the proposition for $\varphi \in \Phi_{n+1}$ (it remains only to note that $\mathrm{Ad}(\mathrm{id}+c)(\mathrm{id}+\varphi) = \mathrm{id}+\varphi\circ(\mathrm{id}+c)$, and $\varphi\circ(\mathrm{id}+c) \to 0$ in $(\Pi^{\forall}, \infty)$ together with φ).

Suppose now that $\varphi \in \Psi_{n+1}$,

$$\varphi = \exp\psi, \qquad \psi \in \mathscr{L}_* \circ \ln^{[n]},\ \mathrm{Re}\,\psi \to -\infty,\ \psi' \to 0 \text{ in } (\Pi^{\forall}, \infty).$$

Then $\mathrm{Ad}(\mathrm{id}+c)(\mathrm{id}+\varphi) = \mathrm{id}+\varphi\circ(\mathrm{id}+c)$, and

$$\varphi\circ(\mathrm{id}+c) = \exp\psi\circ(\mathrm{id}+c).$$

Obviously, $\mathrm{Re}\,\psi\circ(\mathrm{id}+c) \to -\infty$ and $\psi'\circ(\mathrm{id}+c) \to 0$ in $(\Pi^{\forall}, \infty)$.

We prove that $\psi\circ(\mathrm{id}+c) \in \mathscr{L}_* \circ \ln^{[n]}$. Indeed,

$$\mathscr{L}_* = \mathscr{F}^{n-1}_{\mathbf{R},\mathrm{id}} + \mathscr{L}(\mathscr{F}^{n-1}_{++}, \mathscr{F}^n_+).$$

As indicated at the beginning of the proof,

$$\mathrm{id}+c = A^{-n}g_0, \qquad g_0 \in G^{n-1}_{\mathrm{rap}}.$$

What is more, $g_0 \in \mathrm{id} + \mathscr{F}^{n-1}_{\mathbf{R},+\mathrm{id}}$. We have that

$$\mathscr{F}^{n-1}_{\mathbf{R},\mathrm{id}} \circ \ln^{[n]} \circ(\mathrm{id}+c) = \mathscr{F}^{n-1}_{\mathbf{R},\mathrm{id}} \circ g_0 \circ \ln^{[n]} \subset \mathscr{F}^{n-1}_{\mathbf{R},\mathrm{id}} \circ \ln^{[n]},$$

by the third shift lemma, $\mathrm{SL3}_n$, §1.10. Similarly,

$$\mathscr{F}^{n-1}_{++} \circ \ln^{[n]} \circ(\mathrm{id}+c) \subset \mathscr{F}^{n-1}_{++} \circ \ln^{[n]}.$$

Finally,

$$\mathscr{F}^n_{+g} \circ \ln^{[n]} \circ(\mathrm{id}+c) = \mathscr{F}^n_{+g} \circ g_0 \circ \ln^{[n]} = \mathscr{F}^n_{+g\circ g_0} \circ \ln^{[n]}.$$

This proves that $\psi\circ(\mathrm{id}+c) \in \mathscr{L}_* \circ \ln^{[n]}$, and at the same time the whole proposition. ▶

Propositions 4-7 are used in the proof of all assertions concerned with inclusions of standard domains or their images in other standard domains.

F. Inclusion Lemma. *Let* $\{\varphi_j, \psi_j, \kappa_j\}$ *be a tuple of germs of holomorphic functions tending to zero in* $(\widetilde{\Pi}, \infty)$ *such that*:

1. $\varphi_j = \varphi_{sj} + \varphi_{rj}$, $\psi_j = \psi_{sj} + \psi_{rj}$,

$$\sum |\varphi_{rj}| + |\psi_{rj}| + |\kappa_j| \to 0 \quad in\ (\widetilde{\Pi}, \infty);$$

2. *the germs* φ_{sj} *and* ψ_{sj} *belong to a differential algebra ordered in* $(\Pi^\forall, \infty)$ *by the relation* $\operatorname{Re}\cdot \prec 0$, *and for some germ* $\psi \in \{\varphi_{sj}, \psi_{sj}\}$

$$\sum |\varphi'_{rj}| + |\psi'_{rj}| + |\kappa'_j| = o(|\operatorname{Re}\psi'|);$$

3. *there exists a neighborhood* $\mathscr{U}$ *of* 1 *on the line* $\mathbf{R}$ *such that*

$$-\sum \theta_j\varphi_{sj} + \sum \tilde{\theta}_j\tilde{\psi}_{sj} \succ 0 \quad for\ any\ \theta_j, \tilde{\theta}_j \in U. \qquad (***)$$

Then

$$\prod(\mathrm{id}+\psi_j)\circ(\mathrm{id}+\kappa_j)(\Pi, \infty) \subset \prod(\mathrm{id}+\varphi_j)(\Pi, \infty).$$

Remarks. 1. The inequality in the conditions of the lemma shows that the relation $-\sum\varphi_{sj} + \sum\psi_{sj} \succ 0$ holds "with room to spare." In particular,

$$-\sum \varphi_{sj}(1+o(1)) + \sum \psi_{sj}(1+o(1)) \succ 0.$$

2. Some of the germs φ_{sj} and ψ_{sj} can be zero, but not all: by assumption, $\psi \neq 0$.

Proof. Let

$$F = [\Pi(\mathrm{id}+\varphi_j)]^{-1} \circ \Pi(\mathrm{id}+\psi_j)\circ(\mathrm{id}+\kappa_j).$$

The conclusion of the lemma is equivalent to the inequality

$$|\operatorname{Im} F| \le |\eta| \quad \text{in } (\Pi, \infty).$$

It, in turn, follows from the inequality

$$\operatorname{Re} F' - 1 \prec 0 \quad \text{in } (\Pi, \infty).$$

But

$$F' - 1 = \left(-\sum \varphi'_{sj} + \sum \psi'_{sj}\right)(1+o(1)) \quad \text{in } (\Pi, \infty).$$

This follows from assertions 1 and 2. By inequality $(***)$ and Proposition 3 in §5.3, for all θ_j and $\tilde{\theta}_j$

$$-\sum \theta_j\varphi'_{sj} + \sum \tilde{\theta}_j\psi'_{sj} \prec 0 \quad \text{in } (\Pi, \infty).$$

This gives us that $\operatorname{Re} F' - 1 \prec 0$ in (Π, ∞), which proves the lemma. ►

G. The properties of standard domains of class $n+1$: continuation. In this subsection we prove

PROPERTY 2. *For any two standard domains of class* $n+1$ *there exists a domain of the same class belonging to their intersection.*

PROOF. Let Ω_1 and Ω_2 be two standard domains of class $n+1$. Then, by the definition in subsection A,

$$(\ln \Omega_j, \infty) = (\mathrm{id} + \varphi_j)(\Pi, \infty), \qquad \varphi_j \in S_{n+1}, \ j = 1, 2.$$

The domain Ω is sought in the form

$$\Omega = A^{-1}(\mathrm{id} + \psi)(\mathbf{C}^+ \backslash K), \qquad \psi \in S_{n+1} \backslash \mathscr{O}.$$

It suffices to choose a germ ψ such that

$$(\mathrm{id} + \psi)(\Pi, \infty) \subset (\mathrm{id} + \varphi_j)(\Pi, \infty).$$

For this it suffices that the inequality $\operatorname{Re}\psi \succ 2|\operatorname{Re}\varphi_j|$ hold. The decomposition for the germs φ_j and ψ that is required in the inclusion lemma (ψ is taken as ψ_1, and $\psi_2 = \cdots = \mathrm{id}$) is given by Proposition 5. The germ ψ is chosen according to Proposition 6. The preceding inclusion then follows from the inclusion lemma.

In more detail, let $\varphi_{js} \in \widetilde{S}_{n+1}$ be the germs given for φ_j by Proposition 5 in the case when $\varphi_j \notin \mathscr{O}$; $\varphi_{js} = \zeta^{-1} \in \widetilde{S}_{n+1}$ in the case when $\varphi_j \in \mathscr{O}$. Suppose that $\varphi_{1s} \succeq \varphi_{2s}$ in $(\Pi^\forall, \infty)$; since the algebra $\widetilde{S}_{n+1}$ is ordered, this can be achieved by a change in the numbering. By Proposition 6, there exist germs $\psi \in S_{n+1}$ and $\psi_s \in \widetilde{S}_{n+1}$ such that

$$\psi = \psi_s(1 + o(1)), \quad \psi' = \psi_s'(1 + o(1)), \quad \operatorname{Re}\psi \succ 2|\operatorname{Re}\varphi_{1s}| \ \text{ in } (\Pi^\forall, \infty).$$

The germs φ_1, φ_2, and ψ satisfy all the conditions of the inclusion lemma. Then $(\Omega, \infty) \subset (\Omega_j, \infty)$, by the inclusion lemma.

Property 2 is proved for standard domains of class $n+1$. ▶

H. Germs of class $\mathscr{L}^n$. In this subsection we prove the following

LEMMA 5.6_{n+1}. a. $AL^n \subset \mathrm{id} + \mathscr{L}(\mathscr{F}_+^{n-1}) \circ \ln^{[n-1]}$.

b. *Each germ* $h \in L^n$ *is biholomorphic in the germ of any sector* (S_α, ∞), $S_\alpha = \{\zeta \mid |\arg \zeta| < \alpha\}$, $\pi/2 < \alpha < \pi$, *and* $h' \to 1$ *in* (S_α, ∞).

c. *For each germ* $\tilde{h} \in \mathscr{L}^n$ *there exists an equivalent germ* $h \in L^n$ *such that* $\tilde{h} - h = o(|\zeta|^{-N})$ *for each* $N > 0$ *in* (S_α, ∞).

d. *Each germ* $\tilde{h} \in \mathscr{L}^n$ *is biholomorphic in the germ of the sector* (S_α, ∞), *and* $\tilde{h}' \to 1$ *there.*

REMARK. The sector S_α in the statement of the lemma can be considered for arbitrary $\alpha > \pi/2$, but then it must be located on the Riemann surface of the logarithm. For our purposes any angle $\alpha > \pi/2$ arbitrarily close to π works.

PROOF. a. By the definitions of the sets L^n and J^{n-1} (15 in §1.6 and 4^n in §1.7),

$$A^n L^n \subset J^{n-1}.$$

According to Lemma $\mathrm{SL4}_n$ a,

$$J^{n-1} \subset \mathrm{id} + \mathscr{L}(\mathscr{F}_+^{n-1}).$$

Arguing as in the proof of Proposition 7 in §5.2 D, we get that

$$AL^n \subset A^{1-n} J^{n-1} \subset \mathrm{id} + \mathscr{L}(\mathscr{F}_+^{n-1}) \circ \ln^{[n-1]}.$$

The assertion a is proved.

b. It suffices to prove the assertion for generating elements of the group L^n. Let

$$f \in \mathscr{A}^0, \quad g \in G^{n-1}, \quad \mu_n(g) = 0,$$
$$h = A^{-n}(\mathrm{Ad}(g) A^{n-1} f) = \mathrm{Ad}((A^{1-n} g) \circ \ln) f.$$

The germ f^u (the main germ of the tuple f) extends biholomorphically to the germ of the half-strip (Π_a, ∞) for some $a > 0$, where $\Pi_a = \{\zeta \| \eta| \leq a, \xi \geq 0\}$. Further, the germ $\rho = A^{1-n} g \circ \ln$ acts according to the formula

$$(S_\alpha, \infty) \xrightarrow{\ln} (\Pi_\alpha, \infty) \xrightarrow{A^{1-n} g} (\Pi^\forall, \infty).$$

The last assertion follows from the realness of the germ $A^{1-n} g$ and the convergence of its derivative to zero; this requirement appears in the definition of h. This implies that h is biholomorphic in (S_α, ∞).

By assertion a of the lemma,

$$Ah = \mathrm{id} + \varphi, \qquad \varphi \in \mathscr{L}(\mathscr{F}_+^{n-1}) \circ \ln^{[n-1]}.$$

Consequently,

$$h' = (A^{-1}(\mathrm{id} + \varphi))' = (\zeta \cdot \exp \circ \varphi \circ \ln)'$$
$$= (\exp \circ \varphi \circ \ln) \cdot (1 + \varphi' \circ \ln) \to 1 \quad \text{in } (S_\alpha, \infty),$$

because both factors tend to 1: $\varphi \to 0$ and $\varphi' \to 0$ in (Π_α, ∞). This proves assertion b for generating elements of L^n. Using the Cauchy inequality and the arbitrariness of α, we can prove it for an arbitrary element of L^n.

c. We prove the assertion c for generating elements of the group $\mathscr{L}^n$. Let

$$\tilde{h} = A^{-n}(\mathrm{Ad}(g) A^{n-1}(A\tilde{g}) f), \qquad f \in \mathscr{A}^0, \ \tilde{g} \in H^0,$$
$$g \in G^{n-1}, \ \mu_n(g) = 0, \ \tilde{h} \in \mathscr{M}_{\mathbf{R}}.$$

Then $\tilde{h} = h_1 \circ h$, where

$$h = A^{-n}(\mathrm{Ad}(g) A^{n-1} f), \qquad h_1 = A^{-n} \mathrm{Ad}(g) A^n \tilde{g}.$$

We investigate the germ h_1. By the assertion of Example 4 in §1.7,

$$A^n \tilde{g} \in \mathrm{id} + \mathscr{F}^n_{+\,\mathrm{id}}.$$

According to the conjugation lemma CL_n in §4.8,

$$\mathrm{Ad}(g) A^n \tilde{g} \in \mathrm{id} + \mathscr{F}^n_{+g}.$$

Arguing as in the proof of Proposition 7 in §5.2 D, we get that

$$h_1 \in A^{-n}(\mathrm{id} + \mathscr{F}^n_{+g}) \subset \mathrm{id} + \mathscr{F}^n_{+g} \circ \ln^{[n]}.$$

Corollary 4 in §5.2 B shows that $|h_1 - \mathrm{id}| = o(|\zeta|^{-N})$ in (S_α, ∞) for any $N > 0$. This and the condition $h' \to 1$ in (S_α, ∞) give us that

$$|\tilde{h} - h| = o(|\zeta|^{-N}) \quad \text{in } (S_\alpha, \infty).$$

REMARK. The arguments given show that

$$A\tilde{h} = (\mathrm{id} + \varphi), \qquad \varphi \in \mathscr{L}(\mathscr{F}^{n-1}_+, \mathscr{F}^n_+) \circ \ln^{[n-1]}.$$

d. Assertion d follows from assertions b and c. This finishes the proof of Lemma 5.6_{n+1}. ►

I. Admissibility of germs of class $\mathscr{D}^n_{\mathrm{rap}} \circ \mathscr{L}^n$. In this section we prove that the germs in the heading are admissible. We start with a verification of the first requirement in the definition of admissibility; see subsection A.

PROPOSITION 8. *For any germ $\sigma \in \mathscr{D}^n_{\mathrm{rap}} \circ \mathscr{L}^n$ and any standard domain Ω of class $n+1$ there exists a standard domain $\widetilde{\Omega}$ of the same class such that the germ $\rho = \sigma^{-1}$ is biholomorphic in $\widetilde{\Omega}$, and $\rho\widetilde{\Omega} \subset \Omega$.*

PROOF. The biholomorphicity of ρ in the germ of the domain $(\mathbf{C}^+, \infty)$ follows from Lemma 5.6_{n+1} d and Proposition 6 in §5.2 C. We prove that there is a domain $\widetilde{\Omega}$ with the properties in the proposition, or, what is equivalent, we find a germ $\Phi \in S_{n+1}$ such that

$$(\ln \widetilde{\Omega}, \infty) = (\mathrm{id} + \Phi)(\Pi, \infty).$$

By Proposition 4 in subsection E, it can be assumed that

$$(\ln \Omega, \infty) = (\mathrm{id} + \varphi)(\Pi, \infty), \qquad \varphi \in S_{n+1}.$$

Let $\rho = \rho_2 \circ \rho_1$, $\rho_2 \in \mathscr{L}^n$, $\rho_1 \in \mathscr{D}^n_{\mathrm{rap}}$. The inclusion $\rho\widetilde{\Omega} \subset \Omega$ holds if

$$A\rho_2 \circ A\rho_1 \circ (\mathrm{id} + \Phi)(\Pi, \infty) \subset (\mathrm{id} + \varphi)(\Pi, \infty).$$

As shown in the proof of Proposition 4 of subsection E, there exists a germ $\varphi_1 \in \mathbf{\Phi}_{n+1}$ and a $\lambda \in \mathbf{R}$ such that

$$A\rho_1 = (\mathrm{id} + \varphi_1) \circ (\mathrm{id} + \lambda).$$

We find a germ $\Psi \in S_{n+1}$ such that

$$A\rho_2 \circ (\mathrm{id} + \varphi_1) \circ (\mathrm{id} + \Psi)(\Pi, \infty) \subset (\mathrm{id} + \varphi)(\Pi, \infty). \qquad ({}^{*}_{**})$$

When we have done this, we set

$$\mathrm{id} + \Phi = \mathrm{Ad}(\mathrm{id} + \lambda)(\mathrm{id} + \Psi),$$

where $\Phi \in S_{n+1}$ by Proposition 7. For such Φ and Ψ the last and next-to-last inclusions are equivalent.

The germ Ψ will be constructed in a way similar to what was done in subsection G; the inclusion $\binom{*}{**}$ will then be derived from the inclusion lemma.

Let φ_s be the germ given by Proposition 5 for a germ φ in the case when $\varphi \notin \mathscr{O}$; $\varphi_s = 0$ if $\varphi \in \mathscr{O}$. The germ φ_{1s} is determined from the germ φ_1 similarly. Let

$$\psi_0 = \max(\pm\varphi_{1s}, \pm\varphi_s, \zeta^{-1});$$

the maximum is taken in $\widetilde{S}_{n+1}$, and ψ is a corresponding germ in the set $\{\pm\varphi_1, \pm\varphi, \zeta^{-1}\}$: if $\psi_0 = \pm\varphi_{1s}$ (respectively, $\pm\varphi_s$ or ζ^{-1}), then $\psi = \pm\varphi_1$ (respectively, $\pm\varphi$ or ζ^{-1}). According to Proposition 6, there exist germs $\Psi \in S_{n+1}$ and $\Psi_s \in \widetilde{S}_{n+1}$: $\Psi' = \Psi_s'(1+o(1))$ and $\operatorname{Re}\Psi_s \succ 3\operatorname{Re}\psi_0$ in $(\Pi^\forall, \infty)$. By Proposition 5, $\varphi' = \varphi_s'(1+o(1))$ for $\varphi \notin \mathscr{O}$, and $\varphi' = o(\operatorname{Re}\psi_0')$ for $\varphi \in \mathscr{O}$. The analogous equalities are valid for the germ φ_1.

Finally, we let

$$A\rho_2 = \mathrm{id} + \kappa \in A\mathscr{L}^n.$$

By Lemma 5.6_{n+1} d and Proposition 3 in §2.11, $|\kappa| \prec |\zeta|^{-N}$ in $(\Pi^\forall, \infty)$ for arbitrary $N > 0$. By the Cauchy inequality,

$$|\kappa'| \prec |\zeta|^{-3} = o(\operatorname{Re}\psi_0') = o(\operatorname{Re}\Psi_s') \quad \text{in } (\Pi^\forall, \infty).$$

This verifies the first two conditions of the inclusion lemma for the germs $\kappa = \{\kappa_j\}$, $\{\varphi_1, \Psi\} = \{\psi_j\}$, and $\varphi = \{\varphi_j\}$. The third condition is ensured by the choice of Ψ and Ψ_s. The inclusion lemma gives the required inclusion $\binom{*}{**}$.

We verify the remaining requirements in the definition of admissibility.

2°. The derivative ρ' is bounded in $\widetilde{\Omega}$. For germs of class $\mathscr{D}^n_{\mathrm{rap}}$ this follows from Proposition 3, with $n+1$ replaced by n; for germs of class $\mathscr{L}^n$ it follows from Lemma 5.6_{n+1} d; for compositions of them it follows from Proposition 8 and Property 2 of standard domains of class $n+1$.

Requirement 3° for germs of class $\mathscr{D}^n_{\mathrm{rap}}$ is contained in strengthened form in Lemma 5.2_n c, which appears in the induction hypothesis. Indeed, suppose that $\rho \in \mathscr{D}^n_{\mathrm{rap}}$. Then, by Lemma 5.2_n c, there exist an $\varepsilon > 0$ and a standard domain $\widetilde{\Omega}$ of class n (which is in the same time of class $n+1$ by the Inclusion 2° in §4.5C) such that $\varepsilon\xi < \operatorname{Re}\rho$ in $\widetilde{\Omega}$. This is equivalent to the inequality $\varepsilon \operatorname{Re}\sigma < \xi$ in $\rho\widetilde{\Omega}$, where $\rho = \sigma^{-1}$. According to Proposition 8, there exists a standard domain Ω of class $n+1$ such that $\sigma\Omega \subset \widetilde{\Omega}$ for $\sigma \in \mathscr{D}^n_{\mathrm{rap}}$ (since $\mathscr{D}^n_{\mathrm{rap}}$ is a group, σ and $\rho = \sigma^{-1}$ belong to $\mathscr{D}^n_{\mathrm{rap}}$ simultaneously). Consequently, $\rho\widetilde{\Omega} \subset \Omega$. The inequality $\operatorname{Re}\sigma < \varepsilon^{-1}\xi$ holds in Ω.

For $\rho \in \mathscr{L}^n$ the inequalities $\varepsilon\xi < \operatorname{Re}\rho < \varepsilon^{-1}\xi$ hold in some domain of class $n+1$ by Lemma 5.6_{n+1}C and Proposition 1 in §2.11 (by inclusion 2° in §4.5 C, a domain of class n is at the same time of class $n+1$).

The same inequalities hold for compositions in $\mathscr{D}^n_{\mathrm{rap}} \circ \mathscr{L}^n$ in view of Property 2 of domains of class $n+1$ and in view of Proposition 8.

Requirement 4°: $\exp \operatorname{Re} \rho \succ \nu\xi$ in $\widetilde{\Omega}$ for arbitrary $\nu > 0$. This follows from the much stronger inequality $\operatorname{Re} \rho \succ \varepsilon\xi$ in $\widetilde{\Omega}$, which is proved below in subsection J.

This concludes the proof that germs of the class $\mathscr{D}^n_{\mathrm{rap}} \circ \mathscr{L}^n$ are admissible, and Lemma 5.2_{n+1} b is proved.

In the next subsection Lemma 5.2_{n+1} c is proved in a strengthened form.

J. Estimates of the real parts of germs of class $\mathscr{D}^{n+1}_{\mathrm{rap}}$. Here we conclude the proof of Lemma 5.2_{n+1} by proving assertion c.

PROPOSITION 9. *Suppose that $g \in G^n_{\mathrm{rap}}$, and let $\lambda = \lim(A^{-n} g - \mathrm{id})$. Then there exists a standard domain Ω os class $n+1$ such that the inequalities*

$$(1-\delta)\mu\xi \preceq \operatorname{Re} A^{-(n+1)} g \preceq (1+\delta)\mu\xi,$$

hold in Ω for any $\delta > 0$, where $\mu = \exp\lambda$.

PROOF. We prove only the second inequality; the first can be proved similarly, except that Cases 1 and 2 below are interchanged. Both inequalities hold simultaneously in a standard domain of class $n+1$ belonging to the intersection of the standard domains constructed below; such a domain exists in view of Property 2 of standard domains (proved above).

It can be assumed without loss of generality that $\mu = 1$; otherwise, g must be replaced by $\tilde{g} = (A^{n+1}(\mu^{-1})) \circ g \in G^n_{\mathrm{rap}}$. Here μ is the mapping of multiplication by μ; $A^{n+1}\mu \in G^n_{\mathrm{rap}}$, as established in the proof of Proposition 3 in §1.3.

The domain Ω will be sought in the form

$$\Omega = A^{-(n+1)} f(\mathbf{C}^+ \backslash \widetilde{K}), \qquad f \in G^n_{\mathrm{rap}}.$$

The inequality

$$\operatorname{Re} A^{-(n+1)} g \preceq (1+\delta) \operatorname{Re} \zeta \quad \text{in } \Omega$$

in the proposition is equivalent to the inequality

$$\operatorname{Re} \exp A^{-n}(g \circ f) \preceq (1+\delta) \operatorname{Re} \exp A^{-n} f \quad \text{in } \Pi = \ln(\mathbf{C}^+ \backslash \widetilde{K}).$$

Let

$$A^{-n} g = \mathrm{id} + \varphi, \quad A^{-n} f = \mathrm{id} + \psi, \quad A^{-n}(g \circ f) = \mathrm{id} + \tilde{\varphi},$$
$$\tilde{\varphi} = \psi + \varphi \circ (\mathrm{id} + \psi), \quad \tilde{\varphi} \in \mathscr{L}_* \circ \ln^{[n]}, \quad \mathscr{L}_* = \mathscr{L}_*(\mathscr{F}^{n-1}_{\mathbf{R},\mathrm{id}}, \widetilde{\mathscr{F}}^n_+).$$

Since $\mu = 1$, $\varphi \to 0$ in (Π, ∞); below, f will be chosen so that $\psi \to 0$ in (Π, ∞). The last inequality is equivalent to the inequality

$$(\exp \operatorname{Re} \tilde{\varphi}) \cos(\eta + \operatorname{Im} \tilde{\varphi}) \preccurlyeq (1+\delta)(\exp \operatorname{Re} \psi) \cos(\eta + \operatorname{Im} \psi)$$

in (Π, ∞). This inequality, in turn, is a consequence of the following two inequalities:

$$\begin{aligned}&(\operatorname{Re}\exp\tilde{\varphi})\cos\eta \preccurlyeq (1+\delta)(\operatorname{Re}\exp\psi)\cos\eta,\\ &(\operatorname{Im}\exp\tilde{\varphi})\sin\eta \succcurlyeq (1+\delta)(\operatorname{Im}\exp\psi)\sin\eta \quad \text{in } (\Pi, \infty).\end{aligned} \qquad \left({}^{*}_{*}\right)$$

The last inequality follows from the relation

$$\operatorname{sgn}\operatorname{Im}(\exp\tilde{\varphi} - (1+\delta)\exp\psi) = \operatorname{sgn}\eta \quad \text{in } (\Pi, \infty),$$

which, in turn, follows from the inequality

$$\operatorname{Re}(\exp\tilde{\varphi})' \succcurlyeq (1+\delta)\operatorname{Re}(\exp\psi)' \quad \text{in } (\Pi, \infty). \qquad \left({}^{*\,*}_{\;*}\right)$$

The inequality $\left({}^{*}_{*}\right)$ follows from the inequality

$$\operatorname{Re}\exp\tilde{\varphi} \preccurlyeq (1+\delta)\operatorname{Re}\exp\psi \quad \text{in } (\Pi, \infty).$$

The last two inequalities will be ensured by the choice of ψ. The second is trivial: for arbitrary $\psi \to 0$ the left-hand side tends to 1 in $(\Pi^{\forall}, \infty)$, while the right-hand side tends to $1+\delta$, since $\varphi \to 0$. To ensure the inequality $\left({}^{*\,*}_{\;*}\right)$, we consider three cases.

CASE 1: $\varphi \succ 0$, $\varphi \notin \mathscr{O}$. In this case there exists by Proposition 5 a germ $\varphi_s \in \mathscr{K}$ such that $\varphi = \varphi_s(1+o(1))$ and $\varphi' = \varphi_s'(1+o(1))$ in $(\Pi^{\forall}, \infty)$. By Proposition 6, there exist for arbitrary $N > 0$ germs $\psi \in S_{n+1}$ and $\psi_s \in \widetilde{S}_{n+1}$ such that

$$\psi = \psi_s(1+o(1)), \qquad \psi' = \psi_s'(1+o(1)),$$
$$\psi_s \succ N\varphi_s \quad \text{in } (\Pi^{\forall}, \infty).$$

Take $N > 1/\delta$. Since $\exp\tilde{\varphi} = 1+o(1)$ and $\exp\psi = 1+o(1)$, we use assertion 1 in subsection D and get that

$$\begin{aligned}(\exp\tilde{\varphi})' &= \varphi_s'(1+o(1)) + \psi_s'(1+o(1)),\\ (\exp\psi)' &= \psi_s'(1+o(1)) \quad \text{in } (\Pi^{\forall}, \infty).\end{aligned}$$

By Proposition 3 in §5.3 and relation $\varphi \to 0$, the inequalities $\varphi_s \succ 0$ and $\varphi_s \prec \psi_s/N$ give us that $0 \succ \varphi_s' \succ \psi_s'/N$. From this,

$$\varphi_s' \cdot (1+o(1)) \succcurlyeq (\delta + o(1))\psi_s' \quad \text{in } (\Pi^{\forall}, \infty).$$

Together with the two preceding equalities this gives the inequality $\left({}^{*\,*}_{\;*}\right)$.

CASE 2: $\varphi \prec 0$, $\varphi \notin \mathscr{O}$. In this case we can take $\psi \equiv 0$. By Proposition 5 in subsection E, and Proposition 3 in §5.3, the left-hand side of $\left({}^{*\,*}_{\;*}\right)$ is positive in $(\Pi^{\forall}, \infty)$ in this case; by the definition of ψ, the right-hand side is identically equal to zero.

CASE 3: $\varphi \in \mathscr{O}$. In this case we take $\psi = \zeta^{-1}$. Then $\varphi = o(1)\zeta^{-1}$, $\tilde{\varphi} = \zeta^{-1}(1+o(1))$, $\tilde{\varphi}' = \zeta^{-2}(1+o(1))$, $(\exp\tilde{\varphi})' = -\zeta^{2}(1+o(1))$, and $(\exp\psi)' = -\zeta^{-2}(1+o(1))$, and the inequality $\left({}^{*\,*}_{\;*}\right)$ is obvious. ►

We summarize. Assertion a of Lemma 5.2_{n+1} was proved in subsection B, and assertion b in subsection I. Assertion c follows from Proposition 9 (the first inequality) and Proposition 3 (the second inequality): in view of Proposition 3 the derivative σ' is bounded in $(\mathbf{C}^+, \infty)$. Assertion d was proved in subsections C and G. Lemma 5.2_{n+1} is thus proved.

Lemma 5.6_{n+1} was proved in subsection H.

To conclude let us prove a proposition that will be used in the proof of Lemma 5.8_{n+1}.

PROPOSITION 10. *For any germ* $\tilde{h} \in \mathscr{L}^n$ *there is a domain* Ω *of class* $n+1$ *such that for any* $\delta > 0$

$$(1-\delta)\xi \prec \operatorname{Re}\tilde{h} \prec (1+\delta)\xi$$

in Ω.

PROOF. The proof follows the proof of Proposition 9 with much simplification. The domain Ω can be taken in the form

$$\Omega = A^{-(n+1)} f(\mathbf{C}^+ \backslash \widetilde{K}), \qquad A^{-n} f = \operatorname{id} + \psi, \ \psi = \zeta^{-1}$$

(see Case 3 in the previous proof). Only $\widetilde{K}$ depends on $\tilde{h}$.

It suffices to prove the proposition for any $h \in L^n$ instead of $\tilde{h} \in \mathscr{L}^n$; this follows from Lemma 5.6_{n+1} c. Take h instead of $A^{-(n+1)}g$ in Proposition 9 and denote

$$Ah = \operatorname{id} + \varphi, \qquad \tilde{\varphi} = \psi + \varphi \circ (\operatorname{id} + \psi).$$

Repeating the calculations in the proof of Proposition 9, we thus get that it suffices to prove the inequality $({}^*_*{}^*)$. By Lemma 5.6_{n+1} a, $\varphi \in \mathscr{L}(\mathscr{F}_+^{n-1}) \circ \ln^{[n-1]}$. The germ φ' has the same property. By Corollary 4 in 5.2 B (with n replaced by $n-1$) φ' decreases faster than any power in (Π, ∞). Thus, in the inequality $({}^*_*{}^*)$ we have that

$$(\exp\tilde{\varphi})' = -(\exp\tilde{\varphi}) \circ (\zeta^{-2} - \varphi' \circ (\operatorname{id} + \psi) \cdot (1 - \zeta^{-2})),$$
$$(1+\delta)\operatorname{Re}(\exp\psi)' = -(1+\delta)\zeta^{-2}\exp\zeta^{-1}.$$

Since $\exp\tilde{\varphi} = 1 + o(1)$, the inequality $({}^*_*{}^*)$ becomes obvious. ►

§5.5. Admissible germs increasing more rapidly than linear germs

In this section we continue the investigation of germs of class $\mathscr{D}_*^n$ (see Definition 16.b in §1.6) begun in §§5.2 and 5.4. Namely, denote by $\mathscr{D}_{\text{slow}}^{m+1}$ the set

$$\mathscr{D}_{\text{slow}}^{m+1} = \{\sigma = A^{-(m+1)} g \mid g \in G_{\text{slow}}^{m^+}\}.$$

Note that the difference $\mathscr{D}_*^n \backslash \mathscr{D}_{\text{slow}}^n \circ \mathscr{L}^n$ was investigated in §5.4.

A. Formulations. The main result in this section is

LEMMA 5.5_{n+1}. a. *Germs of class* $\mathscr{D}^n_{\text{slow}} \circ \mathscr{L}^n$ *are admissible of class* $n+1$. *If* $g \in G^{n^-}_{\text{slow}}$, *then* $(A^{-n} g\Pi, \infty) \subset (\Pi, \infty)$. *The following assertions hold for any germ* $g \in G^n$.

b. *For arbitrary* $C > 0$ *and* $\varepsilon > 0$

$$(\ln|\zeta|)^c \prec |A^{-(n+1)} g| \prec \exp|\zeta|^\varepsilon \quad in\ (S^\forall, \infty),$$

$$(\ln|\zeta|)^c \prec |A^{-n} g| \prec \exp \xi^\varepsilon \quad in\ (\Pi^\forall, \infty).$$

c. *Let* $\sigma = A^{-(n+1)} g$. *Then* $\sigma'/\sigma \to 0$ *in* $(\mathbf{C}^+, \infty)$.

d. *In addition, let* $g \in G^{n^+}_{\text{slow}}$ *and* $\rho = \sigma^{-1}$. *Then there exists a standard domain of class* $n+1$ *such that* $\operatorname{Re}\rho < \varepsilon\xi$ *in this domain for arbitrary* $\varepsilon > 0$.

The definition of an admissible germ of class $n+1$ was given in §§1.5 and 1.7 and recalled at the beginning of §5.4. The proof of the lemma will be broken up into 5 propositions. Some requirements of the definition of admissible germs are proved at the same time as the assertions of the lemma. We list the corresponding implications so that we can then formulate and prove only the propositions.

It was proved in Proposition 6 in §5.2 C that the germs of class $\mathscr{D}^m_{\text{slow}}$, $m \leq n+1$, are biholomorphic in the germ of the half-plane $(\mathbf{C}^+, \infty)$, and even in the sector S_α: $|\arg\zeta| \leq \alpha < \pi$. The fact that $\rho\tilde{\Omega} \subset \Omega$ for germs of this class will be proved in Proposition 3 of subsection 5.5 D (used for n instead of $n+1$). This verifies the first requirement in the definition of admissible germs for germs of the class $\mathscr{D}^n_{\text{slow}}$.

The second requirement in this definition is verified for germs of the same class in Proposition 2 of §5.5 B. The third requirement follows from Proposition 4 of §5.5 E, which is an explicit strengthening of this requirement.

We note that the first three requirements in the definition of admissible germs were proved in §5.4 for germs of class $\mathscr{L}^n$. Moreover, these requirements have the semigroup property. Consequently, the fact that these requirements hold for germs of class $\mathscr{D}^n_{\text{slow}}$ (proved in the propositions cited above) implies that they hold for germs of class $\mathscr{D}^n_{\text{slow}} \circ \mathscr{L}^n$.

Assertions b and c of the lemma appear in assertion 3° of Proposition 6 in §5.2 C. Requirement 4 in the definition of admissible germs of class $\mathscr{D}^n_{\text{slow}} \circ \mathscr{L}^n$ is derived from the same proposition by the following way. Proposition 10 in §5.4J reduces the requirement to the germs of class $\mathscr{D}^n_{\text{slow}}$. The first requirement in the definition of the admissible germs implies that for any $\sigma \in \mathscr{D}^n_{\text{slow}}$ a standard domain $\tilde{\Omega} \in \mathbf{\Omega}_n$ exists in which $\operatorname{Re}\rho > |\rho|^{1/2}$, $\rho = \sigma^{-1}$; for this the inclusion $\rho\tilde{\Omega} \subset \Omega$ is sufficient, where Ω is a quadratic standard domain. Now by Proposition 6 in §5.2C for any $C > 0$, $\nu > 0$ we have $\exp \operatorname{Re}\rho \succ |\zeta|^c \succ \xi^{c/2} \succ \nu\xi$.

Assertion d of the lemma is proved in Proposition 4, which is based essentially on results in subsection B.

We proceed to the formulation and proof of the propositions. For each of them a corresponding standard domain of class $n+1$ is found separately where necessary. A standard domain in which the statements of these propositions hold simultaneously exists by Property 2° of standard domains of class $n+1$ (proved in §5.4). Everywhere in what follows, the investigation of the case $\mu_{n+1}(g) < 1$ is essentially easier than that of the case $\mu_{n+1}(g) = 1$. Let us begin with a study of this last case. Note that if $\sigma \in \mathscr{D}^{n+1}_{\text{slow}}$, $\rho = \sigma^{-1}$ then $\rho \in A^{-(n+1)} G^{n^-}_{\text{slow}}$. These germs are studied below.

B. Germs of class $\mathscr{D}^{n+1}_{\text{slow}}$ with generalized exponent 1.

In this subsection we investigate germs of the form $A^{-n}g$, $g \in G^n$, $\mu_{n+1}(g) = 1$. Recall the notation:

$$\widetilde{\mathscr{F}}^n_+ = \mathscr{F}^{n-1}_{++} \cup \mathscr{F}^n_+,$$
$$\mathscr{L}_* = \{\varphi_s + \varphi_r | \varphi_s \in \mathscr{F}^{n-1}_{\mathbf{R},\text{id}}, \varphi_r \in \widetilde{\mathscr{F}}^n_+, \varphi_s \neq 0\},$$
$$A^{-n}g = \text{id} + \varphi.$$

Everywhere below, $n \geq 1$, since Lemma 5.5_1 was proved in §2.6.

PROPOSITION 1. *Suppose that* $g \in G^n_{\text{slow}}$, $\mu_{n+1}(g) = 1$, *and* $A^{-n}g = \text{id} + \varphi$. *Then*

1°. $\varphi \in \mathscr{L}_* \circ \ln^{[n]}$.

If in addition $g \preccurlyeq \text{id}$, *then*:

2°. $\operatorname{Re}\varphi \to -\infty$, $\varphi' \to 0$, *and* $\operatorname{Im}\varphi \to 0$ *in* $(\Pi^\forall, \infty)$;

3°. $\arg\varphi \to \pi$ *and* $\arg\varphi' \to \pi$ *in* $(\Pi^\forall, \infty)$.

REMARK. This immediately yields Proposition 5.1_{n+1}; see §3.7.

PROOF. 1°. We prove first that under the conditions of the proposition $k = k(g) \geq n-2$. Indeed, by Proposition 11 in §5.2 F, $\mu_{n+1}(g) = 0$ or ∞ when $k(g) \leq n-3$, and this contradicts the condition.

Consider now the case $k(g) = -1$. Then $n \leq k+2$, and $n = 1$, because $n > 0$, as indicated before the statement of the proposition. By Proposition 10 in §5.2 D, in this case

$$A^{-n}g = A^{-1}g = (\exp\mu + o(1))\zeta^\alpha, \qquad \alpha\zeta + \beta \neq \text{id}, \mu = \exp\beta.$$

This contradicts the convergence $(A^{-1}g)' \to 1 = \mu_{n+1}(g)$ and proves that $k(g) \neq -1$.

Accordingly, $k(g) \geq 0$. In this case one of Propositions 8 or 9 in §5.2 D is applicable, and it immediately yields assertion 1°, as in Corollary 4, §5.2 G.

2°. We get from assertion 1° and Corollary 3 in §5.2 B that the limit $\lambda = \lim\varphi$ exists in $(\tilde{\Pi}, \infty)$. If $\lambda \neq -\infty$, then the germ φ is bounded on $(\mathbf{R}^+, \infty)$. By Corollary 2 in §5.2 G, $g \in G^n_{\text{rap}}$ in this case, which contradicts a condition of the proposition. Therefore, $\varphi \to -\infty$ in $(\tilde{\Pi}, \infty)$. According to Propositions 8 and 9 in §5.2 D, applied to the respective cases $k = n-1$ and $k = n-2$, $\varphi' \to 0$ in $(\Pi^\forall, \infty)$. Consequently, $\operatorname{Im}\varphi \to 0$ in $(\Pi^\forall, \infty)$.

3°. We get from assertion 1°, Propositions 8 and 9 in §5.2 D, and Corollary 2 in §5.2 B that

$$\arg\varphi \to 0 \pmod{\pi} \quad \text{in } (\Pi^\forall, \infty).$$

Since $\varphi \to -\infty$ on $(\mathbf{R}^+, \infty)$, we get that

$$\arg\varphi \to \pi \quad \text{in } (\Pi^\forall, \infty).$$

It now follows from Proposition 5 in §5.3 D that

$$\arg\varphi' \to \pi \quad \text{in } (\Pi^\forall, \infty).$$

Since $|\varphi| \to \infty$ and $\arg\varphi \to \pi$ in $(\Pi^\forall, \infty)$, it follows that $\operatorname{Re}\varphi \to -\infty$ in $(\Pi^\forall, \infty)$. ▶

C. Boundedness of derivatives.

PROPOSITION 2. *Suppose that $g \in G^n_{\text{slow}}$, $\rho = A^{-(n+1)}g$. Then there exists a disk K such that in $\mathbf{C}^+\backslash K$ the derivative ρ' is bounded and even tends to zero.*

CASE 1°: $\mu_{n+1}(g) = 1$. In this case there exists by Proposition 1 in B a disk K such that

$$A^{-n}g = \mathrm{id} + \varphi, \qquad \operatorname{Re}\varphi \to -\infty,\ \varphi' \to 0$$

in the half-strip $\Pi = \ln(\mathbf{C}^+\backslash K)$. We have

$$(A^{-(n+1)}g)' = (\exp\circ\varphi\circ\ln)(1 + \varphi'\circ\ln).$$

The second factor tends to 1 and the first to 0 in (Π, ∞), and this proves the proposition in Case 1°.

CASE 2°: $\mu_{n+1}(g) < 1$. In this case

$$(A^{-(n+1)}g)' = (A^{-(n+1)}g)\cdot(A^{-n}g)'\cdot\zeta^{-1}.$$

But

$$A^{-n}g = (\mu_{n+1}(g) + o(1))\zeta \quad \text{in } (\Pi, \infty).$$

Consequently,

$$A^{-(n+1)}g = \zeta^{\mu_{n+1}(g)+o(1)} \quad \text{in } (\mathbf{C}^+, \infty).$$

Therefore,

$$(A^{-(n+1)}g)' = \zeta^{\mu_{n+1}(g)-1+o(1)}(A^{-n}g)'\circ\ln.$$

The first factor tends to zero in $\mathbf{C}^+\backslash K$, and the second to $\mu_{n+1}(g)$. ▶

D. Mappings of standard domains.

PROPOSITION 3. *For an arbitrary germ* $g \in G_{\text{slow}}^{n^-}$, $\rho = A^{-(n+1)}g$, *and an arbitrary standard domain* Ω *of class* $n+1$ *there exists a standard domain* $\tilde{\Omega}$ *of the same class such that the germ* ρ *extends biholomorphically to* $\tilde{\Omega}$, *and* $\rho\tilde{\Omega} \subset \Omega$.

PROOF. The proof is in large part parallel to that of Proposition 6 in §5.4. The biholomorphicity of the germ ρ in $(\mathbf{C}^+, \infty)$ follows from Proposition 6 in §5.2. We prove the existence of the desired domain $\tilde{\Omega}$. Let

$$A^{-n}g = \text{id} + \varphi, \quad \Omega = A^{-1}(\text{id} + \psi)(\mathbf{C}^+ \backslash K), \quad \psi \in S_{n+1}.$$

We prove that $\tilde{\Omega}$ can be taken to be a domain coinciding with Ω outside some compact set:

$$\tilde{\Omega} = A^{-1}(\text{id} + \psi)(\mathbf{C}^+ \backslash \tilde{K}).$$

The inclusion $\rho\tilde{\Omega} \subset \Omega$ holds for sufficiently large $\tilde{K}$ if

$$(\text{id} + \varphi) \circ (\text{id} + \psi)(\Pi, \infty) \subset (\text{id} + \psi)(\Pi, \infty). \qquad (*)$$

Let $\mu = \mu_{n+1}(g) < 1$. Then $(\text{id} + \psi)' \to 1$ in (Π, ∞) (see the remark in §5.4E), $(\text{id} + \varphi)' \to \mu < 1$ in (Π, ∞), and the inclusion $(*)$ is obvious.

Let $\mu = \mu_{n+1}(g) = 1$. Then $\varphi \in \mathscr{L}_* \circ \ln^{[n]}$ by Proposition 1, where $\mathscr{L}_* = [\mathscr{L}(\mathscr{F}_{\mathbf{R},\text{id}}^{n-1}, \widetilde{\mathscr{F}}_+^n) \backslash \mathscr{L}(\widetilde{\mathscr{F}}_+^n)]$, and $\operatorname{Re}\varphi \to -\infty$ and $\varphi' \to 0$ in $(\Pi^\forall, \infty)$. We decompose φ into a sum: $\varphi = \varphi_s + \varphi_r$, $\varphi_s \in \mathscr{L}(\mathscr{F}_{\mathbf{R},\text{id}}^{n-1}) \circ \ln^{[n]}$, $\varphi_s \neq 0$, $\varphi_r \in \mathscr{L}(\widetilde{\mathscr{F}}_+^n) \circ \ln^{[n]}$. Two cases will be considered:

CASE 1: $\psi \in \mathscr{L}(\widetilde{\mathscr{F}}_+^n) \circ \ln^{[n]}$. In this case the inclusion $(*)$ follows from the inclusion lemma in subsection F of §5.4.

CASE 2: $\psi \in S_{n+1} \backslash \mathscr{L}(\widetilde{\mathscr{F}}_+^n) \circ \ln^{[n]}$. Then $\psi' = \psi_s'(1 + o(1))$ in $(\Pi^\forall, \infty)$, and $\psi_s \in \tilde{S}_{n+1}$ (see Proposition 5 in §5.4 E). Further, $\operatorname{Re}(\varphi_s + C\psi_s) \to -\infty$ in $(\Pi^\forall, \infty)$ for any $C \in \mathbf{R}$, because $\varphi_s \to -\infty$ and $\psi_s \to 0$ in $(\Pi^\forall, \infty)$. Consequently, since φ_s and ψ_s belong to an ordered differential algebra on $(\Pi^\forall, \infty)$, we get that $\varphi_s' + C\psi_s' \prec 0$ in $(\Pi^\forall, \infty)$ for every $C \in \mathbf{R}$. The relation $(*)$ now follows from the inclusion lemma in subsection F of §5.4. ►

The analogue of Proposition 3 is valid for the germ $\rho = A^{-n}g$ when $g \in G_{\text{slow}}^{n-1^-}$; the proof is completely analogous. Precisely this fact is needed to prove that germs of class $\mathscr{D}_{\text{slow}}^n \circ \mathscr{L}^n$ are admissible. Moreover, in §5.7 we use

COROLLARY TO PROPOSITION 3. *Suppose that*

$$g \in G_{\text{slow}}^{n^-}, \quad \mu_{n+1}(g) = 1, \quad A^{-n}g = \text{id} + \varphi,$$
$$\rho_\lambda = \text{id} + \lambda\varphi, \quad \lambda > 0.$$

Then the germ of the half-strip $(\rho_\lambda \Pi, \infty)$ *belongs to the germ of the logarithm of any standard domain of class* $n+1$.

PROOF. The proof of Proposition 3 uses only the following requirements on the correction φ of the germ $A^{-n}g$: $\varphi \in \mathscr{L}_*$, $\operatorname{Re}\varphi \to -\infty$, $\varphi' \to 0$ in $(\Pi^\forall, \infty)$. These properties are preserved, of course, under multiplication of φ by a positive constant. ▶

E. An estimate of the real parts of germs of class $(\mathscr{D}^{n+1}_{\text{slow}})^{-1}$ in standard domains of class $n+1$.

PROPOSITION 4. *Suppose that* $g \in G^{n-}_{\text{slow}}$, $\rho = A^{-(n+1)}g$. *Then there exists a standard domain* Ω *of class* $n+1$ *such that in this domain* $\operatorname{Re}\rho \prec \varepsilon\xi$ *for any* $\varepsilon > 0$.

PROOF. We consider two cases:

$$\mu_{n+1}(g) = 1 \quad \text{and} \quad \mu_{n+1}(g) < 1;$$

the second is considerably simpler.

The proof of Proposition 4 is parallel in large part to that of Proposition 9 in §5.4 J.

CASE 1: $\mu_{n+1}(g) = 1$. In this case the domain Ω will be found in the form

$$\Omega \in A^{-1}(\mathrm{id}+\psi)(\mathbf{C}^+ \backslash K), \qquad \psi \in \Psi_{n+1}.$$

We begin with simplifications, parallel to those carried out in §5.4 J, of the inequality to be proved. The inequality

$$\operatorname{Re} A^{-(n+1)}g \prec \varepsilon \operatorname{Re}\zeta \quad \text{in } \Omega$$

in the proposition is equivalent to the inequality

$$\operatorname{Re}\exp(A^{-n}g) \circ (\mathrm{id}+\psi) \prec \varepsilon \operatorname{Re}\exp(\mathrm{id}+\psi)$$

in (Π, ∞). Let:

$$A^{-n}g = \mathrm{id}+\varphi, \quad (\mathrm{id}+\varphi)\circ(\mathrm{id}+\psi) = \mathrm{id}+\tilde\varphi,$$
$$\tilde\varphi = \varphi\circ(\mathrm{id}+\psi) + \psi.$$

By Proposition 1 of B, $\operatorname{Re}\varphi \to -\infty$ and $\varphi' \to 0$ in $(\Pi^\forall, \infty)$. By the remark in §5.4 E, $\psi \to 0$ and $\psi' \to 0$ in $(\Pi^\forall, \infty)$. The last inequality is equivalent to the inequality

$$(\exp\operatorname{Re}\tilde\varphi)\cos(\eta + \operatorname{Im}\tilde\varphi) \prec \varepsilon(\exp\operatorname{Re}\psi)\cos(\eta+\operatorname{Im}\psi).$$

Arguing as in §5.4 J, we reduce the inequality obtained to the following two inequalities:

$$\operatorname{Re}(\exp\tilde\varphi) \preccurlyeq \varepsilon\operatorname{Re}\exp\psi \quad \text{in } (\Pi, \infty);$$
$$\operatorname{Re}(\exp\tilde\varphi)' \succcurlyeq \varepsilon\operatorname{Re}(\exp\psi)' \quad \text{in } (\Pi, \infty). \tag{$*$}$$

The first inequality is trivial: the left-hand side tends to zero, while the right-hand side tends to ε in (Π, ∞). We prove the second.

We note the important fact that $\exp\varphi \in \Psi_{n+1}$. Indeed, all the requirements on φ imposed in the definition of the set Ψ_{n+1} are satisfied: φ is real (on $(\mathbf{R}^+, \infty)$) and by Proposition 1 of subsection B,

$$\varphi \in \mathscr{L}(\mathscr{F}^{n-1}_{\mathbf{R},\mathrm{id}}, \widetilde{\mathscr{F}}^n_+) \circ \ln^{[n]},$$
$$\operatorname{Re}\varphi \to -\infty, \ \varphi' \to 0 \quad \text{in } (\Pi, \infty).$$

We now take the desired germ in the form $\psi = \exp\varphi/2 \in \Psi_{n+1}$ and prove the inequality $(*)$ with this germ. An equivalent inequality is

$$\operatorname{Re}[(\exp\tilde{\varphi})\tilde{\varphi}'] \succcurlyeq \frac{\varepsilon}{2}\operatorname{Re}\left[(\exp\psi)\left(\exp\frac{\varphi}{2}\right)\varphi'\right]. \tag{$**$}$$

Let us prove the following two assertions:

1°. $\tilde{\varphi} = \varphi\cdot(1+o(1))$ and $\tilde{\varphi}' = \varphi'\cdot(1+o(1))$ in (Π, ∞);

2°. the arguments of both the functions whose real parts are taken in the previous inequality tend to π in (Π, ∞).

These assertions are proved below. We derive the inequality $(**)$ from them. In view of assertion 2° it suffices to prove for any $\varepsilon > 0$ the opposite inequality for the moduli of the functions whose real parts are being compared. We prove the equivalent relation

$$|(\exp\tilde{\varphi})\tilde{\varphi}'/(\exp\psi)\psi'| \to 0 \quad \text{in } (\Pi, \infty).$$

Using assertion 1° and the convergence $\psi \to 0$ in (Π, ∞), we get that

$$\begin{aligned}(\exp\tilde{\varphi})\tilde{\varphi}'/(\exp\psi)\psi' &= 2\exp\left(\varphi(1+o(1)) - \frac{\varphi}{2}\right)\cdot\varphi'(1+o(1))/\varphi' \\ &= 2\exp\left(\varphi\left(\frac{1}{2}+o(1)\right)\right) \to 0,\end{aligned}$$

because $\operatorname{Re}\varphi \to -\infty$ in (Π, ∞). This proves the inequality $(**)$, and with it the whole proposition in Case 1, modulo assertions 1° and 2°.

Let us prove assertion 1°. The function $\operatorname{Re}\varphi$ is harmonic and of constant sign in $(\Pi^\varepsilon, \infty)$, and $\psi \to 0$ and $\varphi \to -\infty$ in (Π, ∞). Consequently, $\operatorname{Re}\psi = o(1)\operatorname{Re}\varphi$. Further, by assertion 1 in §5.4 D,

$$\operatorname{Re}\varphi \circ (\mathrm{id}+\psi) = (\operatorname{Re}\varphi)(1+o(1)).$$

Finally, $\arg\varphi \to \pi$ in (Π, ∞); this follows from Proposition 1 of §5.5 B. Therefore, $\varphi = (\operatorname{Re}\varphi)(1+o(1))$. This proves the first equality in assertion 1°.

We prove the second. We have that $\tilde{\varphi}' = \varphi' \circ (\mathrm{id}+\psi)\cdot(1+\psi') + \psi'$. By Proposition 1 in subsection B,

$$\varphi = \varphi_s + \varphi_r, \qquad \varphi_s \in \mathscr{L}(\mathscr{F}^{n-1}_{\mathbf{R},\mathrm{id}}) \circ \ln^{[n]}, \ \varphi_r \in \mathscr{L}(\widetilde{\mathscr{F}}^n) \circ \ln^{[n]}, \ \varphi_s \neq 0.$$

Note that $\varphi_s' \neq 0$. Otherwise the equality $\varphi = C + o(1)$ would hold, and that would contradict the convergence $\operatorname{Re}\varphi \to -\infty$. Further, φ_s belongs to the ordered differential algebra $\mathscr{L}(\mathscr{F}_{\mathbf{R},\mathrm{id}}^{n-1})\circ\ln^{[n]}$ on $(\Pi^{\forall}, \infty)$, and $\operatorname{Re}\varphi_s \to -\infty$. Therefore, $\varphi_s' \prec 0$ in $(\Pi^{\forall}, \infty)$. Next, $\varphi_r' = o(\varphi_s')$, by ${}^{\mathrm{C}}\mathrm{OL2}_n$. From this, $\varphi' = \varphi_s' \cdot (1 + o(1))$ in $(\Pi^{\forall}, \infty)$. Since $\psi \to 0$ in $(\Pi^{\forall}, \infty)$, we get from assertion 1 in §5.4 D that $\varphi' \circ (\mathrm{id} + \psi) = \varphi' \cdot (1 + o(1))$ in (Π, ∞). Further, $\psi' = \frac{\varphi'}{2}\exp\varphi/2$; the second factor tends to zero in (Π, ∞). The second equality of assertion 1° follows from this.

Let us prove assertion 2°. We have that $\tilde{\varphi} = \varphi \circ (\mathrm{id} + \psi) + \psi$. By Proposition 1 in subsection B, $\operatorname{Im}\varphi \to 0$ in $(\Pi^{\varepsilon}, \infty)$. Further, $\psi \to 0$. Consequently, $\arg\exp\tilde{\varphi} \to 0$. Then $\tilde{\varphi}' = \varphi' \cdot (1 + o(1))$. Consequently, $\arg\tilde{\varphi}' = \arg\varphi' + o(1)$. But $\arg\varphi \to \pi$ and $\arg\varphi' \to \pi$ in (Π, ∞), by Proposition 1. Consequently, $\arg(\exp\tilde{\varphi})\tilde{\varphi}' \to \pi$ in (Π, ∞). Similarly,

$$\arg[(\exp\psi)(\exp\varphi/2)\varphi'] = \operatorname{Im}\psi + \operatorname{Im}\varphi/2 + \arg\varphi' = \pi + o(1).$$

This proves assertion 2°, and with it the whole of Proposition 4 in Case 1.

REMARK. The domain Ω in Proposition 4 was chosen to be of type 2. I was not able to choose a standard domain of class $n + 1$ and type 1 in which the inequality of the proposition would be satisfied, although I believe that this can be done, and I have a proof of it for the cases $n = 1$ and 2. Proposition 4 was the only place in the proof where standard domains of type 2 had to be introduced.

To prove Proposition 4 it remains to consider one more case.

CASE 2: $\mu = \mu_{n+1}(g) < 1$. In this case there exists a disk K such that the inequality of the proposition holds in the domain $\mathbf{C}^+ \setminus K$. As above, let $\Pi = \ln(\mathbf{C}^+ \setminus K)$. The inequality of the proposition is equivalent to the following:

$$(\exp\circ\operatorname{Re}A^{-n}g)\cos(\operatorname{Im}A^{-n}g) \prec \varepsilon\exp\xi \qquad (***)$$

in (Π, ∞). Since $\mu = \lim_{(\Pi,\infty)}(A^{-n}g)' < 1$ and $A^{-n}g$ is real on $\mathbf{R}^+$, we get that the function $|\operatorname{Im}A^{-n}g|$ on (Π, ∞) is bounded away from $\pi/2$, and hence the cosine $\cos(\operatorname{Im}A^{-n}g)$ is bounded away from zero. On the other hand, $\operatorname{Re}A^{-n}g - \xi \to -\infty$, because $|A^{-n}g| = |\zeta|(\mu + o(1))$ and $\arg(A^{-n}g - \zeta) \to \pi$ in (Π, ∞). This proves the inequality $(***)$, and with it Proposition 4.

The proof of Lemma 5.5_{n+1} is thereby completed.

§5.6. Admissible germs of class $\mathscr{D}_0^{n+1}$

In this section we prove

LEMMA 5.4_{n+1}. a. *Germs of class $\mathscr{D}_0^{n+1}$ are admissible of class $n+1$.*

b. *Germs of the form $A^{-(n+1)}g\circ\exp$, where $g \in G_{\mathrm{rap}}^n \cup G_{\mathrm{slow}}^{n^+}$ are nonessential.*

c. *The product of two germs of class $\mathscr{D}_0^{n+1}$ is a nonessential germ.*

PROOF. a. Recall that $\sigma \in \mathscr{D}_0^{n+1}$ if and only if there exists a germ $g \in G_{\text{slow}}^{n-}$ such that $\sigma = A^{-(n+1)}g \circ \exp$. The following two expressions will be used below for the inverse germ $\rho = \sigma^{-1}$:

$$\rho = \ln \circ A^{-(n+1)}g^{-1}; \qquad \rho = A^{-n}g^{-1} \circ \ln .$$

The definition of admissible germs of class $n+1$ was given in §§1.5 and 1.7 and recalled at the beginning of §5.4. We verify in turn the requirement of this definition for the germ σ.

1°. We prove that the germ $\rho = \sigma^{-1}$ admits a biholomorphic extension to $(\mathbf{C}^+, \infty)$. This is equivalent to there being a biholomorphic extension of the germ $A^{-n}g^{-1}$ to (Π, ∞), where $\Pi = \{\xi \geq a, |\eta| \leq \pi/2\}$, $a > 0$. But it was proved in Proposition 6 in §5.2 that the germ $A^{-(n+1)}g^{-1}$ has a biholomorphic extension to $(\mathbf{C}^+, \infty)$. This is equivalent to there being a biholomorphic extension of $A^{-n}g^{-1}$ to (Π, ∞). The first part of requirement 1° is verified.

We verify the second part of requirement 1° in the definition of admissible germs. Let Ω be an arbitrary standard domain of class $n+1$. We prove that there exists a standard domain $\tilde{\Omega}$ of class $n+1$ such that $\rho\tilde{\Omega} \subset \Omega$.

Take $\tilde{\Omega} = \mathbf{C}^+ \backslash K$ for a sufficiently large disk K. Then $\rho\tilde{\Omega} = A^{-n}g^{-1}\Pi$, where $\Pi = \ln(\mathbf{C}^+ \backslash K)$. In view of assertion a of Lemma 5.3_{n+1}, the germ $A^{-n}g^{-1}$ carries the germ (Π, ∞) into the germ of any sector, which, in turn, belongs to the germ of any standard domain.

This proves that

$$\rho(\mathbf{C}^+, \infty) \subset \Omega.$$

2°. Let us prove that the derivative ρ' is bounded. We have that

$$\rho' = (A^{-n}g^{-1})' \circ \ln \cdot \zeta^{-1}.$$

It follows from Proposition 6 in §5.2C that in $(\mathbf{C}^+, \infty)$

$$|(A^{-n}g^{-1})'| \circ \ln \prec \exp |\ln \zeta|^{\varepsilon} \prec \exp \varepsilon \ln |\zeta| = |\zeta|^{\varepsilon}$$

for arbitrary $\varepsilon > 0$. Consequently, in $(\mathbf{C}^+, \infty)$

$$|\rho'| \prec |\zeta|^{-1+\varepsilon}.$$

3°. We get an upper estimate for $\operatorname{Re} \rho$. It follows from assertion $*$ in §5.2 C that $\arg A^{-n}g^{-1} \to 0$ in (Π, ∞). Consequently, the functions $\operatorname{Re} \rho$ and $|\rho|$ are comparable in $(\mathbf{C}^+, \infty)$: their quotient tends to 1. It follows from Proposition 6 in §5.2 that in $(\mathbf{C}^+, \infty)$

$$|\rho| = |A^{-n}g^{-1} \circ \ln| \prec \exp |\ln \zeta|^{\varepsilon} \prec |\zeta|^{\varepsilon}$$

for every $\varepsilon > 0$. In a quadratic standard domain, where $|\zeta| \prec (\operatorname{Re} \zeta)^3$, this implies that

$$|\rho| \prec (\operatorname{Re}\zeta)^{3\varepsilon} \prec \mu\xi$$

for any $\mu > 0$.

4°. We get a lower estimate for $\exp \operatorname{Re}\rho$. Writing ρ in the form

$$\rho = \ln \circ A^{-(n+1)} g^{-1},$$

we get that

$$\exp \operatorname{Re}\rho = |A^{-(n+1)} g^{-1}|.$$

Let $A^{-n} g^{-1} = \mathrm{id} + \varphi_n$; $\varphi_n \succ 0$, because $g^{-1} \succ \mathrm{id}$. According to Corollary 4 in §5.2 G, $\operatorname{Re}\varphi_n \to +\infty$ in (Π, ∞). Consequently,

$$|A^{-(n+1)} g^{-1}| = |\zeta \exp \varphi_n \circ \ln| \succ \mu|\zeta|$$

in $(\mathbf{C}^+, \infty)$ for any $\mu > 0$.

This verifies all the requirements in the definition of admissible germs of class $n+1$ and completes the proof of assertion a in Lemma 5.4_{n+1}. Let us prove assertion b. By Definition 3 in §1.5, a germ σ is said to be nonessential of class $n+1$ if it admits a biholomorphic extension to some standard half-strip Π_* and there exists a standard domain of class $n+1$ belonging to $\sigma\Pi_*$.

Here Π_* is a half-strip of the form

$$\Pi_* = \Phi\Pi, \qquad \Phi = \zeta + \zeta^{-2}, \Pi = \{\zeta | \xi \geq a \geq 0, |\eta| < \tfrac{\pi}{2}\}.$$

We note that $h_0 \stackrel{\text{def}}{=} A^{n+1}\Phi \in G^n_{\text{rap}}$. To prove assertion b it suffices to prove that if $\sigma = A^{-(n+1)} g \circ \exp$, where $g \in G^n_{\text{rap}} \cup G^{n+}_{\text{slow}}$, then there exists a standard domain Ω of class $n+1$ such that

$$\rho(\Omega, \infty) \subset (\Pi_*, \infty), \qquad \rho = \sigma^{-1}.$$

The domain Ω will be found in the form

$$\Omega = A^{-(n+1)} f(\mathbf{C}^+ \backslash K), \qquad f \in G^n_{\text{rap}},\ K \text{ a disk}.$$

For such a domain

$$\rho(\Omega, \infty) = \ln \circ A^{-(n+1)} (g^{-1} \circ f)(\mathbf{C}^+, \infty) = A^{-n} \tilde{g}(\Pi_*, \infty),$$

where $\tilde{g} = g^{-1} \circ f \circ h_0^{-1}$. We consider two cases: $g \in G^n_{\text{rap}}$ and $g \in G^{n+}_{\text{slow}}$.

CASE 1: $g \in G^n_{\text{rap}}$. In this case we take $f = gh_0$. We get that $\tilde{g} = \mathrm{id}$. Then $\rho(\Omega, \infty) = (\Pi_*, \infty)$, which is what was to be proved.

CASE 2: $g \in G^{n+}_{\text{slow}}$. In this case it is possible to take the germ $f \in G^n_{\text{rap}}$ to be arbitrary, for example, $f = h_0$. Then $\tilde{g} = g^{-1} \in G^{n-}_{\text{slow}}$. It follows from

Lemma 5.5_{n+1}a that

$$\rho(\Omega, \infty) = (A^{-n} g^{-1})(\Pi, \infty) \subset (\Pi_*, \infty),$$

since $g^{-1} \in G^{n-}_{\text{slow}}$ this time. Assertion b of the lemma is proved.

We prove assertion c. Let $\sigma_1, \sigma_2 \in \mathscr{D}_0^{n+1}$. By Definition 16a in §1.6, there exist germs $g_1, g_2 \in G^{n^-}_{\text{slow}}$, such that

$$\sigma_1 = A^{-(n+1)} g_1 \circ \exp, \qquad \sigma_2 = A^{-(n+1)} g_2 \circ \exp.$$

Let $\sigma = \sigma_1 \circ \sigma_2$, $\rho = \sigma^{-1}$. Our goal is to prove that $\rho(\mathbf{C}^+, \infty) \subset (\Pi_*, \infty)$. The germ ρ has the form

$$\begin{aligned}\rho &= A^{-n} g_2^{-1} \circ \ln \circ A^{-n} g_1^{-1} \circ \ln \\ &= \ln \circ A^{-(n+1)} g_2^{-1} \circ A^{-n} g_1^{-1} \circ \ln.\end{aligned}$$

We transform the third factor in order then to use assertion a in Lemma 5.3_{n+1}. By the multiplicative decomposition theorem MDT_n,

$$g_1^{-1} = g_0 j h, \qquad g_0 \in G^{n-1}, j \in J^{n-1}, h \in H^n. \tag{$*$}$$

From Corollary 3 in §5.2 G,

$$(A^{-n} j h)' \to 1 \quad \text{in } (\Pi, \infty).$$

Since the composition jh is real on $(\mathbf{R}^+, \infty)$, this implies that

$$A^{-n} j h(\Pi, \infty) \subset (\Pi^\varepsilon, \infty)$$

for each $\varepsilon > 0$. Further, ([15]) $A g_0 \in G^n$. Let $g_3 = g_2^{-1} A g_0$. Then

$$\rho = \ln \circ A^{-(n+1)} g_3 \circ A^{-n}(jh) \circ \ln.$$

The germ ρ acts as follows:

$$\begin{aligned}\rho\colon (\mathbf{C}^+, \infty) &\xrightarrow{\ln} (\Pi, \infty) \xrightarrow{A^{-n}(jh)} (\Pi^\varepsilon, \infty) \xrightarrow{A^{-(n+1)} g_3} \\ &(S^\forall, \infty) \xrightarrow{\ln} (\Pi^\forall, \infty).\end{aligned}$$

The action of the germ $A^{-(n+1)} g_3$ was investigated in assertion a of Lemma 5.3_{n+1}. This proves assertion c, and Lemma 5.4_{n+1} is proved. ►

([15]) See §4.5 C.

§5.7. Proof of Lemma 5.7_{n+1}

A. Formulation.

LEMMA 5.7_{n+1} a. *Suppose that* Ω *is an arbitrary standard domain of class* $n+1$, *and* $\sigma_1, \sigma_2 \in \mathscr{D}_0^{n+1}$ *or* $\sigma_1, \sigma_2 \in \mathscr{D}_1^n$, *with* $\sigma_1 \succ \sigma_2$. *Then:*

$$(\sigma_1 \Pi_{\text{main}} \cap \Omega, \infty) \supset (\sigma_2 \Pi_{\text{main}} \cap \Omega, \infty)$$

if the germs σ_1 *and* σ_2 *are not weakly equivalent; the germs* σ_1 *and* σ_2 *can be renumbered so that for arbitrary* ε *and* δ *with* $0 < \delta < \varepsilon < 1$

$$(\sigma_1 \Pi_{\text{main}}^{(\delta)} \cap \Omega, \infty) \subset (\sigma_2 \Pi_{\text{main}}^{(\varepsilon)} \cap \Omega, \infty)$$

if σ_1 *and* σ_2 *are weakly equivalent.*

b. $\rho_\lambda \Pi$ *is a half-strip of type* W *for arbitrary* $\lambda > 0$.

c. *The germ of this half-strip at infinity belongs to the germ at infinity of the logarithm of any standard domain of class* $n+1$.

d. *If* γ_λ *is a boundary function of the half-strip* $\rho_\lambda \Pi$, $\lambda > 0$, $\mu > 0$, *then*

$$\gamma_\lambda \searrow 0, \quad \gamma_\lambda / \gamma_\mu \to \lambda / \mu, \quad |\xi \gamma_\lambda' / \gamma_\lambda| \prec 2$$

and

$$|\xi \gamma_\lambda'' / \gamma_\lambda'| \prec 3 \text{ on } (\mathbf{R}^+, \infty)$$

The formulation uses the definitions in §§3.2 and 3.9. Namely, σ_1 and σ_2 are weakly equivalent if the correction of the quotient $\sigma_1^{-1} \circ \sigma_2$ is bounded on $(\mathbf{R}^+, \infty)$. The definition of the germ ρ_λ is recalled below in C.

B. Connection between the geometric and analytic ordering of special admissible germs.

In this subsection Lemma 5.7_{n+1} a is proved. We consider the same three cases as in the table in §3.10, setting $\sigma = \sigma_1$ and $\sigma_{\text{next}} = \sigma_2$ in it. Let $\rho = \sigma_1^{-1} \circ \sigma_2$. For the case when σ_1 and σ_2 are not weakly equivalent we prove that ([16])

$$(\rho \Pi, \infty) \subset (\Pi, \infty).$$

We begin with the simplest case ② of the table:

$$\sigma_1 = A^{-n} g_1 \circ \exp, \qquad \sigma_2 = A^{-n} g_2 \circ h,$$
$$g_1, g_2 \in G^{n-1}, \; h \in \mathscr{L}^n.$$

Then

$$\rho = \ln \circ A^{-n} g \circ h, \qquad g = g_1^{-1} \circ g_2 \in G^{n-1}.$$

The germ ρ acts according to the scheme

$$\rho \colon (\Pi, \infty) \xrightarrow{h} (\Pi^{\varepsilon}, \infty) \xrightarrow{A^{-n} g} (S^{\forall}, \infty) \xrightarrow{\ln} (\Pi^{\forall}, \infty).$$

([16]) The inclusion $(\rho \Pi_{\text{main}} \cap \Omega, \infty) \subset (\Pi_{\text{main}} \cap \Omega, \infty)$ follows easily from this, the definition of the domain Π_{main} (§1.6 A), the elementary relation $A^n \Phi \subset G_{\text{rap}}^n$, $\Phi = \zeta + \zeta^{-2}$, and the fact that all the arguments of this subsection go for $\Pi^{\forall}$ instead of Π.

Here Π^ε is the ε-neighborhood of the right half-strip Π. The first inclusion on the left-hand side follows from Lemma 5.6_{n+1}: $h = (1+o(1))\zeta$, the second follows from Lemma 5.3_{n+1}, and the third is obvious. This proves the needed inclusion in case ②.

The cases ① and ③ in the table can be reduced to each other. Let us begin with case ①; case ③ is reduced to it at the end of the subsection. It will be assumed that $\sigma_1, \sigma_2 \in \mathscr{D}_0^{n+1}$; the case of germs of class $\mathscr{D}_0^n$ is analyzed similarly (just replace n by $n-1$).

Thus, let

$$\sigma_1 = A^{-(n+1)} g_1 \circ \exp, \qquad \sigma_2 = A^{-(n+1)} g_2 \circ \exp,$$
$$g_1, g_2 \in G^n, \sigma_1 \succ \sigma_2.$$

Then

$$\rho = A^{-n} g, \qquad g = g_1^{-1} \circ g_2 \prec \mathrm{id}, \ g \in G^n.$$

We consider four cases.

CASE 1: $\mu_{n+1}(g) = \mu < 1$. In this case $\mu = \lim_{(\Pi, \infty)} (A^{-n} g)' < 1$, by the definition of a generalized exponent. Consequently, $(\rho\Pi, \infty) \subset ((1-\delta)\Pi, \infty)$ for arbitrary $\delta \in (0, 1-\mu)$.

In the following three cases $\mu_{n+1}(g) = 1$. In these cases, by Proposition 1 in §5.5 B,

$$A^{-n} g = \mathrm{id} + \varphi, \qquad \varphi \in \mathscr{L}_* \circ \ln^{[n]}, \ \mathrm{Re}\, \varphi \to -\infty, \ \varphi' \to 0 \text{ in } (\Pi, \infty).$$

Recall that

$$\mathscr{L}_* = \mathscr{L}_*(\mathscr{F}_{\mathbf{R}, \mathrm{id}}^{n-1}, \mathscr{F}_{++}^{n-1}, \mathscr{F}_+^n), \qquad \widetilde{\mathscr{F}}_+^n = \mathscr{F}_{++}^{n-1} \cup \mathscr{F}_+^n.$$

CASE 2: $g \in G_{\mathrm{slow}}^{n-1^-}$. In this case it suffices to prove that

$$\mathrm{sgn}\, \mathrm{Im}\, \varphi = -\mathrm{sgn}\, \eta,$$

or

$$\mathrm{Re}\, \varphi' \prec 0 \quad \text{in } (\Pi, \infty).$$

To prove the last relation we decompose φ into a sum

$$\varphi = \varphi_s + \varphi_r, \qquad \varphi_s \in \mathscr{L}(\mathscr{F}_{\mathbf{R}, \mathrm{id}}^{n-1}) \circ \ln^{[n]}, \quad \varphi_r \in \mathscr{L}(\widetilde{\mathscr{F}}_+^n) \circ \ln^{[n]}.$$

According to Corollary 1 in §5.2 B, $\varphi_r \to 0$ in (Π, ∞). Therefore, $\mathrm{Re}\, \varphi_s \to -\infty$ in (Π, ∞), because the germ $\mathrm{Re}\, \varphi$ has the same property by Proposition 1 in §5.5 B. From this, $\mathrm{Re}\, \varphi_s' \neq 0$. Consequently, by Lemma ${}^{\mathrm{C}}\mathrm{OL2}_n$,

$$\varphi_r' = o(\mathrm{Re}\, \varphi_s'), \qquad \mathrm{Re}\, \varphi' = (1 + o(1))\, \mathrm{Re}\, \varphi_s' \quad \text{in } (\Pi, \infty).$$

But the differential algebra $\mathscr{L}(\mathscr{F}_{\mathbf{R}, \mathrm{id}}^{n-1}) \circ \ln^{[n]}$ is ordered on (Π, ∞). Consequently, the condition $\mathrm{Re}\, \varphi_s \to -\infty$ implies that $\varphi_s \prec 0$ in (Π, ∞), by Proposition 3 in §5.3. Consequently, $\mathrm{Re}\, \varphi' \prec 0$. This proves Lemma 5.7_{n+1} a in Case 2.

In the next two cases $g \in G^n_{\mathrm{rap}}$. Then by Corollary 2 in §5.2 G,

$$A^{-n} g = \mathrm{id} + \lambda + \tilde{\varphi}, \qquad \lambda \in \mathbf{R}, \quad \tilde{\varphi} \to 0 \text{ in } (\Pi, \infty).$$

It follows from the definition in §3.2 C that the germs σ_1 and σ_2 are weakly equivalent. By Corollary 2 in §5.2 G,

$$\tilde{\varphi} \in \mathscr{L}_* \circ \ln^{[n]} \quad \text{for } k(g) = n-1,$$
$$\tilde{\varphi} \in \mathscr{L}_{\mathbf{R}}(\mathscr{F}^n_+) \circ \ln^{[n]} \quad \text{for } k(g) = n.$$

CASE 3: $g \in G^n_{\mathrm{rap}}$, $\varphi_s \neq 0$, where

$$\tilde{\varphi} = \varphi_s + \varphi_r, \qquad \varphi_s \in \mathscr{F}^{n-1}_{\mathbf{R},\mathrm{id}} \circ \ln^{[n]}, \ \varphi_r \in \mathscr{L}(\widetilde{\mathscr{F}}^n_+) \circ \ln^{[n]}.$$

In this case $\varphi_s' \neq 0$, because $\varphi_s \to 0$ and $\varphi_s \neq 0$. Then

$$\operatorname{Re} \tilde{\varphi}' = (1 + o(1)) \operatorname{Re} \varphi_s' \quad \text{in } (\Pi, \infty).$$

In this case the inclusions

$$(\rho\Pi, \infty) \subset (\Pi, \infty) \quad \text{or} \quad (\rho\Pi, \infty) \supset (\Pi, \infty)$$

follow from the inequalities

$$\operatorname{Re} \varphi_s' \prec 0 \quad \text{or} \quad \operatorname{Re} \varphi_s' \succ 0,$$

of which one is valid, because

$$\varphi_s' \in \mathscr{F}^{n-1}_{\mathbf{R},\mathrm{id}} \circ \ln^{[n]} = \mathscr{F}\mathscr{C}^{n-1}_{\mathbf{R}} \circ \ln, \qquad \varphi_s' \neq 0,$$

and the differential ring $\mathscr{F}\mathscr{C}^{n-1}_{\mathbf{R}} \circ \ln$ is ordered in (Π, ∞).

CASE 4: $g \in G^n_{\mathrm{rap}}$,

$$\tilde{\varphi} \in \mathscr{L}(\widetilde{\mathscr{F}}^n_+) \circ \ln^{[n]}.$$

We prove that in this case

$$(\mathrm{id} + \tilde{\varphi}) \circ \Phi_{1-\delta}(\Pi, \infty) \subset \Phi_{1-\varepsilon}(\Pi, \infty)$$

for arbitrary $0 < \delta < \varepsilon < 1$, where

$$\Phi_c(\zeta) = \zeta + c\zeta^{-2}.$$

As shown above,

$$g_c \overset{\mathrm{def}}{=} A^n \Phi_c \in H^{n-1} \cap \mathscr{M}_{\mathbf{R}} \subset G^{n-1}.$$

We prove an inclusion equivalent to the original one:

$$\tilde{\rho}(\Pi, \infty) \overset{\mathrm{def}}{=} \Phi^{-1}_{1-\varepsilon} \circ (\mathrm{id} + \tilde{\varphi}) \circ \Phi_{1-\delta}(\Pi, \infty) \subset (\Pi, \infty).$$

Indeed, by the conjugation lemma CL_n and the inequality $\varphi_{1-\varepsilon} \succ \mathrm{id}$,

$$\operatorname{Ad}(\Phi_{1-\varepsilon})(\mathrm{id} + \tilde{\varphi}) = \mathrm{id} + \tilde{\tilde{\varphi}} \in \mathrm{id} + \mathscr{L}(\widetilde{\mathscr{F}}^n_+).$$

Further,

$$\Phi_{1-\varepsilon}^{-1} \circ \Phi_{1-\delta} = \mathrm{id} + \varphi_s ,$$
$$\varphi_s \in \mathscr{F}_{\mathbf{R}, +\mathrm{id}}^{n-1} \circ \ln^{[n]} , \ \varphi_s \succ 0 , \ \varphi_s \to 0.$$

Consequently,

$$\operatorname{Re} \tilde{\rho}' - 1 = \operatorname{Re} \varphi_s'(1 + o(1)) \quad \text{in } (\Pi, \infty) ,$$

and $\varphi_s' \prec 0$ in (Π, ∞) by Proposition 3 in §5.2. This proves the needed inclusions and finishes the proof of Lemma 5.7_{n+1} a in case ①.

Finally, let us consider Case ③:

$$\sigma_1 = A^{-n} g_1 \circ h_1 , \qquad \sigma_2 = A^{-n} g_2 \circ h_2 ,$$
$$g_j \in G_{\mathrm{slow}}^{n-1^+} , \ h_j \in \mathscr{L}^n .$$

As above, we set $\rho = \sigma_1^{-1} \circ \sigma_2 \prec \mathrm{id}$. Then

$$\rho = h_1^{-1} \circ A^{-n} (g_1^{-1} \circ g_2) \circ h_2^{-1} .$$

By assertion $*$ in §2.7 C, for any $h \in \mathscr{L}^n$ there exists a $g_h \in G^n$ such that $A^n h = g_h$. Consequently,

$$\rho = A^{-n} g , \qquad g = g_{h_1}^{-1} \circ g_1^{-1} \circ g_2 \circ g_{h_2} \in G^n .$$

For such ρ Lemma 5.7_{n+1} a was proved in the treatment of case ①.

C. Half-strips of type W. We prove Lemma 5.7_{n+1}b: $\rho_\lambda \Pi$ is a half-strip of type W for any $\lambda \in (0, 1)$. Recall that

$$\rho = A^{-n} g = \mathrm{id} + \varphi , \qquad g \in G_{\mathrm{slow}}^{n^-} , \ \mu_{n+1}(g) = 1 , \ \rho_\lambda = \mathrm{id} + \lambda \varphi .$$

By Proposition 1 in §5.5B, $\varphi \in \mathscr{L}_* \circ \ln^{[n]}$ (see the beginning of subsection B, Case 1, for details). By Proposition 6 in §5.3, $\rho_\lambda \Pi$ is a half-strip of type W.

Lemma 5.7_{n+1} c is contained in §5.5 D as the Corollary to Proposition 3.

D. Boundary functions of special half-strips. Here we prove assertion d in Lemma 5.7_{n+1}. The convergence $\gamma_\lambda'/\gamma_\lambda \to 0$ on $(\mathbf{R}^+, \infty)$ follows immediately from the inequality

$$|\xi \gamma_\lambda'/\gamma_\lambda| \prec 2 \quad \text{on } (\mathbf{R}^+, \infty) , \tag{$*$}$$

which also appears in the statement of Lemma 5.7_{n+1} d. We prove it; the remaining assertions of Lemma 5.7_{n+1} d turn out to be proved simultaneously.

It follows from the definition of the function γ_λ that

$$\gamma_\lambda(\xi + \lambda \operatorname{Re} \varphi(\xi + i\tfrac{\pi}{2})) = -\lambda \operatorname{Im} \varphi(\xi + i\tfrac{\pi}{2}). \tag{$**$}$$

For brevity let $f(\xi) = \operatorname{Im} \varphi(\xi + i\pi/2)$. Equation $(**)$ indicates a similarity between the functions γ_λ and $-\lambda f$. We prove the strengthened formula $(*)$ for f instead of γ_λ:

$$|\xi f^{(k)}/f^{(k-1)}| \prec k + 1 \quad \text{on } (\mathbf{R}^+, \infty) , \tag{$***$}$$

and then derive $(*)$ from this by using the similarity noted above. First let $k = 1$. Recall that $\operatorname{Re}\varphi \to -\infty$, $\varphi' \to 0$ in (Π, ∞), and $\varphi \in \mathscr{L}_* \circ \ln^{[n]}$. Consequently, by Proposition 5 in §5.3,

$$\varphi^{(m)} \to 0, \quad \varphi^{(2m-1)} \prec 0, \quad \varphi^{(2m)} \succ 0 \quad \text{in } (\Pi, \infty)$$

for any positive integer m. These inequalities and the Cauchy-Riemann equations give us that

$$f \prec 0, \quad f' \succ 0 \quad \text{on } (\mathbf{R}^+, \infty).$$

Therefore, $\xi f'/f \prec 0$, and it suffices to prove that

$$\xi f'/f \succ -2 \quad \text{on } (\mathbf{R}^+, \infty).$$

This inequality is equivalent to the inequality

$$\xi f' + 2f \prec 0 \quad \text{on } (\mathbf{R}^+, \infty).$$

And it, in turn, follows from the inequality

$$\operatorname{Im}(\zeta\varphi' + 2\varphi) - \eta \operatorname{Re}\varphi' \prec 0 \quad \text{in } (\Pi, \infty) \text{ for } \eta > 0.$$

Below we prove

PROPOSITION 1. *For any real* $\varphi \in \mathscr{L}_* \circ \ln^{[n]}$, $\varphi \to -\infty$, $\varphi' \to 0$ *in* $(\Pi^{\forall}, \infty)$ *and for* $\eta > 0$ *one has*

$$\operatorname{Im}\varphi \prec \eta \operatorname{Re}\varphi'$$

in $(\Pi^{\forall}, \infty)$.

The proof will be given for Π instead of $\Pi^{\forall}$; the arguments go for $\Pi^{\forall}$ word for word.

Proposition 1 implies that it suffices to prove the inequality

$$\operatorname{Im}(\zeta\varphi' + \varphi) \prec 0 \quad \text{in } (\Pi, \infty) \text{ for } \eta > 0,$$

which, in turn, follows from the inequality

$$\operatorname{Re}(\zeta\varphi' + \varphi)' \prec 0 \quad \text{in } (\Pi, \infty).$$

Let us prove it. To do this we use the fact that $\varphi \in \mathscr{L}_* \circ \ln^{[n]}$ and the condition that $\varphi \to -\infty$ on $(\mathbf{R}^+, \infty)$. As usual, we decompose φ into a sum

$$\varphi = \varphi_s + \varphi_r, \qquad \varphi_s \in \mathscr{L}(\mathscr{F}^{n-1}_{\mathbf{R},\mathrm{id}}) \circ \ln^{[n]}, \; \varphi_r \in \mathscr{L}(\widetilde{\mathscr{F}^n_+}) \circ \ln^{[n]}.$$

Then, since $\zeta \in \mathscr{F}^{n-1}_{\mathbf{R},\mathrm{id}} \circ \ln^{[n]}$, and since $\mathscr{L}(\widetilde{\mathscr{F}^n_+}) \circ \ln^{[n]}$ is an ideal in the differential algebra $\mathscr{L}(\mathscr{F}^{n-1}, \mathscr{F}^n_+) \circ \ln^{[n]}$, we get that

$$(\zeta\varphi' + \varphi)' = \tilde{\varphi}_s + \tilde{\varphi}_r, \qquad \tilde{\varphi}_s = (\zeta\varphi_s' + \varphi_s)', \quad \tilde{\varphi}_r \in \mathscr{L}(\widetilde{\mathscr{F}^n_+}) \circ \ln^{[n]}.$$

If $\tilde{\varphi}_s \neq 0$, then $\tilde{\varphi}_r = o(1)\tilde{\varphi}_s$ in (Π, ∞) in view of Lemma ${}^{\mathrm{C}}\mathrm{OL2}_n$. We prove that in this case $\tilde{\varphi}_s \neq 0$, and, what is more, $\tilde{\varphi}_s \prec 0$ in (Π, ∞).

Assume not. Since the differential algebra $\mathscr{L}(\mathscr{F}^{n-1}_{\mathbf{R},\mathrm{id}}) \circ \ln^{[n]}$ is ordered in (Π, ∞), we get from this that

$$\tilde{\varphi}_s \equiv \zeta\varphi_s'' + 2\varphi_s' \succcurlyeq 0 \quad \text{in } (\Pi, \infty).$$

This implies that on $(\mathbf{R}^+, \infty)$

$$\xi\varphi_s'' + 2\varphi_s' \succcurlyeq 0.$$

We reduce this inequality to a contradiction of the requirements

$$\varphi_s \to -\infty, \quad \varphi_s' \to 0, \quad \varphi_s' \prec 0 \quad \text{in } (\Pi, \infty),$$

which follow from the analogous requirements on φ, since $\varphi_r = \varphi - \varphi_s \to 0$ and $\varphi_r' = o(1)\varphi_s'$ in (Π, ∞).

It follows from the assumption made that

$$\ln|\varphi_s'| \preccurlyeq -2/\xi,$$

$$|\varphi_s'| \preccurlyeq C\xi^{-2}, \qquad |\varphi_s| \preccurlyeq C_1 - C\xi^{-1}$$

for some C_1, $C \in \mathbf{R}$. This contradicts the convergence $\varphi_s \to -\infty$ and proves the inequality $(***)$ for $k = 1$, modulo Proposition 1.

Let us derive from this the main inequality, $(*)$, in this subsection. It follows from the convergence $\varphi' \to 0$ in (Π, ∞) that

$$\operatorname{Re}\varphi(\xi + i\tfrac{\pi}{2}) = o(1)\xi.$$

Equation $(**)$ now implies that

$$\gamma_\lambda(\xi) = -\lambda f \circ ((1 + o(1))\xi).$$

Elementary arguments analogous to those in the proof of Proposition 2 in §3.3 enable us to deduce from $(***)$ that

$$f \circ ((1 + o(1))\xi) = (1 + o(1))f(\xi).$$

Therefore,

$$\gamma_\lambda(\xi) = -\lambda f(\xi)(1 + o(1)). \qquad ({}_{*}{}^{*}{}_{*})$$

The proof of the next result is completely analogous.

Proposition 2. *For any positive integer* k

$$f^{(k)} \circ ((1 + o(1))\xi) = (1 + o(1))f^{(k)}(\xi).$$

The proof repeats almost word-for-word the arguments of this subsection and is presented in subsection E with minimal comments. The equation $(**)$ implies that

$$\gamma_\lambda(\xi) = -\lambda f((1 + o(1))\xi).$$

Proposition 2 now gives us that

$$\gamma_\lambda^{(k)}(\xi) = -\lambda f^{(k)}(\xi)(1 + o(1)). \qquad ({}^{*}{}_{*}{}^{*})$$

All the assertions of Lemma 5.7_{n+1} d now follow from the formulas $({}_{*}{}^{*}{}_{*})$, $({}^{*}{}_{*}{}^{*})$ and $(***)$ and the inequality $f \prec 0$.

E. Proof of Proposition 1. Suppose that φ is a real holomorphic function in (Π, ∞), and for each positive integer m

$$\varphi^{(m)} \to 0, \quad \varphi^{(2m-1)} \prec 0, \quad \varphi^{(2m)} \succ 0 \quad \text{in } (\Pi, \infty).$$

These assumptions are fulfilled for φ in Proposition 1 by the Remark in §5.3B, Proposition 3 in §5.3C and Corollary 2 in §5.2B. It is required to prove that $\operatorname{Im}\varphi \prec \eta \operatorname{Re}\varphi'$ in (Π, ∞) for $\eta > 0$. We prove a more general inequality that is used in the proof of Proposition 2.

PROPOSITION 1 BIS. *Under the assumptions in the beginning of the subsection*

$$(-1)^k(\operatorname{Im}\varphi^{(k-1)} - \eta \operatorname{Re}\varphi^{(k)}) \succ 0 \quad \textit{in } (\Pi, \infty) \textit{ for } \eta > 0$$

for any positive integer k.

Proposition 1 is the particular case of Proposition 1 bis for $k = 1$. We have that

$$\operatorname{Im}\varphi^{(2m-2)}(\xi + i\eta) = \int_0^\eta \operatorname{Re}\varphi^{(2m-1)}(\xi + i\theta)\, d\theta,$$

$$\operatorname{Im}\varphi^{(2m-1)}(\xi + i\eta) = \int_0^\eta \operatorname{Re}\varphi^{(2m)}(\xi + i\theta)\, d\theta.$$

The integrands of both integrals are monotone. The germ of the first is increasing in (Π, ∞) for $\eta > 0$, and that of the second is decreasing:

$$\frac{\partial}{\partial\eta}\operatorname{Re}\varphi^{(2m-1)} \succ 0, \quad \frac{\partial}{\partial\eta}\operatorname{Re}\varphi^{(2m)} \prec 0 \quad \text{in } (\Pi, \infty) \text{ for } \eta > 0.$$

Indeed, by the Cauchy-Riemann equations,

$$\frac{\partial}{\partial\eta}\operatorname{Re}\varphi^{(2m-1)} = -\operatorname{Im}\varphi^{(2m)}, \qquad \frac{\partial}{\partial\eta}\operatorname{Re}\varphi^{(2m)} = -\operatorname{Im}\varphi^{(2m+1)}.$$

The inequalities $-\operatorname{Im}\varphi^{(2m)} \succ 0$ and $-\operatorname{Im}\varphi^{(2m+1)} \prec 0$ follow from the fact that φ is real and from the inequalities

$$\operatorname{Re}\varphi^{(2m+1)} \prec 0, \qquad \operatorname{Re}\varphi^{(2m+2)} \succ 0,$$

which appear in the condition of the proposition. Estimating the integrals in terms of the maximal and minimal values of the integrands and taking into account that $\eta > 0$, we get the inequalities in Proposition 1 bis. ▶

F. Proof of Proposition 2. For an arbitrary positive integer k we prove the estimate $(***)$:

$$|\xi f^{(k)}/f^{(k-1)}| \prec k + 1 \text{ on } (\mathbf{R}^+, \infty), \quad \text{where } f(\xi) = \operatorname{Im}\varphi(\xi + i\tfrac{\pi}{2}),$$

under the assumptions at the beginning of subsection E for $\varphi \in \mathscr{L}_* \circ \ln^{[n]}$. It suffices to prove (observing that $f^{(k)}/f^{(k-1)} \prec 0$ and $f^{(k-1)}(-1)^k \succ 0$) that $(-1)^k(\xi f^{(k)} + (k+1)f^{(k-1)}) \succ 0$ on $(\mathbf{R}^+, \infty)$. By the definition of the function f at the beginning of subsection D, the left-hand side is equal to the restriction of the function $\operatorname{Im}(\zeta\varphi^{(k)} + (k+1)\varphi^{(k-1)}) - \eta \operatorname{Re}\varphi^{(k)}$ to $(\{\eta = \frac{\pi}{2}\}, \infty)$. Thus, it suffices to prove that $(-1)^k \operatorname{Im}(\zeta\varphi^{(k)} + (k+1)\varphi^{(k-1)}) - \eta \operatorname{Re}\varphi^{(k)} \succ 0$ in (Π, ∞) for $\eta > 0$. From Proposition 1 bis, it suffices to prove the inequality $(-1)^k \operatorname{Im}(\zeta\varphi^{(k)} + k\varphi^{(k-1)}) \succ 0$ in (Π, ∞) for $\eta > 0$, or

$$(-1)^k \operatorname{Re}(\zeta\varphi^{(k+1)} + (k+1)\varphi^{(k)}) \succ 0 \quad \text{in } (\Pi, \infty).$$

Let the decomposition $\varphi = \varphi_s + \varphi_r$ be the same as in subsection D. Then

$$\zeta\varphi^{(k+1)} + (k+1)\varphi^{(k)} = \tilde{\tilde{\varphi}}_s + \tilde{\tilde{\varphi}}_r,$$
$$\tilde{\tilde{\varphi}}_s = \zeta\varphi_s^{(k+1)} + (k+1)\varphi_s^{(k)}, \qquad \tilde{\tilde{\varphi}}_r \in \mathscr{L}(\widetilde{\mathscr{F}}_+^n) \circ \ln^{[n]}.$$

If $\tilde{\tilde{\varphi}}_s \neq 0$, then $\tilde{\tilde{\varphi}}_r = o(1)\tilde{\tilde{\varphi}}_s$ in (Π, ∞). We prove that $\tilde{\tilde{\varphi}}_s \succ 0$ in (Π, ∞). Assume not. Let $\tilde{\tilde{\varphi}}_s \preccurlyeq 0$ in (Π, ∞). Then on $(\mathbf{R}^+, \infty)$

$$(-1)^k(\xi\varphi_s^{(k+1)} + (k+1)\varphi_s^{(k)}) \preccurlyeq 0.$$

Consequently, $\varphi_s^{(k)} \prec C\xi^{-(k+1)}$. It follows from this and the condition $\varphi_s' \to 0$ on $(\mathbf{R}^+, \infty)$ that the germ φ_s is bounded on $(\mathbf{R}^+, \infty)$, and this contradicts the requirement that $\operatorname{Re}\varphi \to -\infty$ on $(\mathbf{R}^+, \infty)$. The contradiction obtained proves Proposition 2. The proof of Lemma 5.7_{n+1} d is complete together with this. ►►

§5.8. Convexity

The main result of the section is the

CONVEXITY LEMMA. *Suppose that σ is a slow germ of class $\mathscr{D}_0^m$ for $m = n$ or $m = n+1$, or $\sigma \in \mathscr{D}_{\text{slow}}^n \circ \mathscr{L}^n$, $\rho = \sigma^{-1}$, and Ω is a quadratic standard domain. Then the images under the action of σ of horizontal rays located in the germ of the domain $(\rho\Omega, \infty)$ are convex upward in the upper half-plane and convex downward in the lower half-plane.*

In subsections B–E all the arguments are in $(\rho\mathbf{C}^+, \infty)$ instead of $(\rho\Omega, \infty)$. This difference becomes essential only in subsection F. Subsection A is devoted to an auxiliary fact.

A. Comparison of a harmonic function of constant sign with its derivatives.

PROPOSITION 1. *Suppose that φ is a holomorphic function in the "expanding half-strip"*

$$\mathscr{U}: \xi \geq 1, |\eta| \leq \gamma(\xi), \gamma \to \infty \quad on\ (\mathbf{R}^+, \infty),$$

with $\operatorname{Re}\varphi \succ 0$ in $\mathscr{U}$, and let $\widetilde{\mathscr{U}} \subset \mathscr{U}$ be another half-strip such that $d \to \infty$ in $(\widetilde{\mathscr{U}}, \infty)$, where $d(\zeta)$ is the distance from a point ζ to the boundary of the domain $\mathscr{U}$. Then for an arbitrary positive integer k

$$\varphi^{(k)} / \operatorname{Re}\varphi \to 0 \quad in\ (\widetilde{\mathscr{U}}, \infty).$$

REMARK. The analogous statement is valid if $\mathscr{U}$ is replaced by $(\Pi^{\forall}, \infty)$.

This is an elementary and undoubtedly known fact in the theory of holomorphic (more precisely, harmonic) functions. It is used in this section with $k = 1, 2$.

PROOF. It follows from Harnack's inequality that

$$|\operatorname{Re}\varphi - \operatorname{Re}\varphi(\zeta)| \le \tfrac{3}{d}(\zeta)|\operatorname{Re}\varphi(\zeta)|$$

in the 1-neighborhood of each point $\zeta \in \widetilde{\mathscr{U}}$ for sufficiently large $\operatorname{Re}\zeta$. The proposition now follows from the Bernstein estimates on the derivatives of a harmonic function. In greater detail, it follows from the Bernstein estimates that there exists a $C > 0$ such that

$$|\operatorname{grad}\operatorname{Re}\varphi(\zeta)| \le \tfrac{C}{d}(\zeta)|\operatorname{Re}\varphi(\zeta)|.$$

By the Cauchy-Riemann equations, $|\operatorname{grad}\operatorname{Re}\varphi| = |\varphi'|$. Consequently, $|\varphi'| = o(\operatorname{Re}\varphi)$ in $(\widetilde{\mathscr{U}}, \infty)$. Similarly, for an arbitrary positive integer k there exists a $C_k > 0$ such that

$$|\mathscr{D}^{\alpha}\operatorname{Re}\varphi| \le C_k|\operatorname{Re}\varphi(\zeta)|/d(\zeta)$$

for $\alpha = (\alpha_1, \alpha_2) \in \mathbf{Z}_+^2$, $k = |\alpha|$. But the derivative $\varphi^{(k)}$ can be expressed linearly in terms of the derivatives $\mathscr{D}^{\alpha}\operatorname{Re}\varphi$ for $|\alpha| = k$. This yields the proposition. ▶

B. The convexity lemma for slow germs of class $\mathscr{D}_0^m$, $m = n$ or $m = n+1$. Let us prove the lemma for $m = n + 1$; the proof for $m = n$ is obtained by a word-for-word repetition with $n + 1$ replaced by n.

By definition, a slow germ σ in the set $\mathscr{D}_0^{n+1}$ has the form

$$\sigma = \exp\circ A^{-n}g, \qquad g \in G^{n-}_{\text{slow}}, \quad \mu_{n+1}(g) = 0.$$

To prove the lemma it suffices to verify that

$$\begin{aligned} &1^\circ.\quad \operatorname{sgn}(\sigma''/\sigma') = -\operatorname{sgn}\eta \quad \text{in } (\rho\mathbf{C}^+, \infty) \\ &2^\circ.\quad |\arg\sigma'| \prec \pi/2 \quad \text{in } (\rho\Omega, \infty). \end{aligned} \tag{$*$}$$

The second assertion is proved in subsection D.

Let us prove the first. We set $\tilde{\sigma} = A^{-n}g^{-1}$. Then $\rho = \tilde{\sigma}\circ\ln$. In the germ of the domain $(\rho\mathbf{C}^+, \infty)$ the function σ''/σ' takes the same values as the function $(\sigma''/\sigma')\circ\rho$ in $(\mathbf{C}^+, \infty)$, or $(\sigma''/\sigma')\circ\tilde{\sigma}$ in (Π, ∞). A simple computation gives us that

$$(\sigma''/\sigma')\circ\tilde{\sigma} = (\tilde{\sigma}' - \tilde{\sigma}'')/\tilde{\sigma}'^2.$$

From this,

$$\arg(\sigma''/\sigma')\circ\tilde{\sigma} = \operatorname{Im}\Psi,$$

where $\Psi = \ln(\tilde{\sigma}' - \tilde{\sigma}'') - 2\ln\tilde{\sigma}'$. We want to prove that

$$\operatorname{sgn}\operatorname{Im}\Psi = -\operatorname{sgn}\eta \quad \text{in } (\Pi, \infty).$$

To do this it suffices to prove that

$$\operatorname{Re}\Psi' \prec 0 \quad \text{in } (\Pi, \infty).$$

We have that

$$\Psi' = \frac{\tilde{\sigma}'' - \tilde{\sigma}'''}{\tilde{\sigma}' - \tilde{\sigma}''} - 2\frac{\tilde{\sigma}''}{\tilde{\sigma}'}.$$

In subsection C below we prove

PROPOSITION 2. *Let* $C > 1$ *be arbitrary; as above, let* $\tilde{\Pi} = A^{-n}C\Pi$. *Assume that* $g \in G^{n^+}_{\text{slow}}$, $\mu_{n+1}(g) = \infty$, *and* $\tilde{\sigma} = A^{-n}g$. *Then* $\arg\tilde{\sigma}' \to 0$ *and* $\arg\tilde{\sigma}'' \to 0$ *in* $(\tilde{\Pi}, \infty)$.

It follows from this proposition and Proposition 1 in subsection A that in (Π, ∞)

$$\tilde{\sigma}'' = o(\tilde{\sigma}'), \qquad \tilde{\sigma}''' = o(\tilde{\sigma}'').$$

Therefore,

$$\Psi' = -\frac{\tilde{\sigma}''}{\tilde{\sigma}'}(1 + o(1)).$$

Again applying Proposition 2, we get that

$$\arg\Psi' \to \pi \quad \text{in } (\Pi, \infty).$$

The equality $(*)$ follows from this.

REMARK. Proposition 2_{n+1} in §3.10 D follows immediately from Proposition 2. Indeed, by Lemma 5.6_{n+1}, $h' = 1 + o(1)$ and $\arg h' \to 0$ in $(\Pi^{\forall}, \infty)$ for any $h \in \mathscr{L}^n$. Therefore, it suffices to prove that $\arg\sigma' \to 0$ in $(\Pi^{\forall}, \infty)$ for any $\sigma = A^{-n}g$, $g \in G^{n-1^+}_{\text{slow}}$.

But $G^{n-1} \subset G^n$. By definition, the germ $A^{-n}g$ increases on $(\mathbf{R}^+, \infty)$ more rapidly than any linear germ. According to Lemma 5.3_{n+1} b, the limit

$$\mu_{n+1}(g) = \lim_{(\Pi^{\forall}, \infty)} (A^{-n}g)'$$

exists. Consequently, it is equal to infinity.

Proposition 2 now yields Proposition 2_{n+1} in §3.10 D.

C. Proof of Proposition 2.

PROOF. *Step* 1. Suppose first that $g = g_1 \in G^{n-1}$. The requirement $\mu_{n+1}(g) = \infty$ means that $g_1 \in G^{n-1^+}_{\text{slow}}$. We consider three cases: $k(g) = n-2$, $k(g) = n-3$; and $k(g) = n-4$. The case $k(g) = n-1$ is impossible, since $\mu_{n+1}(g) = 1$; see §5.2 E. The propositions in §5.2 referred to below are used for $n-1$ instead of n. The estimates given by these propositions are valid in the domain $\tilde{\Pi}' = A^{1-n}C\Pi$ for any $C > 0$. Let $A^{1-n}g_1 = \text{id} + \varphi_{n-1}$. Then

$$\begin{aligned} \tilde{\sigma}' &= \tilde{\sigma} \cdot (1 + \varphi'_{n-1}) \circ \ln \cdot \zeta^{-1}, \\ \tilde{\sigma}'' &= \tilde{\sigma} \cdot [\varphi'_{n-1}(1 + \varphi'_{n-1}) + \varphi''_{n-1}] \circ \ln \cdot \zeta^{-2}. \end{aligned} \qquad (**)$$

CASE 1: $k(g) = n - 2$. Then by Proposition 8 in §5.2 D,

$$\varphi_{n-1} \in \mathscr{L}_{\mathbf{R}}(\mathscr{F}_{\mathrm{id}}^{n-2}, \mathscr{F}_{+}^{n-1}) \circ \ln^{[n-1]};$$

$$\begin{aligned}
&\varphi_{n-1} = \varphi_s + \varphi_r, \qquad \varphi_s \in \mathscr{F}_{\mathbf{R},\mathrm{id}}^{n-2} \circ \ln^{[n-1]} = \mathscr{F}\mathscr{C}_{\mathbf{R}}^{n-2} \circ \ln, \\
&\varphi_s \neq 0, \; \varphi_r \in \mathscr{L}(\mathscr{F}_{++}^{n-2}, \mathscr{F}_{+}^{n-1}) \circ \ln^{[n-1]}, \\
&\varphi_{n-1}^{(l)} \to 0 \text{ in } (\tilde{\Pi}, \infty)
\end{aligned} \qquad \binom{*}{*}$$

for an arbitrary positive integer l. Further, $\operatorname{Re}\varphi_{n-1} \to +\infty$ in $(\tilde{\Pi}', \infty)$ according to the assumption that $(A^{-n}g_1)' \to \infty$. Consequently, by Proposition 1 in subsection A, $\varphi_{n-1}'' = o(\operatorname{Re}\varphi_{n-1}')$ in $(\tilde{\Pi}', \infty)$. Further,

$$\varphi_{n-1}' = \varphi_s'(1 + o(1)), \qquad \varphi_s' \neq 0,$$

because $\operatorname{Re}\varphi_{n-1} \to +\infty$ and $\varphi_r \to 0$ in $(\tilde{\Pi}', \infty)$. Consequently,

$$\arg\varphi_{n-1}' = \arg\varphi_s' + o(1) \to 0 \quad \text{in } (\tilde{\Pi}, \infty),$$

since $\arg\varphi_s' \to 0 \pmod{\pi}$ and $\varphi_s' \succ 0$ in $(\tilde{\Pi}', \infty)$. Consequently,

$$\arg(1 + \varphi_{n-1}') \to 0 \quad \text{in } (\tilde{\Pi}, \infty),$$

$$\begin{aligned}
&\arg[\varphi_{n-1}'(1 + \varphi_{n-1}') + \varphi_{n-1}''] \\
&\qquad = (\arg\varphi_{n-1}')(1 + o(1)) \to 0 \quad \text{in } (\tilde{\Pi}', \infty).
\end{aligned}$$

The arguments of all three factors in each of the formulas $(**)$ thus tend to zero. For the third factor this follows from Proposition 4 in §5.2, and for the second it was just proved (note that $(\tilde{\Pi}, \infty) \subset \exp(\tilde{\Pi}', \infty)$). For the first factor this is a consequence of Lemma 5.3_{n+1} a and the observation that $\arg A^{-n}g_1$ takes in $\tilde{\Pi} = (A^{-n}C)\Pi$ the same values that $\arg A^{-n}(g_1 \circ C)$ takes in Π, and $g_1 \circ C \in G^{n-1}$.

CASE 2: $k(g) = n - 3$. Then by Proposition 9 in §5.2, the decomposition $\binom{*}{*}$ holds, and $\arg\varphi_{n-1}' \to 0$ in $(\tilde{\Pi}', \infty)$. The only difference between this and Case 1 is that now $\lim_{(\tilde{\Pi}, \infty)} \varphi_{n-1}' = \lambda$, and λ is not necessarily equal to 0: the cases $\lambda \in \mathbf{R}^+$ and $\lambda = +\infty$ are also possible. Proposition 2 is proved as in Case 1 for $\lambda = 0$. For $\lambda \in \mathbf{R}^+$

$$\varphi_{n-1}'(1 + \varphi_{n-1}') + \varphi_{n-1}'' = \lambda + \lambda^2 + o(1) \quad \text{in } (\tilde{\Pi}', \infty).$$

For $\lambda = \infty$

$$\varphi_{n-1}'(1 + \varphi_{n-1}') + \varphi_{n-1}'' = \varphi_{n-1}'^2(1 + o(1)) \quad \text{in } (\tilde{\Pi}', \infty).$$

In both cases

$$\arg[\varphi_{n-1}'(1 + \varphi_{n-1}') + \varphi_{n-1}''] \to 0 \quad \text{in } (\tilde{\Pi}', \infty).$$

The subsequent argument is as in Case 1.

CASE 3: $k(g) \leq n-4$. In this case

$$\operatorname{Re} \varphi'_{n-1} \to +\infty, \quad \arg \varphi'_{n-1} \to 0 \quad \text{in } (\tilde{\Pi}', \infty),$$

by Proposition 11 in §5.2 F. The subsequent argument goes as in Case 2.

This completes the proof of Proposition 2 for $g = g_1 \in G^{n-1}_{\text{slow}}$.

Step 2. Suppose now that $g \in G^{n^+}_{\text{slow}}$. Then by the multiplicative decomposition theorem, MDT_n,

$$g = g_1 \circ g_2, \qquad g_1 \in G^{n-1}, \; g_2 = j \circ h, \; j \in J^{n-1}, \; h \in H^n.$$

According to Lemma SL4_n a, $k(g_2) \geq n-1$. Proposition 7 in §5.2 D gives us that

$$A^{1-n} g_2 \overset{\text{def}}{=} \text{id} + \psi, \qquad \psi \in \mathscr{L}_{\mathbf{R}}(\mathscr{F}^{n-1}_+, \mathscr{F}^n_+) \circ \ln^{[n-1]}.$$

Assume now that

$$A^{1-n} g_1 = \text{id} + \tilde{\varphi}_{n-1}, \qquad A^{1-n} g = \text{id} + \varphi_{n-1},$$
$$\varphi_{n-1} = \tilde{\varphi}_{n-1} \circ (\text{id} + \psi) + \psi.$$

REMARK. For $g_2 = \text{id}$ one has that $\tilde{\varphi}_{n-1} = \varphi_{n-1}$. Thus, Step 1 proves that in $(\tilde{\Pi}', \infty)$

$$\tilde{\varphi}_{n-1} = \varphi_s(1 + o(1)), \qquad \tilde{\varphi}'_{n-1} = \varphi'_s(1 + o(1)),$$
$$\varphi_s \in \mathscr{F}^{n-2}_{\mathbf{R}, \text{id}} \circ \ln^{[n-1]}$$

in Cases 1 and 2;

$$\operatorname{Re} \tilde{\varphi}'_{n-1} \to +\infty$$

in Case 3; and

$$\arg \tilde{\varphi}'_{n-1} \to 0$$

in all the three cases.

ASSERTION $*$. *In the germ of the domain* $(\ln \tilde{\Pi}, \infty)$

1°. $\varphi'_{n-1} = \tilde{\varphi}'_{n-1}(1 + o(1))$;
2°. $\varphi''_{n-1} = o(\operatorname{Re} \tilde{\varphi}'_{n-1})$;
3°. $\arg(1 + \varphi'_{n-1}) \to 0$;
4°. $\arg[\varphi'_{n-1}(1 + \varphi'_{n-1}) + \varphi''_{n-1}] \to 0$.

PROOF. In the case when $\psi = 0$ we have that $\varphi_{n-1} = \tilde{\varphi}_{n-1}$, and Assertion $*$ follows from the properties of $\tilde{\varphi}_{n-1}$ proved in Step 1 and Proposition 1 of subsection A.

Now prove that the appearance of the nonzero term ψ is nonessential. It was for this purpose that Step 1 was taken and the factor g_1 was singled out.

1°. We have that

$$\varphi'_{n-1} = \tilde{\varphi}'_{n-1} \circ (\text{id} + \psi) \cdot (1 + \psi') + \psi'.$$

The relation $\psi \in \mathscr{L}(\mathscr{F}_{+}^{n-1}, \mathscr{F}_{+}^{n}) \circ \ln^{[n-1]}$ implies that for some $C_0 > 0$

$$\psi \to 0 \quad \text{and} \quad \psi' \to 0 \quad \text{in } (A^{1-n}C_0 \circ \ln \Omega, \infty)$$

for some standard domain Ω of class $n-1$. Moreover, for any $C > 0$

$$(\ln \tilde{\Pi}, \infty) \subset (A^{1-n}C_0 \circ \ln \Omega, \infty),$$

where $\tilde{\Pi} = A^{-n}C\Pi$. Indeed, $A^{-n}(CC_0^{-1})\Pi \subset (\Omega, \infty)$ for any $C > 0$.

Further, assertion 1 in §5.4 D gives us that

$$\operatorname{Re} \varphi'_{n-1} \circ (\mathrm{id} + \psi) = \operatorname{Re} \tilde{\varphi}'_{n-1} \cdot (1 + o(1))$$

in $(\ln \tilde{\Pi}, \infty)$. Finally, in Cases 1 and 2

$$\operatorname{Re} \tilde{\varphi}'_{n-1} = \operatorname{Re} \tilde{\varphi}_s \cdot (1 + o(1)), \qquad \varphi'_s \in \mathscr{F}_{\mathbf{R}, \mathrm{id}}^{n-2} \circ \ln^{[n-1]}$$

by the remark at the beginning of Step 2. Thus, $\psi' = o(\operatorname{Re} \tilde{\varphi}'_s)$ by Lemma ${}^{\mathrm{C}}\mathrm{OL2}_n$. In Case 3 the same is given by the same remark and the relation $\psi \to 0$ in $(\ln \tilde{\Pi}, \infty)$. This proves 1°.

2°. We have

$$\varphi''_{n-1} = \tilde{\varphi}''_{n-1} \circ (\mathrm{id} + \psi) \circ (1 + \psi')^2 + \tilde{\varphi}_{n-1} \circ (\mathrm{id} + \psi) \cdot \psi'' + \psi''.$$

Assertion 2° is now a consequence of the following equalities, which are valid in $(\ln \tilde{\Pi}, \infty)$:

$$\psi' = o(1), \quad \psi'' = o(1), \quad \psi'' = o(\operatorname{Re} \varphi'_{n-1}),$$
$$\tilde{\varphi}'_{n-1} \circ (\mathrm{id} + \psi) = \tilde{\varphi}'_{n-1} \cdot (1 + o(1)).$$

3°. Assertion 3° is a direct corollary to assertion 1° and the remark cited.

4°. Assertion 4° is derived from the previous ones like the analogous statement in Step 1. ►

The arguments of all the factors in (**) tend to zero in $(\tilde{\Pi}, \infty)$. For the second factor it is given by assertion *; for the first and third factors it was proved in Step 1. ►

D. Arguments of derivatives of germs in the class $\mathscr{D}_0^{n+1}$. We prove assertion 2° in subsection B, strengthening it: under the conditions of subsection B,

$$0 \prec \arg \sigma' \prec \frac{\pi}{2} \quad \text{in } (\rho\mathbf{C}^+, \infty) \text{ for } \eta > 0, \tag{***}$$
$$0 \succ \arg \sigma' \succ -\frac{\pi}{2} \quad \text{in } (\rho\mathbf{C}^+, \infty) \text{ for } \eta < 0.$$

PROOF. The function σ' takes the same values in $(\rho\mathbf{C}^+, \infty)$ that $\sigma' \circ \rho = 1/\rho'$ takes in $(\mathbf{C}^+, \infty)$. Further,

$$\rho' = \tilde{\sigma}' \circ \ln \cdot \zeta^{-1}.$$

The function ρ' takes in $(\mathbf{C}^+, \infty)$ the same values that the function

$$\rho' \circ \exp = \tilde{\sigma}' \cdot \exp(-\zeta)$$

takes in (Π, ∞). Furthermore,

$$\arg(\tilde{\sigma}' \cdot \exp(-\zeta)) = \arg \tilde{\sigma}' - \eta.$$

It suffices to prove that in (Π, ∞)

$$\operatorname{sgn}(\arg \tilde{\sigma}' - \eta) = -\operatorname{sgn} \eta \quad \text{and} \quad \operatorname{sgn} \arg \tilde{\sigma}' = \operatorname{sgn} \eta.$$

But $\arg \tilde{\sigma}' = \operatorname{Im} \ln \tilde{\sigma}'$. Therefore, it suffices to prove that

$$\operatorname{Re} \tilde{\sigma}''/\tilde{\sigma}' - 1 \prec 0 \quad \text{and} \quad \tilde{\sigma}''/\tilde{\sigma}' \succ 0 \quad \text{in } (\Pi, \infty).$$

The first inequality follows from the fact that $(\sigma''/\sigma') \to 0$ in (Π, ∞), which follows from Proposition 1 in subsection A and Proposition 2, and the second follows from Proposition 2. ►

E. The convexity lemma for germs of class $\mathscr{D}^n_{\text{slow}} \circ \mathscr{L}^n$. Here we prove the lemma in the heading. Let $\sigma \in \mathscr{D}^n_{\text{slow}} \circ \mathscr{L}^n$. As in subsection B, it suffices to prove two assertions:

1°. $\operatorname{sgn} \arg \sigma''/\sigma' = -\operatorname{sgn} \eta$ in $(\rho\mathbf{C}^+, \infty)$;
2°. $|\arg \sigma'| \prec \pi/2$ in $(\rho\Omega, \infty)$.

We begin with assertion 1°. By definition,

$$\sigma = A^{-n} g_1 \circ \tilde{h}, \qquad g_1 \in G^{n-1^+}_{\text{slow}}, \ \tilde{h} \in \mathscr{L}^n.$$

In the case $\tilde{h} = \mathrm{id}$, assertion 1° can easily be derived from assertion $*$ in section C (the reduction is below). The following preparations are in order to prove that the appearance of a nonidentity $\tilde{h}$ is nonessential from the point of view of assertion 1°.

By Lemma 5.6_{n+1} c, there is a germ $h \in L^n$ such that $\tilde{h} - h = o(|\zeta|^{-N})$ for any $N > 0$ in (S_α, ∞), $S_\alpha = \{\zeta \,|\, |\arg \zeta| < \alpha\}$, $\alpha \in (\pi/2, \pi)$. Lemma 5.6_{n+1} a gives us that

$$Ah = \mathrm{id} + \psi_1, \qquad \psi_1 \in \mathscr{L}(\mathscr{F}^{n-1}_+) \circ \ln^{[n-1]}.$$

Further, $A\tilde{h} = \mathrm{id} + \psi$, $\psi = \psi_1 + \psi_2$ (this defines ψ_2).

Let $\tilde{h} - h = \kappa$. Then

$$\psi_2 = \ln(1 + \tfrac{\kappa}{h} \circ \exp) = o(\exp(-N\xi))$$

for any $N > 0$ in (Π_α, ∞), $\Pi_\alpha = \{|\eta| < \alpha, \xi \geq 0\}$, $\alpha \in (\pi/2, \pi)$.

Now we prove that ψ is "nonessential." Let

$$A\sigma = \mathrm{id} + \varphi_{n-1}, \qquad Ag_1 = \mathrm{id} + \tilde{\varphi}_{n-1}.$$

Then

$$\varphi_{n-1} = \tilde{\varphi}_{n-1} \circ (\mathrm{id} + \psi) + \psi.$$

ASSERTION $**$. *Assertion* $*$ *in section* C *holds for* φ_{n-1} *and* $\tilde{\varphi}_{n-1}$ *defined above, not only in* $(\ln\tilde{\Pi}, \infty)$, *but also in* (Π_α, ∞), $\alpha \in (\pi/2, \pi)$. *Moreover,* $\varphi'''_{n-1} = o(\operatorname{Re}\tilde{\varphi}'_{n-1})$ *in* (Π_α, ∞).

PROOF. The function $\tilde{\varphi}_{n-1}$ is the same as in assertion $*$. In the case when $\psi = 0$ and $\varphi_{n-1} = \tilde{\varphi}_{n-1}$, assertion $**$ follows from the remark in section C and the fact that $(\Pi^\forall, \infty) \subset (\tilde{\Pi}', \infty)$.

The foregoing proves that ψ_1 and ψ_2 and all their derivatives (and thus ψ and all its derivatives) are small compared with $\operatorname{Re}\tilde{\varphi}'_{n-1}$ in (Π_α, ∞). After this, assertion $**$ is proved just like assertion $*$. ►

Now let us prove assertion 1°. By the formulas $(**)$,

$$\arg\sigma''/\sigma' \circ \exp = \arg[\varphi'_{n-1}(1+\varphi'_{n-1}) + \varphi''_{n-1}] - \arg(1+\varphi'_{n-1}) - \eta \overset{\text{def}}{=} \operatorname{Im}\Psi,$$

$$\Psi = \ln[\varphi'_{n-1}(1+\varphi'_{n-1}) + \varphi''_{n-1}] - \ln(1+\varphi'_{n-1}) - \zeta.$$

By the Cauchy-Riemann condition, it suffices to prove that

$$\operatorname{Re}\Psi' \prec 0 \quad \text{in } (\ln\rho\mathbf{C}^+, \infty) \subset (\Pi_\alpha\Omega, \infty)$$

for any $\alpha \in (\pi/2, \pi)$. We have that

$$\Psi' = \frac{2\varphi'_{n-1}\varphi''_{n-1} + \varphi''_{n-1} + \varphi'''_{n-1}}{\varphi'_{n-1}(1+\varphi'_{n-1}) + \varphi''_{n-1}} - \frac{\varphi''_{n-1}}{1+\varphi'_{n-1}} - 1.$$

By assertion $**$, the first two terms tend to zero in (Π_α, ∞). This proves assertion 1°.

Assertion 2° follows from Proposition 3 below.

F. Estimate of the argument of the derivative of a germ of class $\mathscr{D}^n_{\text{slow}} \circ \mathscr{L}^n$.

PROPOSITION 3. *For any arbitrary germ* $\sigma \in \mathscr{D}^n_{\text{slow}} \circ \mathscr{L}^n$, $\rho = \sigma^{-1}$, *there exists a quadratic domain* Ω *such that in* $\rho\Omega$

$$0 < \arg\sigma' < \tfrac{\pi}{2} \quad \textit{for } \eta > 0,$$
$$0 > \arg\sigma' > -\tfrac{\pi}{2} \quad \textit{for } \eta < 0.$$

PROOF. The first of these inequalities is equivalent to the second in view of the realness of σ and the symmetry principle; therefore, we prove only the first. We prove first that for an arbitrary quadratic domain Ω and any $\varepsilon > 0$

$$-\varepsilon < \arg\sigma' < \tfrac{\pi}{2} + \varepsilon \quad \text{in } (\rho\Omega, \infty) \text{ for } \eta > 0. \tag{$^{\,*}_{*\,*}$}$$

An equivalent inequality is

$$\varepsilon > \arg\rho' > -(\tfrac{\pi}{2} + \varepsilon) \quad \text{in } (\Omega, \infty) \text{ for } \eta > 0.$$

It suffices to prove that

$$-(\tfrac{\pi}{2} + \varepsilon) < \arg\rho' \circ \exp < \varepsilon \quad \text{in } (\Pi, \infty) \text{ for } \eta > 0.$$

Let $\tilde{\rho} = A\rho$. Then

$$\arg \rho' \circ \exp = \operatorname{Im} \tilde{\rho} + \arg \tilde{\rho}' - \eta.$$

By Lemma 5.6_{n+1} d, Lemma 5.3_{n+1} b, and Proposition 2 in subsection B, $\arg \tilde{\rho}' \to 0$ in $(\Pi^{\forall}, \infty)$. Further, for each $\varepsilon > 0$

$$|\tilde{\rho}'| < 1 + \varepsilon \quad \text{in } (\Pi^{\forall}, \infty).$$

Indeed, by Lemmas 5.3_n b and 5.6_{n+1} d, the limit $\mu = \lim \tilde{\rho}'$ exists in $(\Pi^{\forall}, \infty)$ and $\mu \leq 1$, because $\tilde{\rho} \prec \mathrm{id}$. Therefore, for any $\varepsilon > 0$

$$\operatorname{Im} \tilde{\rho} - \eta \prec \varepsilon \quad \text{in } (\Pi, \infty) \text{ for } \eta > 0.$$

Finally, $\operatorname{Im} \tilde{\rho} \succ 0$ in $(\Pi^{\forall}, \infty)$ for $\eta > 0$. Therefore, $\operatorname{Im} \tilde{\rho} - \eta \succ -(\frac{\pi}{2} + \varepsilon)$ in (Π, ∞) for arbitrary $\varepsilon > 0$. This proves the estimate $({}_{*}{}^{*}{}_{*})$.

We now derive assertion 2° in subsection E from assertion 1° in the same subsection. It follows from assertion 1° that the image under σ of the intersection of a horizontal ray $L \subset \mathbf{C}^+$ and the domain $\rho\Omega$ is located locally to the right of its tangent for a quadratic standard domain Ω "sufficiently far to the right." By Lemmas 5.3_n b and 5.6_{n+1} d, $(\sigma\Pi^{\forall}, \infty) \subset (S^{\forall}, \infty)$. Consequently, $(\Pi^{\forall}, \infty) \subset (\rho S^{\forall}, \infty)$. Therefore, each horizontal ray $L \subset \mathbf{C}^+$ enters $\rho\Omega$ and stays there; a priori, this can happen after several entrances and exits. We prove that an arbitrary sufficiently high ray intersects $\rho\partial\Omega$ at exactly one point. Let Γ^+ be the part of the boundary of Ω lying in the upper half-plane and convex downwards. It can be assumed that the domain Ω belongs to $\mathbf{C}^+$ and that the estimates $({}_{*}{}^{*}{}_{*})$ hold in it. It follows from these estimates that outside some compact set the tangent vector to the curve $\rho\Gamma^+$ lies in the cone $\arg\zeta \in (-\varepsilon, \frac{\pi}{2} + \varepsilon)$. Consequently, the points of the intersection $\rho\Gamma^+ \cap L$ follow in the same order on L and $\rho\Gamma^+$: the so-called nontrivial meandering permutations are not possible (see Figure 22). Let $\eta_0 > 0$ be such that $\rho\Gamma^+ \supset \rho\partial\Omega \cap \{\eta > \eta_0\}$. Assume that the ray L with vertex $i\eta$, $\eta > \eta_0$, intersects the curve $\rho\Gamma^+$ at more than one point. Then the first and second (along L) points of intersection lie in the same order on $\rho\Gamma^+$. But this is impossible, because Ω can be chosen so that

$$\operatorname{sgn} \arg(\sigma''/\sigma') = -\operatorname{sgn} \eta \quad \text{in } \rho\Omega. \qquad ({}^{*}{}_{*}{}^{*})$$

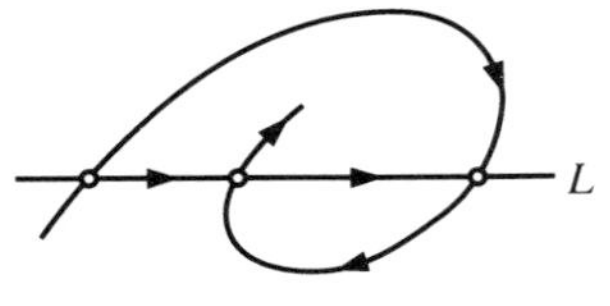

FIGURE 22

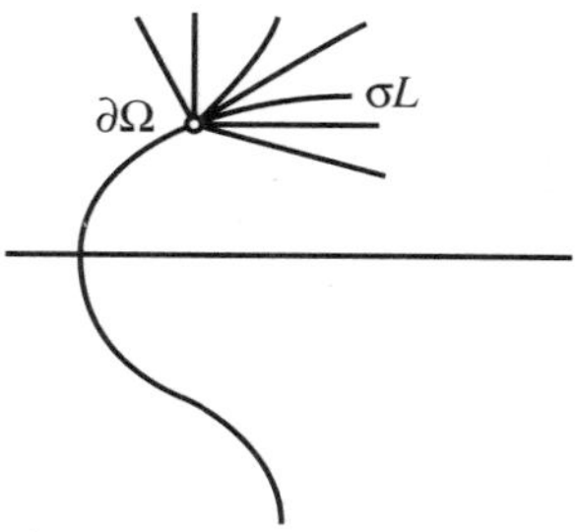

FIGURE 23

Indeed, let γ and S be the arc and segment on $\rho\Gamma^+$ and L between the points of intersection. The curve $\sigma\gamma$ lies locally to the left of any of its tangents, while the curve σS lies locally to the right. Both curves lie in the cone $\arg(\zeta - \zeta_0) \in (-\varepsilon, \frac{\pi}{2} + \varepsilon)$, where ζ_0 is their common initial point. Therefore, the curves lie on different sides of their tangents at ζ_0, and cannot have a second point of intersection (Figure 23).

Accordingly, the curve $\rho\Gamma^+$ intersects each horizontal ray $L \subset \mathbf{C}^+ \cap \{\eta > \eta_0\}$ in a single point. We prove that the curve $\sigma(L \cap \rho\Omega)$ has at all points a tangent vector lying in the open first quadrant. Indeed, the slope of the curve σL decreases monotonically in view of the equality $({}^*{}_*{}^*)$, and cannot be made less than $-\varepsilon$, according to what was proved at the beginning of the subsection. Therefore, if at some point the slope of σL is negative and equal to $-\alpha$, then the slope of the subsequent part of this curve will be included in the interval $(-\alpha, -\varepsilon)$, and the curve intersects the real axis; this contradicts the fact that σ is bijective and real. Together with each point of zero slope on σL there is always a point of negative slope in view of the equality $({}^*{}_*{}^*)$. Therefore, the slope of σL can only be positive; it is less than the slope of the curve Γ^+ at the point of the intersection $\Gamma^+ \cap \sigma L$, that is, less than $\pi/2$. This proves that $0 < \arg\sigma' < \pi/2$ in the domain $\rho\Omega \cap \{\eta > \eta_0\}$. Similarly, by the symmetry principle, $0 > \arg\sigma' > -\frac{\pi}{2}$ in the domain $\rho\Omega \cap \{\eta < -\eta_0\}$.

Suppose that $\Pi_0 = \rho\Omega \cap \{|\eta| \leq \eta_0\}$. By Proposition 2 and Lemma 5.3_n b, $\arg\sigma' \to 0$ in (Π_0, ∞). Similarly,

$$\operatorname{sgn}\arg\sigma' = \operatorname{sgn}\eta \quad \text{in } (\Pi_0, \infty).$$

Therefore, for a standard domain $\tilde{\Omega} \subset \Omega$ sufficiently far to the right the inequalities of Proposition 3 hold in the domain $\rho\tilde{\Omega}$. ►

§5.9. Distortion theorems for special admissible germs

In this section we prove Lemma 5.8_{n+1}. It follows in a relatively easy way from the convexity lemma in §5.8 for all germs except for rapid germs and sectorial germs, for which it is proved in subsection D.

A. Formulations. We recall two definitions. The standard partition Ξ_{st} is the partition by the horizontal rays $\eta = \pi k$.

The total slope of the lines of the partition $\sigma_* \Xi_{\mathrm{st}}$ in a domain Ω is the sum of the secants of the slope angles of the lines of the partition, taken at the points of intersection of all these lines with the segment $\{\operatorname{Re} \zeta = \xi\} \cap \Omega$ (Definition 1 in §3.4).

LEMMA 5.8_{n+1}. *Suppose that* σ *is an admissible germ of class* $\mathscr{D}_0^{n+1}$ *or* $\mathscr{D}_1^n$. *Then there exists a standard domain* Ω *of class* $n+1$ *such that*:

a. *The boundary lines of the partition* $\sigma_* \Xi_{\mathrm{st}}$ *of* Ω *are the graphs of the functions* $\eta = \eta(\xi)$.

b. *The total slope of the lines of the partition* $\sigma_* \Xi_{\mathrm{st}}$ *in* Ω *does not exceed* $C\xi^4$ *for some* $C > 0$.

c. *The real part of the germ* $\rho = \sigma^{-1}$ *does not decrease upon moving away from the real axis along a vertical in* (Ω, ∞) *if*

$$\sigma \in \mathscr{D}_0^{n+1} \cup \mathscr{D}_0^n \cup \mathscr{D}_{\mathrm{slow}}^n \circ \mathscr{L}^n,$$

and for any $\delta > 0$

$$\operatorname{Re} \rho \succ (1-\delta)\rho \circ \operatorname{Re},$$

if $\sigma \in \mathscr{D}_{\mathrm{rap}}^n \circ \mathscr{L}^n$.

It suffices to consider the following three cases.

CASE 1: σ is a slow germ that is not comparable with a linear germ.

CASE 2: $\sigma \in \mathscr{D}_{\mathrm{rap}}^n \circ \mathscr{L}^n$.

CASE 3: σ is a rapid or sectorial germ.

In Case 2, Lemma 5.8_{n+1} c follows from Propositions 9 and 10 in §5.4 J.

B. Slow germs that are not comparable with linear germs. For these germs assertions a and c of the lemma follow directly from the estimates on the arguments of their derivatives—the estimates (***) in §5.8 D—and from Proposition 3 in §5.8 F.

We prove assertion b. Let us begin with a general proposition that helps in estimating the total slope of the lines of the partition $\sigma_* \Xi_{\mathrm{st}}$.

PROPOSITION 1. *Suppose that* ρ *is a holomorphic diffeomorphism of a quadratic standard domain* Ω *into* $\mathbf{C}$ *that is real on* $(\mathbf{R}^+, \infty)$ *and has bounded derivative* ρ', *and let* $\sigma = \rho^{-1}$ *be the inverse mapping. Then the number of points in the intersection of the segment* $\{\operatorname{Re} \zeta = \xi\} \cap \Omega$ *with the lines of the partition* $\sigma_* \Xi_{\mathrm{st}}$ *does not exceed* $C\xi^2$ *for some* $C > 0$.

PROOF. The distances between two successive points of intersection of a vertical line with the lines of the partition $\sigma_* \Xi_{\mathrm{st}}$ are bounded below by a positive constant. Indeed, under the mapping ρ the vertical segment between these two points is carried into a curve joining two points on different edges of a half-strip of width π. The length of this curve is at least π and is at most the product of the distance under investigation by the Lipschitz constant

L of the mapping ρ; it is finite, since the derivative ρ' is bounded. This implies that the distance under study is not less than π/L. But the length of the vertical segment cut out on the line $\operatorname{Re}\zeta = \xi$ by the quadratic standard domain does not exceed $C\xi^2$ for some $C > 0$. This yields the assertion.

Let us return to the investigation of Case 1. In this case the number of points of intersection of the lines of the partition $\sigma_* \Xi_{\text{st}}$ with the verticals $\operatorname{Re}\zeta = \xi$ inside a quadratic standard domain does not exceed $C\xi^2$ for some $C > 0$ (for rapid and sectorial germs this number can be estimated by a constant independent of ξ).

We now investigate the slope of the lines of the partition $\sigma_* \Xi_{\text{st}}$. These lines are convex upward in the upper half-plane and downward in the lower half-plane, by the convexity lemma in §5.8. If Ω is a quadratic standard domain sufficiently far to the right, then this convexity holds everywhere in it, and not just in a neighborhood of infinity. Then the slope angle of the kth line of the partition, counted upward from the real axis, does not exceed the slope angle of the boundary line of the quadratic standard domain at the point where it intersects this line of the partition; see §5.8 F. In view of Proposition 1, this point does not lie below the line $\eta = Ck$ for some $C > 0$. The slope angle at this point of the boundary of the quadratic standard domain does not exceed $\pi/2 - C_1/\sqrt{k}$ for some $C_1 > 0$. Consequently, the secant of the slope angle of this line does not exceed $C_2 k$. The slopes of the lines of the partition in the lower half-plane are estimated similarly. The number of points of intersection of the lines of the partition with the vertical $\operatorname{Re}\zeta = \xi$ in the domain Ω does not exceed $C\xi^2$ by Proposition 1. This yields Lemma 5.8_{n+1} in Case 1.

C. Almost linear germs. We prove a and b of Lemma 5.8_{n+1} in Case 2 (c is already proved). Let

$$\sigma \in A^{-n} g \circ h, \qquad g \in G^{n-1}_{\text{rap}}, \ h \in \mathscr{L}^n.$$

In this case σ' has a real finite limit in a quadratic standard domain. This follows from Corollary 2 in §5.2G and Lemma 5.6_{n+1} a in §5.4H. Consequently, the slope angle of the lines of the partition $\sigma_* \Xi_{\text{st}}$ can be made arbitrarily close to zero in a quadratic standard domain sufficiently far to the right. This proves a of Lemma 5.8_{n+1} in Case 2. The number of points of intersection of the segment $\{\operatorname{Re}\zeta = \xi\} \cap \Omega$ with the lines of the partition was estimated from above in Proposition 1. Consequently, the total slope of the lines of the partition can be estimated from above by the function $C\xi^2$ for sufficiently large C. This completes the proof of Lemma 5.8_{n+1} in Case 2.

D. Proof of Lemma 5.8_{n+1} in the case of sectorial and rapid germs. We consider the germs of the class $\mathscr{D}_0^{n+1}$:

$$\sigma = \exp \circ A^{-n} g, \qquad g \in G^{n^-}_{\text{slow}}, \ \mu_{n+1}(g) \in (0, 1]. \qquad (*)$$

The argument is analogous for germs of class $\mathscr{D}_0^n$; just replace $n+1$ by n everywhere. For germs of the form $(*)$

$$\rho(\mathbf{C}^+, \infty) = \tilde{\sigma} \circ \ln(\mathbf{C}^+, \infty) = (\tilde{\sigma}\Pi, \infty),$$

where $\tilde{\sigma} = A^{-n} g^{-1}$, $\tilde{\sigma}' \to \mu = \mu_{n+1}(g^{-1})$ in (Π, ∞), and $\mu \in [1, \infty)$. Then the germ of the domain $\rho(\mathbf{C}^+, \infty)$ belongs to the germ of the half-strip $((\mu+\varepsilon)\Pi, \infty)$ for arbitrary $\varepsilon > 0$. Consequently, the domain $\rho\mathbf{C}^+$ intersects only finitely many rays of the standard partition $\Xi_{\text{st}}: \eta = \pi j$, $j \in \mathbf{Z}$. If $|\pi j| > \mu\pi/2$, then the intersection $\rho\mathbf{C}^+ \cap \{\eta = \pi j\}$ is empty or compact. Choosing a quadratic standard domain Ω sufficiently far to the right, we can make all the intersections $\rho\Omega \cap \{\eta = \pi j\}$ with $|\pi j| > \mu\pi/2$ empty. It is assumed below that this has already been done.

We proceed directly to the proof of Lemma 5.8_{n+1}. Let $\tilde{\sigma} = \mu \circ (\mathrm{id} + \tilde{\varphi}_n)$. Since $A^n \mu \overset{\text{def}}{=} g_0 \in G^n$,

$$g_0^{-1} \circ g = \tilde{g} \in G^n, \quad \mathrm{id} + \tilde{\varphi}_n = A^{-n}\tilde{g}, \quad \mu_{n+1}(\tilde{g}) = 1.$$

We consider three cases: $\tilde{g} \in G_{\text{slow}}^{n^-}$, $\tilde{g} \in G_{\text{rap}}^n$, $\tilde{g} \in G_{\text{slow}}^{n^+}$. We prove that in the first two cases the standard domain Ω can be chosen so that the intersection $\rho\Omega \cap L_\pm$ is empty; here $L_\pm \subset \mathbf{C}_-^+$ is the ray $\eta = \pm\pi\mu/2$. In the third case we verify that the line $\sigma L_\pm$ satisfies the requirements yielding Lemma 5.8_{n+1}.

CASE 1: $\tilde{g} \in G_{\text{slow}}^{n^-}$. In this case the inclusion

$$(A^{-n}\tilde{g}\Pi, \infty) \subset (\Pi, \infty)$$

follows from the corollary to Proposition 3 in §5.5 D, or, equivalently, from Lemma 5.7_{n+1} c. Thus,

$$(\rho\mathbf{C}^+, \infty) = (\mu \circ A^{-n}\tilde{g}\Pi, \infty) \subset \mu(\Pi, \infty).$$

Consequently, $\rho\Omega \cap L_\pm = \varnothing$ for an arbitrary standard domain $\Omega \subset \mathbf{C}^+$ in which a representative of ρ is defined.

CASE 2: $\tilde{g} \in G_{\text{rap}}^n$. We take a standard domain Ω of class $n+1$ such that $(\ln\Omega, \infty) \subset (\Pi, \infty)$. Then by Lemma 5.2_{n+1} a, there exists a domain $\tilde{\Omega}$ of the same class such that $\tilde{\rho}\tilde{\Omega} \subset \Omega$, where $\tilde{\rho} = A^{-(n+1)}\tilde{g}$. Consequently,

$$A^{-n}\tilde{g} \circ \ln\tilde{\Omega} \subset \ln\Omega \subset \Pi;$$

$$\rho\tilde{\Omega} = \mu \circ A^{-n}\tilde{g} \circ \ln\tilde{\Omega} \subset \mu\Pi.$$

This again implies that $\rho\Omega \cap L_\pm = \varnothing$.

CASE 3: $\tilde{g} \in G_{\text{slow}}^{n^+}$. For $\mu = 1$ and $\tilde{g} = g^{-1}$ this inclusion follows from relation $(*)$; for $\mu > 1$ it does not follow.

The number of lines of the partition $\sigma_*\Xi_{\text{st}}$ is finite, as indicated above; the lines of the form σL, $L \subset \{\eta = \pi j\}$ are easy to investigate when

$|\pi j| < \pi\mu/2$, but this is somewhat more difficult when $|\pi j| = \pi\mu/2$. We first consider the first ones. On the ray $L\colon \eta = \pi j$,

$$|\arg\sigma'| = |\operatorname{Im}\tilde{\sigma}^{-1}| = |(\mu^{-1} + o(1))\pi j| < \alpha_0 < \pi/2.$$

Consequently, the line σL is the graph of a function. Further, $|\cos\arg\sigma'| > \cos\alpha_0 > 0$. In the absence of rays of the form $L_\pm\colon \eta = \pi j = \pm\mu\pi/2$ this proves assertions a and b of Lemma 5.8_{n+1}.

Suppose now that $\mu = \pm 2j$ for some $j \in \mathbf{Z}$; then the rays $L_\pm$ exist. We investigate the curve σL_+; the curve σL_- is symmetric to it with respect to the real axis and is investigated similarly. Let us prove that the inequalities

$$0 \prec 1/\cos\arg\sigma' \prec C(\operatorname{Re}\sigma)^4 \tag{$**$}$$

hold on L_+. This yields assertions a and b of Lemma 5.8_{n+1}: assertion b at once, and assertion a from the inequality $\cos\arg\sigma' \succ 0$.

Let $A^{-n}g = \mu^{-1}\circ(\mathrm{id} + \varphi_n)$. As above, $\mu_{n+1}(\mathrm{id} + \varphi_n) = 1$. We have that

$$\sigma' = \sigma\cdot(A^{-n}g)' = \sigma\cdot\mu^{-1}\cdot(1 + \varphi_n'),$$
$$\arg\sigma' = \operatorname{Im}\mu^{-1}(\zeta + \varphi_n) + \arg(1 + \varphi_n').$$

On L_+

$$\cos\arg\sigma' = \cos(\tfrac{\pi}{2} + \alpha) = -\sin\alpha,$$

where

$$\alpha = \mu^{-1}\operatorname{Im}\varphi_n + \arg(1 + \varphi_n');$$
$$\operatorname{Re}\sigma = [\exp\mu^{-1}(\xi + \operatorname{Re}\varphi_n)]\cos\mu^{-1}(\eta + \operatorname{Im}\varphi_n)$$
$$= -[\exp\mu^{-1}(\xi + \operatorname{Re}\varphi_n)]\sin\mu^{-1}\operatorname{Im}\varphi_n.$$

By Proposition 1 in §5.5 B,

$$\varphi_n \in \mathscr{L}_*\circ\ln^{[n]}, \quad \operatorname{Re}\varphi_n \to -\infty,\ \varphi_n' \to 0 \quad \text{in } (\Pi^\forall, \infty).$$

According to Proposition 5 in §5.3 D, $\varphi_n' \prec 0$ in $(\Pi^\forall, \infty)$. We prove that

$$\arg(1 + \varphi_n') = o(\operatorname{Im}\varphi_n).$$

To do this it suffices to prove that for arbitrary $\varepsilon > 0$

$$0 \preccurlyeq \operatorname{sgn}\eta\cdot\arg(1 + \varphi_n') \prec -\varepsilon\operatorname{sgn}\eta\cdot\operatorname{Im}\varphi_n \tag{$***$}$$

in $(\Pi^\forall, \infty)$. Let us prove this. We have that $\varphi_n' \to 0$; consequently, in $(\Pi^\forall, \infty)$

$$\arg(1 + \varphi_n') = (\operatorname{Im}\varphi_n')(1 + o(1)).$$

The equalities

$$\operatorname{sgn}\eta = -\operatorname{sgn}\operatorname{Im}\varphi_n = \operatorname{sgn}\operatorname{Im}\varphi_n'$$

follow from the inequalities $\operatorname{Re}\varphi_n' \prec 0$ and $\operatorname{Re}\varphi_n'' \succ 0$, which, in turn, follow from Proposition 5 in §5.3 D. Further,

$$\operatorname{Im}\varphi_n^{(k)}(\xi + i\eta) = \int_0^\eta \operatorname{Re}\varphi_n^{(k+1)}(\xi + i\theta)\,d\theta\,, \qquad k = 0\,,\,1.$$

Therefore, to prove the second of the inequalities in $(***)$ it suffices to establish that $|\operatorname{Re}\varphi_n''| \prec \varepsilon|\operatorname{Re}\varphi_n'|$ in $(\Pi^\forall, \infty)$ for any $\varepsilon > 0$. This follows from the fact that the function $\operatorname{Re}\varphi_n'$ is of definite sign in $(\Pi^\forall, \infty)$, the fact that the half-strip $\Pi^\forall$ is arbitrary, and the Remark in §5.8 A. Thus, it has been proved that on L_+

$$\begin{aligned}\sin\alpha &= \sin[\mu^{-1}\operatorname{Im}\varphi_n + \arg(1 + \varphi_n')] \\ &= (1 + o(1))\sin\mu^{-1}\operatorname{Im}\varphi_n.\end{aligned}$$

It follows from the equality $\operatorname{sgn}\operatorname{Im}\varphi_n = -\operatorname{sgn}\eta$ that the first of the inequalities in $(**)$ holds. We prove the second. It is equivalent to the following inequality: on L_+

$$\exp\circ 4\mu^{-1}(\xi + \operatorname{Re}\varphi_n) \succ (1 + o(1))|\sin\mu^{-1}\operatorname{Im}\varphi_n|^{-5}.$$

To prove this it is sufficient to prove that for any $\varepsilon > 0$

$$|\operatorname{Im}\varphi_n| \succ \exp(-\varepsilon\xi).$$

Since φ_n is real, it suffices to prove that in (Π, ∞)

$$|\operatorname{Re}\varphi_n'| \succ \exp(-\varepsilon\xi)$$

for any $\varepsilon > 0$.

We mentioned above that $\varphi_n \in \mathscr{L}_* \circ \ln^{[n]}$, $\operatorname{Re}\varphi_n \to -\infty$ and $\varphi_n' \to 0$ in $(\Pi^\forall, \infty)$. Consequently, in $(\Pi^\forall, \infty)$

$$\varphi_n' = \varphi_s'(1 + o(1))\,, \qquad \varphi_s' \in \mathscr{F}_{\mathbf{R},\mathrm{id}}^{n-1} \circ \ln^{[n]} = \mathscr{F}\mathscr{C}_{\mathbf{R}}^{n-1} \circ \ln.$$

By Lemma ${}^{\mathbf{C}}\mathrm{OL}1_n$,

$$\varphi_s' = F \circ \ln\,, \qquad F \in \mathscr{F}\mathscr{C}_{\mathbf{R}}^{n-1}\,,\ |\operatorname{Re}F| \succ \exp(-\alpha\xi)$$

for some α in the domain $T = \ln A^{1-n}C\Pi$ for any $C > 0$. Consequently,

$$|\operatorname{Re}\varphi_s'| \succ |\zeta|^{-\alpha} \succ \exp(-\varepsilon\xi)$$

in $(\Pi^\forall, \infty)$ for any ε. This proves the second inequality in $(**)$. This proves assertions a and b of Lemma 5.8_{n+1} for rapid and sectorial germs.

Let us now prove Lemma 5.8_{n+1} c for these germs. We have that in (Ω, ∞)

$$\begin{aligned}&\rho = \tilde\sigma \circ \ln\,, \quad \tilde\sigma = A^{-n}g^{-1}\,, \quad \mu_{n+1}(g^{-1}) = \mu \in [1\,, \infty)\,, \\ &\rho' = \tilde\sigma' \circ \ln \cdot \zeta^{-1}\,, \quad \arg\rho' = \arg\tilde\sigma' \circ \ln - \arg\zeta.\end{aligned}$$

The function $\arg \rho'$ takes the same values in (Ω, ∞) that $\arg \rho' \circ \exp$ takes in $(\ln \Omega, \infty)$. Further, $\tilde{\sigma} = \mu \circ (\mathrm{id} + \tilde{\varphi}_n)$ and $\tilde{\sigma}' = \mu(1 + \varphi_n')$. We prove that in the germ of the domain (Ω, ∞)

$$\operatorname{sgn} \arg \rho' = -\operatorname{sgn} \eta \quad \text{and} \quad |\arg \rho'| < \tfrac{\pi}{2} + \varepsilon$$

for arbitrary $\varepsilon > 0$. We have

$$\begin{aligned} \arg \rho' \circ \exp &= \arg(1 + \varphi_n') - \eta \\ &= (1 + o(1)) \operatorname{Im} \varphi_n' - \operatorname{Im} \zeta. \end{aligned}$$

But $|\operatorname{Im} \varphi_n'| \prec \varepsilon |\operatorname{Im} \zeta|$ in $(\ln \Omega, \infty)$, because $|\varphi_n''| \prec \varepsilon$, since $\varphi_n'' \to 0$ in $(\Pi^{\forall}, \infty)$. This implies the equality to be proved.

Further,

$$\begin{aligned} |\arg \rho' \circ \exp| &\leq |(1 + o(1)) \operatorname{Im} \varphi_n'| + \tfrac{\pi}{2} \\ &\prec \tfrac{\pi}{2} + \varepsilon \end{aligned}$$

in $(\ln \Omega, \infty)$ for arbitrary $\varepsilon > 0$.

This proves Lemma 5.8_{n+1} c for rapid and sectorial germs. With this the proof of Lemma 5.8_{n+1} is complete.

The proof of all of Lemmas 5.1_{n+1}–5.8_{n+1} are thereby complete, and with them the proof of the identity theorem.

Bibliography

1. V. I. Arnol′d, *Supplementary chapters to the theory of ordinary differential equations*, "Nauka", Moscow, 1978; English transl., *Geometric methods in the theory of ordinary differential equations*, Springer-Verlag, 1982.

2. V. I. Arnol′d and Yu. S. Il′yashenko, *Ordinary differential equations*, Itogi Nauki i Tekhniki: Sovremennye Problemy Mat.: Fundamental. Napravleniya, vol. 1, VINITI, Moscow, 1985, pp. 7–149; English transl., Encyclopedia of Math. Sci., vol. 1 (*Dynamical systems*, I), Springer-Verlag, 1988, pp. 1–148.

3. R. Bamón, *Quadratic vector fields in the plane have a finite number of limit cycles*, Inst. Hautes Études Sci. Publ. Math. No. 64 (1986), 111–142.

4. R. Bamón, J. C. Martin-Rivas, and R. Moussu, *Sur le problème de Dulac*, C. R. Acad. Sci. Paris Sér. I Math. **303** (1986), 737–739.

5. Ivar Bendixson, *Sur les courbes définies par des équations différentielles*, Acta Math. **24** (1901), 1–88.

6. R. I. Bogdanov, *Local orbital normal forms of vector fields on the plane*, Trudy Sem. Petrovsk. Vyp. **5** (1979), 51–84; English transl., *Topics in Modern Math.*, Plenum Press, New York, 1985, pp. 59–106.

7. A. D. Bryuno, *Analytical form of differential equations*. I, II, Trudy Moskov. Mat. Obshch. **25** (1971), 119–262; **26** (1972), 199–239; English transl. in Trans. Moscow Math. Soc. **25** (1971); **26** (1972).

8. Kuo-Tsai Chen, *Equivalence and decomposition of vector fields about an elementary critical point*, Amer. J. Math. **85** (1963), 693–722.

9. Henri Dulac, *Recherches sur les points singuliers des équations différentielles*, J. École Polytech. (2) **9** (1904), 1–125.

10. ——, *Sur les cycles limites*, Bull. Soc. Math. France **51** (1923), 45–188.

11. Freddy Dumortier, *Singularities of vector fields*, Inst. Mat. Pura Apl., Conselho Nac. Desenvohvimento Ci. Tecn., Rio de Janeiro, 1978.

12. Jean Écalle, *Les fonctions résurgentes*. Vols. I, II, Dép. Math., Univ. Paris-Sud, Orsay, 1981.

13. Jean Écalle, J. Martinet, R. Moussu, and J.-P. Ramis, *Non-accumulation des cycles-limites*. I, C. R. Acad. Sci. Paris Sér. I Math. **304** (1987), 375–377.

14. M. G. Golitsyna, *Nonproper polycycles of quadratic vector fields on the plane*, Methods of the Qualitative Theory of Differential Equations, Gor′kov. Gos. Univ., Gorki, 1987, pp. 51–67. (Russian); English transl. in Selecta Math. Soviet. **10** (1991).

15. Masuo Hukuhara, Tosihusa Kimura, and Tizuko Matuda, *Équations différentielles ordinaires du premier ordre dans le champ complexe*, Math. Soc. Japan, Tokyo, 1961.

16. Yu. S. Il′yashenko, *In the theory of normal forms of analytic differential equations violating the conditions of A. D. Bryuno, divergence is the rule and convergence the exception*, Vestnik Moskov. Univ. Ser. I Mat. Mekh. **1981**, no. 2, 10–16; English transl. in Moscow Univ. Math. Bull. **36** (1981).

17. ——, *On the problem of finiteness of the number of limit cycles of polynomial vector fields on the plane*, Uspekhi Mat. Nauk **37** (1982), no. 4, (226), 127. (Russian)

18. ____, *Singular points and limit cycles of differential equations on the real and complex plane*, Preprint, Sci. Res. Computing Center, Acad. Sci. USSR, Pushchino, 1982. (Russian)

19. ____, *Limit cycles of polynomial vector fields with nondegenerate singular points on the real plane*, Funktsional. Anal. i Prilozhen. **18** (1984), no. 3, 32–42; English transl. in Functional Anal. Appl. **18** (1984).

20. ____, *The finiteness problem for limit cycles of polynomial vector fields on the plane, germs of saddle resonant vector fields and non-Hausdorff Riemann surfaces*, Topology (Leningrad, 1982), Lecture Notes in Math., vol. 1060, Springer-Verlag, 1984, pp. 290–305.

21. ____, *Dulac's memoir "Sur les cycles limites" and related questions in the local theory of differential equations*, Uspekhi Mat. Nauk **40** (1985), no. 6 (246), 41–78; English transl. in Russian Math. Surveys **40** (1985).

22. ____, *Separatrix lunes of analytic vector fields on the plane*, Vestnik Moskov. Univ. Ser. I Mat. Mekh. **1986**, no. 4, 25–31; English transl. in Moscow Univ. Math. Bull. **41** (1986).

23. ____, *Finiteness theorems for limit cycles*, Uspekhi Mat. Nauk **42** (1987), no. 3 (255), 223. (Russian)

24. ____, *Finiteness theorems for limit cycles*, Uspekhi Mat. Nauk **45** (1990), no. 2 (272), 143–200; English transl. in Russian Math. Surveys **45** (1990).

25. A. Yu. Kotova, *Finiteness theorem for limit cycles of quadratic systems*, Methods in the Qualitative Theory of Differential Equations, Gor′kov. Gos. Univ., Gorki, 1987, pp. 74–89. (Russian); English transl. in Selecta Math. Soviet. **10** (1991).

26. R. Courant, *Geometrische Funktionentheorie*, Part III in A. Hurwitz and R. Courant, *Vorlesungen über allgemeinen Funktionentheorie and elliptische funktionen*, 3rd ed., Springer-Verlag, 1929.

27. Bernard Malgrange, *Travaux d'Écalle et de Martinet-Ramis sur les systèmes dynamiques*, Séminaire Bourbaki 1981/82, Exposé 582, Astérisque, no. 92–93, Soc. Math. France, Paris, 1982, pp. 59–73.

28. Jean Martinet and Jean-Pierre Ramis, *Problèmes de modules pour des équations différentielles non linéaires du premier ordre*, Inst. Hautes Études Sci. Publ. Math. No. 55 (1982), 63–164.

29. J.-F. Mattei and R. Moussu, *Holonomie et intégrales premières*, Ann. Sci. École Norm. Sup. (4) **13** (1980), 469–523.

30. Robert Moussu, *Le problème de la finitude du nombre de cycles limites* (d'après R. Bamón et Yu. S. Il′yashenko), Séminaire Bourbaki 1985/86, Exposé 655, Astérisque, no. 145–146, Soc. Math. France, Paris, 1987, pp. 89–101.

31. A. Seidenberg, *Reduction of singularities of the differential equation* $A\,dy = B\,dx$, Amer. J. Math. **90** (1968), 248–269.

32. A. N. Shoshitaĭshvili, *Bifurcation of topological type of singular points of vector fields depending on parameters*, Trudy Sem. Petrovsk. Vyp. 1 (1975) 279–309; English transl. in Amer. Math. Soc. Transl. (2) **118** (1982).

33. S. Smale, *Differentiable dynamical systems*, Bull. Amer. Math. Soc. **73** (1967), 747–817.

34. Floris Takens, *Normal forms for certain singularities of vector fields*, Ann. Inst. Fourier (Grenoble) **23** (1973), fasc. 2, 163–195.

35. E. C. Titchmarsh, *The theory of functions*, 2nd ed., Oxford Univ. Press, 1939.

36. S. M. Voronin, *Analytic classification of germs of conformal mappings* $(\mathbf{C}, 0) \to (\mathbf{C}, 0)$ *with linear part the identity*, Funktsional. Anal. i Prilozhen. **15** (1981), no. 1, 1–17; English transl. in Functional Anal. Appl. **15** (1981).

37. S. E. Warschawski, *On conformal mapping of infinite strips*, Trans. Amer. Math. Soc. **51** (1942), 280–335.

38. Jean-Christophe Yoccoz, *Non-accumulation de cycles limites*, Séminaire Bourbaki 1987/88, Exposé 690, Astérisque no. 161–162, Soc. Math. France, Paris, 1988, pp. 87–103.